D0709558

SYMPOSIA OF THE ZOOLOGICAL SOCIETY OF LONDON NO. 66

Marine Mammals: Advances in Behavioural and Population Biology

The track of a female southern elephant seal, which travelled 2500 km from South Georgia (bottom left) to the Antarctic Peninsula (top), shown in three dimensions together with the topography of the Southern Ocean. Each spike in the track shows a dive. Land is shown in yellow and depth is contoured at 500 m intervals. The seal swam within the upper 1000 m during its traverse of the Southern Ocean and then dived to the bottom when over the continental shelf of Antarctica. The data were obtained from a satellite-linked data logger produced by the Sea Mammal Research Unit, NERC. The display was produced by Evans and Sutherland Computer Ltd, UK, using AVS.

Marine Mammals: Advances in Behavioural and Population Biology

The Proceedings of a Symposium held at The Zoological Society of London on 9th and 10th April 1992

Edited by I. L. BOYD

British Antarctic Survey,
Cambridge

Published for THE ZOOLOGICAL SOCIETY OF LONDON

by CLARENDON PRESS · OXFORD

1993

ARLIS
Alaska Resources
Library & Information Services
Anchorage, Alaska

Oxford University Press, Walton Street, Oxford OX2 6DP
Oxford New York Toronto
Delhi Bombay Calcutta Madras Karachi
Kuala Lumpur Singapore Hong Kong Tokyo
Nairobi Dar es Salaam Cape Town
Melbourne Auckland Madrid
and associated companies in
Berlin Ibadan

Oxford is a trade mark of Oxford University Press

Published in the United States
by Oxford University Press Inc., New York

© The Zoological Society of London, 1993

All rights reserved. No part of this publication may be
reproduced, stored in a retrieval system, or transmitted, in any
form or by any means, without the prior permission in writing of Oxford
University Press. Within the UK, exceptions are allowed in respect of any
fair dealing for the purpose of research or private study, or criticism or
review, as permitted under the Copyright, Designs and Patents Act, 1988, or
in the case of reprographic reproduction in accordance with the terms of
licences issued by the Copyright Licensing Agency. Enquiries concerning
reproduction outside those terms and in other countries should be sent to
the Rights Department, Oxford University Press, at the address above.

This book is sold subject to the condition that it shall not,
by way of trade or otherwise, be lent, re-sold, hired out, or otherwise
circulated without the publisher's prior consent in any form of binding
or cover other than that in which it is published and without a similar
condition including this condition being imposed
on the subsequent purchaser.

A catalogue record for this book is available from the British Library

Library of Congress Cataloging in Publication Data
Marine mammals: advances in behavioural and population biology: the
    proceedings of a symposium held at the Zoological Society of London
    on 9th and 10th April 1992/edited by I.L. Boyd.
    (Symposia of the Zoological Society of London: no. 66)
    Includes bibliographical references and index.
    1. Marine mammals—Congresses.   2. Marine mammals—Behavior—
Congresses.   3. Mammal populations—Congresses.   I. Boyd, I. L.
II. Zoological Society of London.   III. Series.
QL1.Z733 no. 66
[QL713.2]   591 s—dc20   [599.5'0451]   93–19398
ISBN 0–19–854069–8

Typeset by Cambrian Typesetters, Frimley, Surrey
Printed in Great Britain by
Bookcraft (Bath) Ltd.
Midsomer Norton, Avon

Preface

The purpose of this symposium was to gather together representative examples of current marine mammal research, to illustrate the breadth and depth of the subject and to draw attention to new techniques and lines of research which offer great possibilities for future development. From the beginning, it was obvious that not every subject could be accommodated, because of lack of both time during the symposium and space in this book. In addition, restrictions on funding limited the number of contributions from outside Britain. However, although the balance has consequently been weighted heavily towards contributions from marine mammal researchers from Britain, the range of work being carried out overseas is well represented; the great majority of those invited from overseas were able to attend the meeting. Many of them successfully obtained their own travel funding, and we are grateful for the enthusiasm and help that have extended further the scope and interest of the book.

This book follows closely the sequence of papers given at the Symposium which was divided into four main topics: (1) genetical identity of stocks and influences on gene flow, (2) ranges and movements, (3) foraging behaviour, diet and reproduction and (4) physiology and bioenergetics. Pinnipeds have a significantly higher profile in the book than cetaceans. This partly reflects the balance of interest amongst marine mammalogists within Britain but it also reflects some differences between the ways in which the science of marine mammalogy has developed in pinnipeds and cetaceans during the last ten years. In general, scientists working on pinnipeds have enormous advantages over those working on cetaceans because their study animals are more accessible and, as a result, progress has been easier in pinniped research. One of the great challenges of the field is to apply to cetaceans many of the techniques which have been developed for pinnipeds.

Financial assistance and encouragement from the Zoological Society of London made this symposium possible. Additional financial assistance with the travel costs of overseas participants was provided by a grant from The Royal Society and the Carnegie Trust for the Universities of Scotland provided a grant for the costs of publishing the colour frontispiece. I am grateful to all the authors whose diligence and prompt submission of manuscripts has made my job as editor much easier than it might otherwise have been. I am also grateful to Miss Unity McDonnell for her assistance with editing the manuscripts and organizing the symposium.

Cambridge, September 1992   I. L. BOYD

Contents

Comparative levels of genetic variability in harbour seals and northern elephant seals as determined by genetic fingerprinting
NILES LEHMAN, ROBERT K. WAYNE & BRENT S. STEWART

Implications of DNA fingerprinting for mating systems and reproductive strategies of pinnipeds
D. J. BONESS, W. D. BOWEN & J. M. FRANCIS

Influence of rare ecological events on pinniped social structure and population dynamics
FRITZ TRILLMICH

Influence of maternal characteristics and environmental variation on reproduction in Antarctic fur seals
N. J. LUNN & I. L. BOYD

Social organization of humpback whales on a North Atlantic feeding ground

PHILLIP J. CLAPHAM

Part II Foraging ecology of marine mammals

Sex differences in diving and foraging behaviour of northern elephant seals

B. J. LE BOEUF, D. E. CROCKER, S. B. BLACKWELL, P. A. MORRIS & P. H. THORSON

Contents

Seasonal dispersion and habitat use of foraging northern elephant seals

BRENT S. STEWART & ROBERT L. DELONG

Studying the behaviour and movements of high Arctic belugas with satellite telemetry

A. R. MARTIN, T. G. SMITH & O. P. COX

Grey seals off the east coast of Britain: distribution and movements at sea
P. S. HAMMOND, B. J. McCONNELL & M. A. FEDAK

Harbour seal movement patterns
PAUL M. THOMPSON

Recent advances in diet analysis of marine mammals
G. J. PIERCE, P. R. BOYLE, J. WATT & M. GRISLEY

Contents

Milk secretion in marine mammals in relation to foraging: can milk fatty acids predict diet?
SARA J. IVERSON

The relationship between reproductive and foraging energetics and the evolution of the Pinnipedia
DANIEL P. COSTA

Part III Physiology and bioenergetics

To what extent can heart rate be used as an indicator of metabolic rate in free-living marine mammals?
P. J. BUTLER

Behavioural and physiological options in diving seals
M. A. FEDAK & D. THOMPSON

How fast should I swim? Behavioural implications of diving physiology

D. THOMPSON, A. R. HIBY & M. A. FEDAK

Role of plasma and tissue lipids in the energy metabolism of the harbour seal

R. W. DAVIS, W. F. BELTZ, F. PERALTA & J. L. WITZTUM

Balancing power and speed in bottlenose dolphins (*Tursiops truncatus*)

T. M. WILLIAMS, W. A. FRIEDL, J. E. HAUN & N. K. CHUN

Contributors

AMOS, W., Department of Genetics, University of Cambridge, Downing Street, Cambridge CB2 3EH, UK.

BELTZ, W. F., Department of Medicine, University of California, La Jolla, CA 92097, USA.

BLACKWELL, S. B., Department of Biology, University of California, Santa Cruz, CA 95064, USA.

BONESS, D. J., Department of Zoological Research, National Zoological Park, Smithsonian Institution, Washington, DC 20008, USA.

BOWEN, W. D., Marine Fish Division, Bedford Institute of Oceanography, Dartmouth, Nova Scotia, B2Y 4A2 Canada.

BOYD, I. L., British Antarctic Survey, Natural Environment Research Council, High Cross, Madingley Road, Cambridge CB3 0ET, UK.

BOYLE, P. R., Department of Zoology, University of Aberdeen, Tillydrone Avenue, Aberdeen AB9 2TN, UK.

BUTLER, P. J., School of Biological Sciences, University of Birmingham, Edgbaston, Birmingham B15 2TT, UK.

CHUN, N. K., Naval Ocean Systems Center, Kailua, HI 96734, USA.

CLAPHAM, P. J., Department of Zoology, University of Aberdeen, Tillydrone Avenue, Aberdeen AB9 2TN, UK; and Cetacean Research Program, Center for Coastal Studies, 59 Commercial Street, PO Box 1036, Provincetown, MA 02657, USA.

COSTA, D. P., Department of Biology and the Institute of Marine Science, University of California, Santa Cruz, CA 95064, USA.

COX, O. P., Sea Mammal Research Unit, Natural Environment Research Council, High Cross, Madingley Road, Cambridge CB3 0ET, UK.

CROCKER, D. E., Department of Biology, University of California, Santa Cruz, CA 95064, USA.

DAVIS, R. W., Department of Marine Biology, Texas A & M University, PO Box 1675, Galveston, TX 77553, USA.

DELONG, R. L., National Marine Mammal Laboratory, NMFS, NOAA, 7600 Sand Point Way, Seattle, Washington 92115, USA.

FEDAK, M. A., Sea Mammal Research Unit, Natural Environment Research Council, High Cross, Madingley Road, Cambridge CB3 0ET, UK.

FRANCIS, J. M., Department of Zoological Research, National Zoological Park, Smithsonian Institution, Washington, DC 20008, USA.

FRIEDL, W. A., Naval Ocean Systems Center, Kailua, HI 96734, USA.

GRISLEY, M., Department of Zoology, University of Aberdeen, Tillydrone Avenue, Aberdeen AB9 2TN, UK.

HAMMOND, P. S., Sea Mammal Research Unit, Natural Environment Research Council, High Cross, Madingley Road, Cambridge CB3 0ET, UK.

HAUN, J. E., Naval Ocean Systems Center, Kailua, HI 96734, USA.

HIBY, A. R., Sea Mammal Research Unit, Natural Environment Research Council, High Cross, Madingley Road, Cambridge CB3 0ET, UK.

HOELZEL, A. R., Department of Genetics, University of Cambridge, Cambridge CB2 3EH, UK: *present address* LVC, National Cancer Institute, Frederick, MD 21702, USA.

IVERSON, S. J., Canadian Institute of Fisheries Technology, Technical University of Nova Scotia, 1360 Barrington Street, PO Box 1000, Halifax, Nova Scotia, B3J 2X4 Canada.

LE BOEUF, B. J., Department of Biology, University of California, Santa Cruz, CA 95064, USA.

LEHMAN, N., Department of Molecular Biology, The Scripps Research Institute, 10666 North Torrey Pines Road, La Jolla, CA 92037, USA; *and* Department of Biology, University of California, Los Angeles, CA 90024–1606, USA.

LUNN, N. J., British Antarctic Survey, Natural Environment Research Council, High Cross, Madingley Road, Cambridge CB3 0ET, UK.

McCONNELL, B. J., Sea Mammal Research Unit, Natural Environment Research Council, High Cross, Madingley Road, Cambridge CB3 0ET, UK.

MARTIN, A. R., Sea Mammal Research Unit, Natural Environment Research Council, High Cross, Madingley Road, Cambridge CB3 0ET, UK.

MORRIS, P. A., Department of Biology, University of California, Santa Cruz, CA 95064, USA.

PERALTA, F., Department of Medicine, University of California, La Jolla, CA 92097, USA.

PIERCE, G. J., Department of Zoology, University of Aberdeen, Tillydrone Avenue, Aberdeen AB9 2TN, UK.

SMITH, T. G., Department of Fisheries & Oceans, Pacific Biological Station, 3190 Hammond Bay Road, Nanaimo, B.C. V9R 5K6, Canada.

STEWART, B. S., Hubbs-Sea World Research Institute, 1700 South Shores Road, San Diego, CA 92109, USA.

THOMPSON, D., Sea Mammal Research Unit, Natural Environment Research Council, High Cross, Madingley Road, Cambridge CB3 0ET, UK; *and* Department of Environmental and Evolutionary Biology, University of Liverpool, PO Box 147, Liverpool L69 3BX, UK.

THOMPSON, P. M., Department of Zoology, University of Aberdeen, Lighthouse Field Station, Cromarty IV11 8YJ, UK.

THORSON, P. H., Department of Biology, University of California, Santa Cruz, CA 95064, USA.

TRILLMICH, F., University of Bielefeld, Faculty of Biology, Behavioural Ecology, PO Box 10 01 31, D-4800 Bielefeld 1, FRG.

WATT, J., Department of Zoology, University of Aberdeen, Tillydrone Avenue, Aberdeen AB9 2TN, UK.

WAYNE, R. K., Department of Biology, University of California, Los Angeles, CA 90024–1606, USA; *and* Institute of Zoology, The Zoological Society of London, Regent's Park, London NW1 4RY, UK.

WILLIAMS, T. M., Naval Ocean Systems Center, Kailua, HI 96734, USA.

WITZTUM, J. L., Department of Medicine, University of California, La Jolla, CA 92097, USA.

Organizer of symposium

DR I. L. BOYD, British Antarctic Survey, Natural Environment Research Council, High Cross, Madingley Road, Cambridge CB3 0ET, UK.

Chairmen of sessions

PROFESSOR P. J. BUTLER, School of Biological Sciences, University of Birmingham, Edgbaston, Birmingham B15 2TT, UK.

DR J. P. CROXALL, British Antarctic Survey, Natural Environment Research Council, High Cross, Madingley Road, Cambridge CB3 0ET, UK.

DR R. W. DAVIS, Department of Marine Biology, Texas A & M University, PO Box 1675, Galveston, TX 77553, USA.

DR P. S. HAMMOND, Sea Mammal Research Unit, Natural Environment Research Council, High Cross, Madingley Road, Cambridge CB3 0ET, UK.

DR J. HARWOOD, Sea Mammal Research Unit, Natural Environment Research Council, High Cross, Madingley Road, Cambridge CB3 0ET, UK.

PROFESSOR B. J. LE BOEUF, Department of Biology, University of California, Santa Cruz, CA 95064, USA.

PROFESSOR P. A. RACEY, Department of Zoology, University of Aberdeen, Tillydrone Avenue, Aberdeen AB9 2TN, UK.

Symp. zool. Soc. Lond. (1993) No. 66: 1–12

Introduction: trends in marine mammal science

I. L. BOYD

British Antarctic Survey
Natural Environment Research Council
Madingley Road
Cambridge CB3 0ET, UK

In the past 20–30 years marine mammalogy has emerged as a discipline in its own right. The willingness of agencies to fund marine mammal research is evidence of its importance both in solving practical problems and in the more esoteric realms of pure science. To an outsider, the study of marine mammals might appear at first to be one of those narrow scientific specializations of little general interest. However, there is probably no other group of animals for which there is such specific legislation, enshrined in international treaties or national laws around the world, to regulate exploitation or conservation, and for which there are such polarized views of how we should manage populations. Often, marine mammals are the stalking horse for public opinion on the wide-ranging principles of environmental conservation. The purpose of this chapter is to examine the development of marine mammal science from its roots in the control of commercial exploitation of seals and whales to its role in providing information essential for the administration of national and international laws and as a subject which is providing new insights into several aspects of mammalian ecology and physiology.

A look back at the history of man's interactions with marine mammals illustrates why we require information about the biology of many species (Bonner 1982; Royal Commission on Seals and the Sealing Industry in Canada 1986). Marine mammals have been subject to subsistence hunting and then, following the industrial revolution, commercial exploitation. However, the time-scale over which industrialized economies sought a return on financial investments in whaling and sealing was out of step with the regeneration time of the marine mammal populations and, as a result, many species were severely depleted in numbers or driven close to extinction. The momentum of technological advance which led the way to the widespread over-exploitation of marine mammals is, today, helping to develop our understanding of these animals and making us more aware of

ZOOLOGICAL SYMPOSIUM No. 66
ISBN 0–19–854069–8

Copyright © 1993 The Zoological Society of London
All rights of reproduction in any form reserved

their vulnerability to pollution, exploitation or disease. However, we still know little about the biology of many species of marine mammals. In a recent report from the International Union for the Conservation of Nature (IUCN in press), lack of knowledge about some species was considered, amongst other things, to be an indirect threat to some of these species. This is acknowledgement that processes exist which could be having detrimental effects on marine mammals but, because there are insufficient data, we have no way of recognizing that any change is occurring.

The seas were once viewed as containing large, untapped and inexhaustible resources as well as having a capacity to absorb pollutants. This view has changed. No longer does anybody have the right to exploit marine mammals when and where they like, partly as a result of international agreements but mostly as a consequence of the unwritten laws of public opinion. Unfortunately, the change in public attitudes to whaling came too late to save many populations of baleen whales from becoming severely depleted, even though to the few scientists involved in whale research at the time it was obvious that those levels of exploitation were unsustainable. Inevitably, the Antarctic whaling industry collapsed in the early 1960s.

In the first two decades of this century, there was a realization that stocks of marine mammals were limited and that exploitation had to be regulated. There was concern as early as 1895 over dwindling stocks of harp seals (*Phoca groenlandica*) and legislation was enacted to reduce the impact of hunting. In 1911, the North Pacific Fur Seal Convention stopped pelagic sealing of northern fur seals (*Callorhinus ursinus*) in order to conserve stocks. In 1919, reduced catches of whales at South Georgia prompted action to reduce the impact of whaling and increase the efficiency with which carcasses were used. This decline in whale stocks at South Georgia was probably the starting point for modern marine mammal science when, as a result, the *Discovery* Investigations (1925–1951) were established to examine the biology of whale stocks in the Southern Ocean. In the 1920s ecology was emerging as an individual discipline and this resulted in the *Discovery* Investigations being founded on sound ecological principles. Not only was the biology of the whales themselves examined, but also their food supply and its distribution and abundance in relation to oceanographic conditions. This research was funded by licence fees levied by the British Government on the whaling industry, mainly at South Georgia. Ironically, the *Discovery* Investigations did nothing to extend the life of the whaling industry or to conserve the whale stocks in the South Georgia region, probably because of the competitive nature of the whaling industry and the economics of whaling. Gross over-exploitation continued unabated and, eventually, the focus of the industry's attention switched to pelagic whaling in international waters.

In the 1950s, the theme of the *Discovery* Investigations was continued by the Falkland Islands Dependencies Survey (later British Antarctic Survey) when it established a programme of research on southern elephant seals at South Georgia under R. M. Laws. At that time elephant seals were being exploited for their oil. Later Laws was to serve as a whaling inspector and scientist aboard a whale factory ship during the final years of the pelagic whaling industry and his studies of reproduction and growth remain as classics. He was also the first scientist to propose a management strategy for a marine mammal population, based on the data he had collected and on the principle of maximum sustainable yield. This was implemented in 1952 to manage the sealing industry at South Georgia, and operated successfully until sealing ended in 1964.

At around the same time as Laws was conducting his pioneering work on southern elephant seals, similar work was beginning under the auspices of the North Pacific Fur Seal Convention to collect biological information on northern fur seals breeding in the Pribilof and Aleutian Islands, and a similar approach was being taken for the north-west Atlantic harp seals by the Canadian Government. Again, the aim was to obtain accurate data about the biology of these seals and then use this to establish an effective management policy for sustainable exploitation.

In parallel with these studies, which basically concentrated on population ecology, there was a growing interest in the unusual anatomy and physiology of marine mammals (Ridgway 1972; Harrison 1972–1977). There was considerable interest in the cognitive capacity of many species (Andersen 1969), partly because of the use of marine mammals as show animals and partly for military reasons (trained seals or dolphins performed tasks on command).

Other major processes which contributed to the birth of marine mammalogy were the establishment of the International Whaling Commission's (IWC) Scientific Committee in 1946 to provide advice to the Commission about the status of whale populations and potential sustainable yields. This committee required data and scientific personnel with a knowledge of both the data and the general biology of the species concerned and it has been largely responsible for the development of the mathematics of marine mammal population dynamics. A more recent development has been the introduction in the United States of legislation specifically to protect marine mammals. This move was, in itself, not new because other nations had enacted legislation to protect marine mammals, but the underlying philosophy of the more recent US legislation was different. Previously, marine mammals had been viewed at best as mysterious animals inhabiting an inhospitable environment and at worst as pests to be eradicated for denying fishermen their livelihood. The US Marine Mammal Protection Act of 1974 recognized the potential for modern methods of fishing to affect marine mammal populations and for all marine industries to accidentally kill marine mammals, through entanglement in nets and other debris, destruction of habitat, disturbance of behaviour patterns or through various forms of pollution. Potential impacts of industrial developments on marine mammals therefore needed to be assessed before these developments took place. In addition, it recognized the need for better information on the biology, ecology and management of marine mammals and, consequently, assigned research responsibilities and funding to various Federal agencies to obtain such information. Between 1976 and 1990, Federal expenditure on marine mammal research increased from $7.7 million in 1976 to a peak of $21.5 million in 1989 (Waring 1991). This supported from 113 (1976) to 177 (1987) research projects on marine mammals.

These projects, together with other statutory bodies with more restricted ranges of influence in various parts of the world, have grown out of a sense of the vulnerability of marine mammals to anthropogenic changes in the marine environment. This includes the vulnerability to disease of populations already depleted through exploitation or sensitized through the effects of pollution. As top predators in marine food chains, seals and whales provide a complex signal, in terms of dispersion, abundance, reproduction and survival, describing the state of the environment they inhabit. This is an integration of the dynamics of marine biogeochemistry, primary production and several layers of secondary production. Marine mammals probably make a trivial contribution to the dynamics of marine systems in terms of their impact on wide-scale energy or carbon flux (although see Huntley, Lopez & Karl 1991), but the effect which these dynamics can have on them is far from trivial and, given sufficient insight, may contain much useful information about long-term trends and acute changes occurring within ecosystems. This is illustrated by the effects of two natural events: that of El Niño on marine mammal populations in the Pacific (Trillmich this volume) and that of periodic reductions in food supply on Antarctic fur seal reproduction and growth (Lunn & Boyd this volume).

Our ability to interpret many of these changes in marine mammal populations or to manage the populations for 'sustainable use' (as defined by IUCN) is founded on knowledge of the fundamental biology of species. For example, it is difficult to enact a management strategy on a population which has been poorly defined, and recently developed methods of determining the genetic relationships between populations, using DNA analysis, have been applied in this context (Hoelzel this volume; Lehman, Wayne & Stewart this volume). In terrestrial mammals, population segregation can often be easily inferred because of geographical isolation; but such inferences are problematical in the case of many marine mammals because geographical isolation is often less obvious and isolating mechanisms are poorly understood. DNA analyses have also allowed us to look back at the effects which past heavy exploitation of marine mammals has had on the genetic diversity of modern populations, although we still understand little about the consequences of genetic depletion, caused by such exploitation, for the future resilience of the population to disease or other unforeseen stress.

It is often the case that observations which can be made routinely and with ease on terrestrial mammals pose difficult technical problems when attempted on marine mammals, mainly because it is difficult to track and observe the behaviour of individuals. It has long been recognized that many cetaceans have a complex social structure with stable groups of individuals or migratory patterns which involve different behavioural interactions at different stages of the migratory process. Clapham (this volume) provides information about social interactions and foraging economics in a large species of baleen whale, the humpback (*Megaptera novaeangliae*), accumulated from many years of observation and the identification of individuals from coloration patterns on tail flukes. Photo-identification is becoming a highly developed method for following individuals over several years and long distances and this study illustrates the large-scale effort which must be invested consistently over many years to achieve significant advances in our knowledge of the behaviour of these animals. Such an approach is more difficult for cetaceans which range more widely or which cannot be identified individually and Amos (this volume) has demonstrated how it is possible to derive information about the social structure of pilot whale (*Globicephala melas*) pods from their genetic composition. In contrast, social behaviour in pinnipeds is well studied (Bartholomew 1970; Renout 1991), at least during the reproductive phase of the annual cycle when most of these species are ashore or on ice, and has provided textbook examples of, for example, extreme forms of defence polygyny. The opportunity now exists to advance this potentially fertile field of pinniped research, using DNA fingerprinting to identify parentage and to measure the genetic basis for the evolution of different mating strategies (Boness, Bowen & Francis this volume).

The most significant advance in marine mammalogy in recent years has probably been the move towards studying animals under wild, unrestrained conditions when they are at sea. Studying cetaceans in this way has always been a problem although, as demonstrated by Williams *et al.* (this volume), the capacity of dolphins to be trained to perform complex tasks can be used to study their physiology and behaviour. In the case of pinnipeds, there are advantages in having access to individuals while they are ashore and the focus has changed towards studying their free-ranging behaviour and physiology. This change has largely resulted from the application of micro-electronics to the problem of collecting and storing information remotely from the observer. In its simplest form, a device contains a microprocessor controller which samples a pre-calibrated pressure transducer (or some other form of transducer, e.g. measurement of heart rate) at a pre-determined interval. The readings are then stored in a memory until the device is retrieved from the animal and the data can then be recovered by the researcher. Many of the initial technical problems have now been overcome; these included designing a watertight housing and achieving low power consumption to reduce battery size while still maintaining potential for long-term deployments. These devices have given us a fundamentally new perspective on the foraging capabilities of almost all the pinniped species on which they have been deployed; ten years ago few people would have predicted the diving capabilities of these animals and the distances over which some of them migrate. More significant constraints are presently limiting the usefulness of these devices for deployment on cetaceans. One of these is attachment of the device, since it is still necessary to capture animals to do this, and another is that many designs of device require to be recovered in order to recover the data and this necessitates having to recapture the animal. Both of these constraints remain to be overcome completely and they have held back the development of behavioural studies of cetaceans compared with pinnipeds. However, Martin & Smith (this volume) have carried out the first major tracking and diving study of a small cetacean using instruments which transmit data to satellites in polar orbits and these devices also provide an estimate of the position of the animals at the time of transmission. It is only a matter of time before similar studies are carried out on other cetaceans.

Perhaps, however, the consummate example of how this technology has revolutionized our view of the behaviour of marine mammals comes from the elephant seals. As shown by Le Boeuf et al. (this volume) and Stewart & DeLong (this volume), northern elephant seals range widely over the North Pacific Basin with apparent segregation of the main foraging areas between males and females. There is, however, disagreement over the foraging strategies of male elephant seals. Stewart & DeLong found little evidence that males dive consistently to the bottom to feed while Le Boeuf et al. found evidence that they do. This discrepancy is interesting; it may be a genuine difference in foraging between males originating from the two colonies in which each research group is working, or there may be different biases in the two sets of measurements.

Similar data now exist for southern elephant seals, largely resulting from the studies at Macquarie Island carried out by Hindell and others (Hindell, Slip & Burton 1991) and at South Georgia by the Sea Mammal Research Unit in collaboration with the Marine Life Sciences Division of the British Antarctic Survey. Much of this work is now carried out using satellite-linked dive recorders and an example of a three-dimensional image of a southern elephant seal track, obtained from such a recorder, is shown in the frontispiece of this book. An intriguing feature of elephant seal foraging patterns, which is also emerging for grey seals (Hammond, McConnell & Fedak this volume) and harbour seals (P. M. Thompson this volume), is the apparent individual variation of ranging behaviour. Some individual grey seals appear to have favoured localities where feeding is likely to be taking place and a similar pattern may be emerging for southern elephant seals. The question is how does this behaviour develop, especially for an elephant seal which may have to locate foraging patches many hundreds of kilometres from its breeding or moulting site? Do these animals know, in a geographical sense, the location of rich feeding grounds or do they locate food by association with particular oceanographic features such as oceanic frontal zones? Stewart & DeLong (this volume) suggest that female northern elephant seals forage in a different water mass from males and Le Boeuf et al. (this volume) suggest that this difference may reflect the different nutritional requirements of males and females. Le Boeuf et al. also suggest that this sexual segregation develops from an early age. To what extent is this a selective process, resulting from those which do not segregate having low survival? How much of it is instinctive and how much learned? Survival rates are probably lowest in pinnipeds in their first year and it seems probable that this is due largely to many newly weaned juveniles never learning to feed effectively. Perhaps locating a rich patch of food early on in life, which is then returned to in successive years, is a prerequisite for survival, or specific feeding specializations may develop early in life when foraging tactics become imprinted behaviour. It may be that specialization on particular types of prey or intimate knowledge of a particular geographical region are alternative strategies followed by different individuals depending on their experiences early in life.

These distinctions are important because they have implications for how individuals will respond to changes in fisheries management. It is also a significant challenge both to document the process of dietary development and foraging specializations from long-term studies of individuals, and to measure the individual specializations in diet which may exist within a population. New or more sophisticated methods of diet analysis may be required before progress can be made. Serology, to identify proteins in the gut or faeces of marine mammals from specific types of prey, or multivariate statistical methods for identifying prey hard parts from gut or faecal samples (Pierce & Boyle this volume) have an important place in the measurement of the breadth of the diet in populations, but methods of identifying marker fatty acids, specific to different types of prey (Iverson this volume), perhaps have more future for measuring diet specialization in individuals. This method works on the principle that fatty acids synthesized by the prey, which are characteristic of that type of prey in terms of chain length, degree of saturation and branching, are sequestered unaltered into the fat reserves of the seal or whale. Thus, by taking a biopsy of fat it is possible to say whether or not the seal or whale has fed on that particular prey type. Future studies will help to develop this method so that it may be possible to quantify the relative contribution of each prey item to the diet. The method also has the advantages that it can integrate the diet of an individual over a relatively long period and that measurements can be obtained from large numbers of individuals with relative ease.

The recent discoveries about the behaviour of pinnipeds when free-ranging at sea have raised a number of other interesting questions for physiologists. Through the efforts of Kooyman and co-workers, the Weddell seal continues to provide some of the best opportunities for studying natural diving in a marine mammal. However, the greatest challenge to conventional thinking has come from elephant seals and numerous studies have now demonstrated the amazing capacity of these animals to live at great depths in the oceans and to remain submerged for periods which could not have been predicted before the measurements were made (e.g. Le Boeuf *et al.* 1988; Boyd & Arnbom 1991; Hindell *et al.* 1991; McConnell, Chambers & Fedak 1992; Le Boeuf *et al.* this volume; Stewart & DeLong this volume). In brief, when at sea adult elephant seals spend 85–90% of their time submerged. Each dive they make lasts 15 to 45 min (although several dives of 1 h or more have been recorded and one has been recorded to last 2 h). Most dives take them to depths of 250–450 m but dives regularly exceed 1000 m and one of over 1500 m has been recorded. Seals normally spend 1–2 min at the surface between dives. The aerobic dive limit (ADL), that is the maximum time an animal can remain submerged without resorting to anaerobic metabolism (Kooyman 1989), is 18–25 min for most average-sized female elephant seals. This is calculated from the estimated oxygen storage capacity of the tissues and the estimated metabolic rate. This means that on something like 50% of dives elephant seals exceed their predicted ADL. This has important implications for how we view metabolism in these animals because, since anaerobic metabolism is one-eighteenth as efficient (in terms of molecules of ATP produced per molecule of glucose metabolized) as aerobic metabolism, then elephant seals would appear to be behaving very inefficiently. Most people would agree that this scenario is very improbable, especially since other species of divers also regularly exceed their ADLs. Birds provide an even more extreme case, especially the penguins of the genus *Aptenodytes* (Kooyman *et al.* 1992). Most diving physiologists would now accept that simple calculations of ADLs are wrong because we do not understand the biochemical pathways used by these animals to maximize the efficiency of oxygen utilization, the nature and size of the oxygen stores in the tissues or how best to measure metabolic rate.

The unique feature of elephant seals is the continuous nature of their diving with consistently long dive durations. Nobody doubts that elephant seals abide by the laws of physics, but it is unclear how they do this. Fedak & Thompson (this volume) explore some ways in which these seals could achieve an oxygen balance. A favoured theory is that seals experience some form of hypometabolism when they dive. Measurements of heart rate would appear to confirm that there is at least reduced circulation of the blood during diving (Kooyman 1989; Thompson & Fedak 1993; Butler this volume) but a normal metabolic rate could be maintained by oxygen stored within the tissues. If hypometabolism does occur, then how is it controlled, what are the costs and why can a female elephant seal, after eight months of being in a state of almost perpetual hypometabolism, produce a precocial pup which, as a foetus, had grown at a rate at least as great as in terrestrial mammals? Logic suggests there must be a cost associated with hypo-metabolism otherwise it would be life on the cheap and all mammals would be doing it. A problem is that most measurements of metabolism are averaged over many dive cycles, so, if seals were to enter a state of hypermetabolism when they surface (when oxygen consumption measurements are made), we would not be able to measure hypometabolism during the dive. Our measurements of metabolism are also too indirect. At best they are a measure of the oxygen used by the whole animal over a dive. What is needed is a measure of individual tissues within a dive. The cost associated with hypometabolism could be induction of sluggishness as in torpid bats, although there appears to be no evidence for this from records of swim velocity, or that, in order to induce hypometabolism, heat has to be lost from the body and then regenerated on return to the surface. Studies of the partitioning of oxygen and fuel reserves, in particular lipids which form the main metabolic fuel for diving marine mammals, at the tissue level (Davis et al. this volume), may provide an important key to understanding the diving metabolism of marine mammals.

Perhaps the most significant implication of studies of diving metabolism in marine mammals is that our understanding of mammalian metabolic physiology is flawed. Marine mammals, despite their incredible feats of endurance, are an extreme modification of the physiological continuum of which man is a part and we should be wary of corralling marine mammals as a special case with little relevance in a wider context. In the longer term, marine mammals may provide revelations about our own physiological capabilities.

However spectacular the physiological performance of many marine mammals appears to be, their physiology still limits their ability to search for and capture prey because they are always constrained by the necessity to return to the surface to breathe. Williams et al. (this volume) have demonstrated how bottlenose dolphins (*Tursiops truncatus*) maximize submergence time, and therefore potential foraging time, through cost-efficient travel. A similar theme has been pursued by D. Thompson, Hiby & Fedak (this volume) who have examined theoretically the economics of specific foraging tactics for marine mammals. One interesting conclusion of this theoretical approach is that, under certain circumstances when prey densities and swimming speeds are greater than a certain threshold, it is more economical for a marine mammal to sit and wait for prey in a particular part of the water column than actively to pursue prey. An important task is now to test these models through the observation of foraging behaviour in relation to prey abundance.

The wider implications of studies of metabolic rate are that it then becomes possible to build a picture of the energy budgets of individuals at various times of year and stages in their life cycles. In the past, doubly labelled water has been used to estimate metabolic rates of pinnipeds at sea (e.g. by Costa, Croxall & Duck 1989) and there have also been numerous studies which have used single-labelled isotopes during fasting. The main problems with measuring metabolism in this way are, first, that this method is restricted to animals which can be recaptured reliably within a certain time after the start of the experiment and, second, the cost. Also, the measure of metabolic rate obtained by this method is an average over several days. Considered together, these are severe restrictions which cannot be easily overcome. However, as argued by Butler (this volume), it should be possible to use heart rate as an indirect measure of metabolic rate when averaged over several dive cycles and recent work has tended to confirm this (Butler *et al.* 1992). The advantages of this method are obvious since it allows the metabolic cost of particular activities to be measured. The only restriction on the period over which measurements are made is the memory capacity of the heart rate recorder (Woakes 1992), and the device is sufficiently inexpensive to allow large numbers to be deployed. However, broad acceptance of this method will only come with its development for a wide range of species, including cetaceans.

As our knowledge of the energetics of marine mammals, in particular the pinnipeds, has developed, patterns have begun to emerge which begin to place each taxonomic group within its ecological context, providing important clues as to why there has been a divergence in pinniped evolution between the otariids (eared seals) and the phocids (earless seals). In his latest of a series of papers on this subject, Costa (this volume) has placed a large number of the measured traits of pinnipeds, from body size, metabolic rate and oxygen storage capacity to dive duration, fasting duration and duration of lactation, into a wide-ranging analysis of the economics of being a predator in the marine environment. This has suggested how particular constraints have led to adoption of divergent predation strategies by the two main taxonomic groups of pinnipeds. It appears that phocids adopt an economical life-style which is geared to exploitation of patchy, dispersed, distant prey resources while the life-style of the otariids is expensive, relying on rich and predictable prey resources, which are often associated with upwelling regions of the oceans (see also Trillmich this volume). Although we can see parallels with this amongst the cetaceans, for example the large baleen versus the small odontocete whales, the information base for the various species is much less well developed.

Broad overviews of this type are important because they place much of the current knowledge within a descriptive model and, therefore, provide a context within which other research can be developed. Marine mammal science arose as an aid to management of marine mammal populations for sustainable exploitation and this provided much of the platform for scientists with interests beyond the immediate applied goals to help the subject to become a discipline in its own right. In recognition of this, the Society for Marine Mammalogy was founded in 1981. This now has over 1200 members and produces a scientific journal. The emphasis has changed from studying marine mammals in order to control exploitation to enhancing our knowledge of their biology and their environment, promoting their conservation, and establishing their role in marine food chains including their interactions with commercial fisheries.

The future is bright and challenging. High on the list of priorities must be the application to cetaceans of technology developed for studies of pinnipeds. The technological advances, in both micro-electronics and biochemistry, provide unparalleled opportunities to examine the interactions of populations and the genetical and social role of individuals within those populations. Together with long-term studies, which are an essential part of any programme of study of long-lived vertebrates, these techniques will help to define how and why marine mammals are dispersed and forage in the way we are now beginning to observe and what consequences different foraging strategies have for the survival and reproductive success of individuals.

References

Andersen, H. T. (Ed.) (1969). *The biology of marine mammals*. Academic Press, New York & London.

Bartholomew, G. A. (1970). A model for the evolution of pinniped polygyny. *Evolution, Lancaster Pa* 24: 546–559.

Bonner, W. N. (1982). *Seals and man: a study of interactions*. University of Washington Press, Seattle & London.

Boyd, I. L. & Arnbom, T. (1991). Diving behaviour in relation to water temperature in the southern elephant seal: foraging implications. *Polar Biol.* 11: 259–266.

Butler, P. J., Woakes, A. J., Boyd, I. L. & Kanatous, S. (1992). Relationship between heart rate and oxygen consumption during steady-state swimming in California sea lions. *J. exp. Biol.* 170: 35–42.

Costa, D. P., Croxall, J. P. & Duck, C. D. (1989). Foraging energetics of Antarctic fur seals in relation to changes in prey availability. *Ecology* 70: 596–606.

Harrison, R. J. (Ed.) (1972–1977). *Functional anatomy of marine mammals* 1–3. Academic Press, London.

Hindell, M. A., Slip, D. J. & Burton, H. R. (1991). The diving behaviour of adult male and female southern elephant seals, *Mirounga leonina* (Pinnipedia: Phocidae). *Aust. J. Zool.* 39: 595–619.

Huntley, M. E., Lopez, M. D. G. & Karl, D. M. (1991). Top predators in the Southern Ocean: a major leak in the biological carbon pump. *Science* 253: 64–66.

IUCN (In press). *Seals, fur seals, sea lions and walruses. Status of pinnipeds and conservation action plan.* International Union for the Conservation of Nature, Gland, Switzerland.

Kooyman, G. L. (1989). *Diverse divers: physiology and behaviour.* Springer-Verlag, Berlin & London. (*Zoophysiology* 23.)

Kooyman, G. L., Cherel, Y., LeMaho, Y., Croxall, J. P., Thorson, P. H., Ridoux, V. & Kooyman, C. A. (1992). Diving behaviour and energetics during foraging cycles in king penguins. *Ecol. Monogr.* 62: 143–163.

Le Boeuf, B. J., Costa, D. P., Huntley, A. C. & Feldkamp, S. D. (1988). Continuous, deep diving in female northern elephant seals, *Mirounga angustirostris. Can. J. Zool.* 66: 446–458.

McConnell, B. J., Chambers, C. & Fedak, M. A. (1992). Foraging ecology of southern elephant seals in relation to the bathymetry and productivity of the Southern Ocean. *Antarct. Sci.* 4: 393–398.

Renouf, D. (Ed.) (1991). *The behaviour of pinnipeds.* Chapman & Hall, London, New York etc.

Ridgway, S. H. (Ed.) (1972). *Mammals of the sea: biology and medicine.* Charles C. Thomas Publishers, Springfield, Illinois.

Royal Commission on Seals and the Sealing Industry in Canada (1986). *Seals and sealing in Canada. Report of the Royal Commission* 2. Ministry of Supply and Services, Ottawa.

Thompson, D. & Fedak, M. A. (1993). Cardiac responses of grey seals during diving at sea. *J. exp. Biol.* 174: 139–164.

Waring, G. H. (1991). *Survey of federally-funded marine mammal research and studies FY74–FY90. Final report to U.S. Marine Mammal Commission in fulfilment of Contract T-75133766.* U.S. Marine Mammal Commission, Washington D.C. 20009.

Woakes, A. J. (1992). A implantable data logging system for heart rate and body temperature. In *Wildlife telemetry: remote monitoring and tracking of animals:* 120–127. (Eds Priede, I. G. & Swift, S. M.). Ellis Horwood Ltd., New York etc.

Genetic identity of stocks and influences on gene flow

Symp. zool. Soc. Lond. (1993) No. 66: 15–32

Genetic ecology of marine mammals

A. R. HOELZEL[1]

Department of Genetics
University of Cambridge
Cambridge CB2 3EH, UK

Synopsis

Recent advances in the field of molecular genetics have facilitated the application of genetic techniques to the analysis of population structure. This paper presents a brief review of the results from studies of genetic variation within and between marine mammal populations. There are at least three categories of applications: (1) paternity testing and kinship analysis to assess aspects of behaviour that limit variation or introduce genetic structure into local populations; (2) the estimation of effective population size, and the interpretation of historical events (such as population bottlenecks) through the analysis of present-day molecular variation; and (3) the assessment of genetic distance between populations. The identification of genetically differentiated populations, and the processes that limit or generate patterns of variation, will be important to the effective conservation of genetic diversity in these species.

Introduction

Genetic ecology is the application of molecular genetic techniques to problems in population biology and behavioural ecology. Since the 1960s and the advent of gel electrophoresis the analysis of proteins has provided a wealth of information on the level of genetic variation within and between populations. The sequence of amino acids in a protein is encoded by the sequence of nucleic acids in genes. Some amino acids are charged, and their configuration in the protein affects the migration of the molecule through an electrical field. Therefore, a small proportion of changes in the DNA can be detected by this kind of analysis. The resolution is low, however, because only coding regions of DNA are represented (and these evolve relatively slowly), and only a small proportion of the changes in those regions will cause a detectable change in the mobility of the protein.

DNA changes over time in a finite number of ways. These include the mutation of one of the four bases into another, the deletion of one or more

---

[1] Current address: LVC, National Cancer Institute, Frederick, MD 21702, USA

ZOOLOGICAL SYMPOSIUM No. 66
ISBN 0–19–854069–8

Copyright © 1993 The Zoological Society of London
All rights of reproduction in any form reserved

bases, the insertion of one or more bases, and the repetition of a segment of DNA. These changes can now be quantified through the direct analysis of DNA, which both increases resolution compared to protein analysis and permits the analysis of markers whose properties best fit a particular application. These analyses can provide information on the genetic structure of populations, given certain assumptions about the rate of change and the mechanisms by which variants spread through a population. There are at least three levels of analysis. (1) If genetic markers in a sexual population segregate independently according to Mendelian expectations, then the pattern of alleles in one parent and offspring can in some cases identify the other parent. Using a highly variable marker, such as minisatellite loci, it is possible to exclude the possibility of an alternative parent with a very small chance of error. (2) The level of variation in a given population is dependent on the effective population size. Therefore, measuring variance should allow the estimation of effective population size, and can in some cases be used to investigate fluctuations in population size over the evolutionary history of the species. (3) If the rate of change is constant over the relevant time frame, then the amount of change can be used to estimate the time over which two sequences have been diverging. This gives a relative indication of the 'genetic distance' between individuals within a local population, in comparison with that between populations.

This short review will describe the methodology used in some of the relevant DNA analyses, give an overview of the published applications involving marine mammal populations, and discuss the potential of this approach and the implications of the results to date. Marine mammals are a diverse group of species all sharing in common a habitat that makes life difficult for field researchers. For this reason at least, a new method of analysis is welcome. However, genetic ecology also promises a great deal more. Early results suggest some surprising answers to previously intractable questions.

Methodology

The first step is to collect tissue or blood samples from animals in the field for analysis. Pinnipeds can often be approached on rookeries and either sedated so that blood can be collected, or sampled for a small clip of skin from the rear flippers. The latter can be obtained by using a pair of ear-clippers designed for marking cattle. A quarter-inch-square sample is generally ample for DNA analyses.

Sampling from cetaceans in the wild is more of a challenge. Two principal methods have been used. One, live capture of dolphins (Duffield & Wells 1991), is limited in application to those species that can practicably be manipulated in this way. The other, remote biopsy sampling (Winn,

Bischoff & Taruski 1973), is more generally applicable. This entails firing a small (approximately 1 cm × 3 cm) cylindrical dart, mounted at the head of an arrow, into the body of the whale. A number of large-scale projects sampling a variety of large and small cetacean species have demonstrated the feasibility of this method (see Hoelzel 1991a). Although the subjects sometimes respond to being sampled, there is no indication that long-term behaviour is affected, and strong reactions are rare (Brown, Kraus & Gaskin 1991; Weinrich *et al.* 1991). The tissue sample collected with a biopsy dart can then be stored at ambient temperature for an extended period of time (months) in a salt/DMSO preservative (Amos & Hoelzel 1991). The dermal portion of the sample can also be preserved in the field for subsequent tissue culture (Lambertsen 1987).

The next step is to extract DNA from the tissue or blood sample. This is usually accomplished in three steps. First the tissue is broken down and the cells lysed in a buffer which includes detergents and the protolytic enzyme, proteinase k. Then the DNA is separated from protein and other contaminants by mixing with phenol and chloroform. Water and phenol are immiscible, and the protein enters the phenol phase, while the DNA remains in the aqueous phase. Finally the DNA is precipitated out of solution with salt and ethanol, and re-dissolved in buffered water.

At this stage, the DNA can be digested with 'restriction enzymes' which recognize specific sequences, usually four or six base-pairs long. The enzyme cuts the DNA at or near the recognition sequence. A given enzyme will always produce the same set of fragments from a given sequence of DNA. Changes can be detected when mutation eliminates old or creates new restriction sites. These are referred to as restriction fragment length polymorphisms (RFLPs). In order to visualize the patterns, the digested DNA must be separated by size in a gel matrix. This is done by casting a gel of agarose or polyacrylamide, and running the DNA through the gel under buffer in an electric current (gel electrophoresis). The DNA migrates in the anodal direction (from − to +). Size separation depends on the concentration of the gel medium, but is roughly exponential with large fragments migrating disproportionately slowly.

DNA in the gel can then be transferred to a membrane by the southern blotting method (Southern 1975). The DNA in the gel is denatured (separated into single strands), and drawn up with the buffer through the gel and onto the membrane by capillary action, after which the DNA is fixed to the membrane with heat or UV light. Single-stranded DNA bound to the membrane in this way can then be 'probed' with the region of DNA under investigation (Fig. 1). For example, a clone of an *Adh* gene could be radioactively labelled by one of several methods, denatured, and annealed in solution to the DNA bound to the membrane. The result is a membrane that is selectively radioactive. When exposed to X-ray film, fragments of the

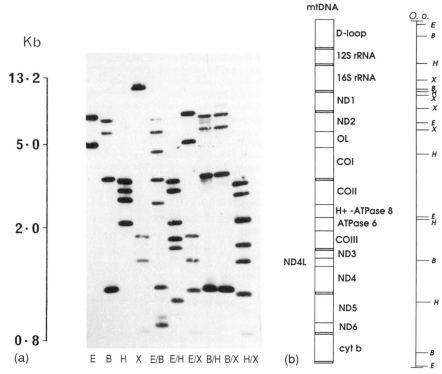

Fig. 1. (a) Killer whale (*Orcinus orca*) whole cell DNA was probed with the mtDNA genome of the dolphin (*Cephalorhynchus commersonii*). This shows a series of digests and double digests with *Eco*RI (E), *Bam*HI (B), *Hind*III (H) and *Xho*I (X). (b) A map of the killer whale (*O.o.*) mitochondrial genome derived from the series of digests shown in (a), compared with a map of the organization of the mammalian mitochondrial genome. The circular genomes are shown as linear to facilitate the comparison. (After Hoelzel 1989.)

sequence that comprise the *Adh* gene will create dark bands on the film. RFLP variation between individuals for this gene would be seen as the varying positions of the bands.

An alternative method of analysis is to amplify a specific sequence of DNA. This can be done by the method known as the polymerase chain reaction (PCR). It is necessary to know the sequence of DNA that flanks the sequence of interest (often achieved by comparing sequences from related species). Short (15–30 bp) sequences of DNA (primers) are then designed to anneal to one strand in a forward direction, and to the other strand in the reverse direction. PCR proceeds as follows in a reaction mixture that includes buffer, nucleotides and a heat-stable polymerase (which is the key to the automated amplification of DNA). First the template DNA is denatured at a high temperature (around 95 °C). Then the temperature is

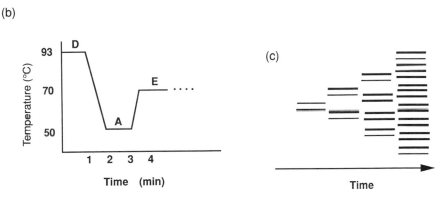

Fig. 2. The polymerase chain reaction (PCR). (a) Primers anneal to denatured template DNA. (b) The reaction is cycled (usually 20–40 times) between denaturing (D), annealing (A) and extension (E) temperatures. (c) Repeated cycling results in a roughly exponential amplification of the target sequence.

reduced to a level appropriate for annealing the primers to the template (usually between 45 °C and 65 °C). Finally the temperature is brought to an optimal level for the polymerase (70 °C to 75 °C) and the cycle is repeated 20–40 times (Fig. 2). This typically produces enough of a specific sequence of DNA to permit 5–10 RFLP analyses, cloning or direct sequencing. The latter is a particularly informative method for population studies (Hoelzel & Green 1992).

Population studies

The characteristics of change in specific genomic regions can be exploited to increase the efficiency of these applications, and fitting protocols to particular questions becomes increasingly practicable as the technology develops. For example, highly variable regions can be used to discern close relationships such as paternity, while the matrilineally inherited mito-chondrial genome is sensitive to the dispersal patterns of females and to fluctuations in population size. In general, an appropriate marker can be found for each hierarchical level of analysis. Repetitive regions such as minisatellites and microsatellites are most appropriate for comparing close relatives and for testing paternity, because they display a very high rate of change. Microsatellites are generally less variable than minisatellites and are appropriate for comparing subpopulation structure, though care must be taken in the interpretation of these patterns (see below). Population level differentiation can be compared using mtDNA, which evolves one or two orders of magnitude faster than the nuclear genomic regions that encode proteins. This is fast enough to give a high level of resolution, but not so fast as to saturate variation between populations (as would be the case for most minisatellite probes). Relatively variable nuclear regions, such as the MHC loci, are also used to compare populations, although considerations related to natural selection will affect the interpretation of variation in some of these markers (and in the MHC genes in particular). Other characteristics that make a specific marker best suited to a given application are related to the kind of change that is being investigated (especially single base-pair mutations as against variation in the number of repeats in an array), the mode of transmission and the way that a region of DNA is spread through a population. These factors will be discussed in more detail in the following sections.

Paternity testing and kinship

The technique of 'DNA fingerprinting' (Jeffreys, Wilson & Thein 1985) is now well established as a method for testing paternity. This method investigates variation in minisatellite loci. Minisatellite DNA regions are repetitive and highly variable in length, but generally inherited according to Mendelian expectations. The core of the repeated sequence is conserved between loci, so a single probe can pick up numerous loci. This produces a ladder of bands, most of which are highly polymorphic in the population. In theory it should be possible to test for kinship by comparing bandsharing between multilocus DNA fingerprints; however, this is not possible in practice beyond the identification of first- or second-order relatives (Lynch 1988; see below).

An individual should derive half of its bands from its mother, and half from its father. In practice a number of bands will often be shared by both parents and the offspring. Paternity is determined by matching bands in the offspring, not present in the mother, with bands in the father (Fig. 3). This process is simpler when single minisatellite or microsatellite loci are analysed. In this case each individual has only one (homozygote) or two (heterozygote) bands, and the level of polymorphism can be as high as 99% for minisatellite loci (Wong et al. 1986). In all cases it is preferable to test that the inheritance of alleles follows Mendelian expectations, and that there is no evidence for linkage effects (see Bruford et al. 1992).

To date, relatively few studies have used these techniques to analyse marine mammal behaviour and population structure. The potential is considerable, but in some cases so are the logistic difficulties. To test paternity it is usually necessary to collect samples from both mother and

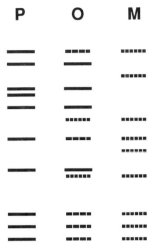

Fig. 3. Schematic representation of paternity testing by DNA fingerprinting. Bands in the offspring (O) were derived from either the mother (M), father (P) or both.

offspring, and the set of potential sires. However, even the identification of potential sires requires the long-term recognition of individuals, and observations of associations between the female and potential mates at the time of conception. One cetacean study where the field observational data are this detailed is the bottlenose dolphin (*Tursiops truncatus*) study on the Gulf coast of Florida (Duffield & Wells 1991). Preliminary results from this study suggest that males moving through the home ranges of matrifocal groups achieve matings during temporary associations. Exclusive male access to females or groups within the community was not observed.

In a study investigating paternity in pods of long-finned pilot whales (*Globicephala melas*) captured in the Faroese drive fishery, Amos, Barrett & Dover (1991) found within pod paternities for only four out of 34 paternity tests. It was assumed that in most cases the entire pod had been captured. Paternal exclusions were achieved using the multi-locus DNA fingerprint probe, 33.15 (Jeffreys *et al.* 1985). A pair of bands apparently representing a single locus within the multi-locus pattern was used to assess the probability that mating was random. On the basis of the bias of some rare paternal alleles inferred from female/foetus pairs, Amos *et al.* (1991) concluded that males often successfully mate with two or more females, and generally not within their own pod.

Results from each of these studies are consistent with observational data for a number of social odontocete species. For example, social groups are apparently also matrifocal, and sub-adult animals maintain long-term associations with the pod, in short-finned pilot whales (*G. macrorhynchus*: J. & S. Heimlich-Boran pers. comm.), killer whales (*Orcinus orca*: Bigg *et al.* 1990) and sperm whales (*Physeter catodon*: Rice 1989). In the killer whale and short-finned pilot whale temporary associations have been observed between matrifocal groups which could serve to facilitate outcrossing between groups. Male associations within the natal pod sometimes continue longer than expected (e.g. in some killer whale populations: Bigg *et al.* 1990), but the basic pattern of female philopatry and male dispersal (actual or effective dispersal through outcrossing between social groups) is typical of the pattern seen for mammalian species (Greenwood 1980).

Varying levels of polygyny have been recognized in pinniped species, from the temporally and spatially plastic territories of male Australian sea lions (*Neophoca cinerea*), to the extreme polygyny of the northern elephant seal (*Mirounga angustirostris*). Other species are apparently promiscuous (e.g. *Phoca vitulina*) or serially monogamous (see Riedman 1990). Molecular analysis will greatly increase our understanding of these breeding systems, though few such studies have been completed to date. Preliminary results from grey seals (*Halichoerus grypus*: Amos 1991) and northern elephant seals (A. R. Hoelzel unpubl.) suggest less polygny than expected.

Assessing kinship using DNA fingerprints is limited by a number of practical considerations (Lynch 1988). First, allelic pairs usually cannot be identified. Second, some proportion of alleles will be run off the end of the gel, and third, co-migrating alleles cannot be distinguished. These problems can be corrected by using hypervariable single locus probes, but problems remain associated with the saturation of variation beyond close relatives and the high mutation rate in these regions. Variable single-locus minisatellite and microsatellite markers can be treated in the same way as allozyme loci to identify kin-groups (see Wilkinson 1985). An assumption is made that kin-groups should show a non-random distribution of allele frequencies.

Hoelzel (1989) used multi-locus DNA fingerprints to show that a pair of adult male killer whales that shared prey and provisioned one another were closely related (Hoelzel 1991b). Amos (this volume) also used microsatellite loci to demonstrate that pilot whales travel in matrifocal kin-groups.

Genetic variation

The analysis of genetic variation can provide important information on effective population size (N_e), and on demographic history. N_e can be defined as the size of an ideal population that would show the same genetic characteristics as the real population. The concept of effective population size provides a way to summarize the various effects of demographics on genetic variation in populations of finite size. The rate of loss of variation in a population is dependent on N_e, and N_e is therefore of central interest to conservation biologists working with small or exploited populations. There are a number of ways that genetic data can be used to estimate N_e, and these are reviewed for application to cetacean populations by Waples (1991; the same analyses would be appropriate for pinnipeds). They can be divided into two general categories: methods for estimating long-term N_e, and methods for estimating current N_e.

Long-term estimates of N_e provide an indication of the effective size of the population (or species) in an evolutionary time frame. These estimates are based on equilibrium levels of diversity and are dependent on accurate estimates and interpretation of mutation rates. Hoelzel & Dover (1991a) estimated a long-term N_e of 400 000 for minke whales, based on mtDNA variation in three populations. They determined a mutation rate for the region of mtDNA under analysis (the control region) by sequencing from three cetacean species. The mutation rate determined in this way was more than an order of magnitude slower than for several other mammalian taxa (see Wilson et al. 1985), which emphasizes the importance of independent assessment of mutation rates for a given DNA region and taxa.

Estimates of current N_e are based on the stochastic effects of genetic drift over a few generations. One method measures temporal changes in allele frequencies (Waples 1989). Allele frequency change is a function of N_e and elapsed time in generations. Therefore, N_e can be estimated from short-term changes in allele frequency, given that there are no mutations, that the loci under investigation are selectively neutral and unlinked, that samples can be chosen at random and that there has been no immigration from neighbouring populations. With whales and seals an additional problem exists related to the fact that generations are overlapping and individuals can reproduce numerous times over the course of a lifetime. Another method limited by similar assumptions is based on the comparison of gametic (linkage) disequilibrium (Hill 1981). Random genetic drift in finite populations is one of the factors that will increase gametic disequilibrium. Both these methods will require large samples of individuals and loci to produce estimates with usefully small standard errors.

A population bottleneck will diminish variation through the effects of sampling and inbreeding. Therefore, genetic variation can give a rough indication of how long ago a bottleneck event may have occurred. For example, the human species has an anomalously low level of mtDNA variability, suggesting a bottleneck event within the last 200 000 years (e.g. Cann, Stoneking & Wilson 1987). To reconstruct a bottleneck event in more detail requires more than genetic data alone. The principal unknown variables are the time, duration and size of the bottleneck. If the population is still in a recovery phase (exhibiting density-independent growth) and data on demographic growth are available, then a most probable event can be statistically described, given historical information on one of the three variables. For example, Hoelzel et al. (in press) used mtDNA variation and census data to back-calculate the size and duration of the bottleneck

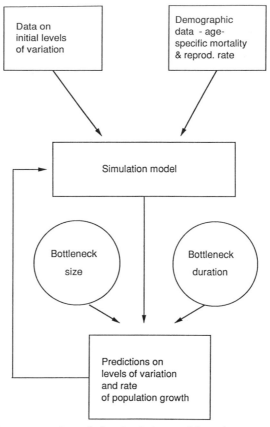

Fig. 4. Schematic representation of the simulation model used to reconstruct historical population bottlenecks. The model determines population growth and loss of variation on the basis of data on initial genetic diversity (often estimated from related species) and species-specific demographic parameters (especially age-specific mortality and reproductive rate), for a chosen bottleneck size and duration. Repeated runs with random numerical seeds provide a statistical simulation of demographic stochasticity. (After Hoelzel *et al.* in press.)

experienced by the northern elephant seal (*Mirounga angustirostris*). A probable date of the event could be inferred from the historical data, and census data were consistent with a population in exponential growth. The reconstruction was treated statistically by using a simulation approach (Fig. 4). Ninety five percent confidence results indicated a bottleneck event of fewer than 30 seals for less than 20 years (or a single-year bottleneck of fewer than 20 seals).

Genetic distance and stock identity

Genetic change accumulates over time, and to the extent that this occurs at a uniform and constant rate, the genetic difference between two populations can indicate how long they have been apart. This 'molecular clock' is notorious for its imperfections, yet molecular phylogenies are often in good agreement with phylogenies derived by other means (Lewin 1990). The question of rate has been investigated extensively for point-mutational changes, especially in genes (see Li, Luo & Wu 1985). Although the rate of change clearly varies between different genomic regions (Vawter & Brown 1986), when the approximate rate for a given region has been established, this can be used to assess the extent of time-dependent differentiation between populations.

Recently, the analysis of short repetitive regions of DNA (known as microsatellites) has been facilitated by the application of PCR (Tautz 1989). These regions change primarily by the mechanism of DNA slippage (expansion and contraction of repetitive regions by mis-match and repair, see Hoelzel & Dover 1991b), and the variability measured is in the length of the repetitive region. Microsatellites evolve quickly and therefore could provide a 'clock' with high resolution. At the same time, many of these loci evolve slowly enough so that the problem of saturating variation between populations could be avoided. For these reasons microsatellite loci have been considered for investigations of genetic distance at the population level. The problem lies in the fact that too little is known about the mechanism of DNA slippage to be sure that rates are unbiased and constant. It is known that the rate of slippage-like mechanisms can change dramatically over evolutionary time (Dover 1987). For this reason population level analyses should include a genetic marker that is known to evolve primarily by point-mutational change, such as mtDNA or allozyme variation. MtDNA is a good choice because it evolves 5–10 times faster than the nuclear genomic DNA that codes for proteins, and it is inherited only through matrilines. Matrilineal inheritance means that the pattern of variation in a population is unaffected by sexual recombination, and this simplifies the interpretation of genetic distance.

A number of studies have investigated mtDNA or allozyme variation between cetacean species and populations. The most extensive allozyme study involved four balaenopterid species (Fig. 5; Wada & Numachi 1991). A total of 17 925 whales from 15 locations were investigated for allozyme variation at up to 45 loci. Three of the four species (fin: *Balaenoptera physalus*; sei: *B. borealis*; and minke whale: *B. acutorostrata*) are distributed in all major oceans and in both hemispheres, while the Bryde's whale (*B. edeni*) has a more restricted range, primarily in tropical and subtropical waters. Fin, sei and minke whales all show some seasonal movement between polar feeding grounds and more temperate regions in the winter. A clear distinction between feeding and breeding grounds (as seen in the humpback whale, *Megaptera novaeangliae*) has not been established. However, to the extent that such a distinction does exist, populations of

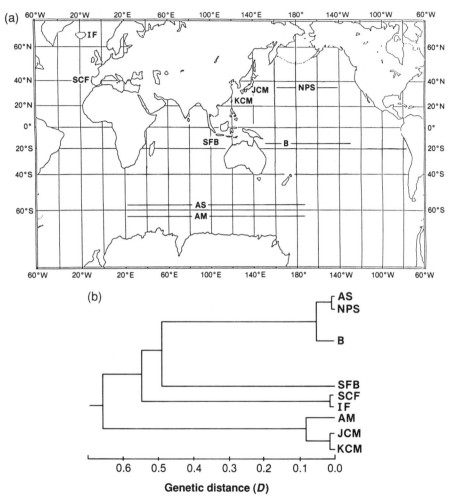

Fig. 5. (a) Map of region from which samples were collected. AS = Antarctic sei, NPS = North Pacific sei, B = Bryde's whale, SFB = small-form Bryde's whale, SCF = Spanish coastal fin whale, IF = Icelandic fin whale, AM = Antarctic minke whale, JCM = Japanese coastal minke whale, KCM = Korean coastal minke whale. (b) Dendrogram showing genetic distance (*D*) between species and populations (after Wada & Numachi 1991 and Danielsdottir *et al.* 1991). This is a phenetic tree based on Nei genetic distance measures and no inference about the real topology is intended.

each species could be expected to be isolated to some degree on either side of the equator. This is because the seasonal difference would put their breeding seasons out of phase. There was some evidence for such a population division in sei whales, but not in fin whales. In fact fin whales showed very little genetic differentiation between regions as geographically isolated as the Antarctic, the North Pacific and the North Atlantic. However, a separate study, also using allozyme diversity, found a population division evident between fin whales off the coasts of Spain and Iceland (Fig. 5; Danielsdottir *et al.* 1991).

Minke whales showed considerable variation between the Antarctic and the North Pacific (Fig. 5), and these results were supported by data for mtDNA diversity (Wada, Kobayashi & Numachi 1991; Hoelzel & Dover 1991a). Hoelzel & Dover (1991a) investigating mtDNA diversity showed a similarly high genetic distance between minke whales in the North Atlantic and those in either the Antarctic or the North Pacific. The genetic distinction was as high as that between some balaenopterid species. These three populations of minke whales had previously been distinguished at the subspecific level on the basis of variation in the pigmentation pattern of the pectoral flippers and the baleen plates (Stewart & Leatherwood 1985). However, two other examples indicate that cetacean morphological variation cannot be interpreted quantitatively. At one extreme, fairly extensive morphological variation between populations of spinner dolphins (*Stenella longirostris*) was not correlated with mtDNA variation (Dizon, Southern & Perrin 1991). At the other extreme, a baleen whale previously considered to be a small form of the Bryde's whale, being similar in morphology except for its size at maturity, was found to have a very high genetic distance from all four balaenopterids investigated (Fig. 5; Wada & Numachi 1991). This genetic distance easily distinguishes the small-form Bryde's whale as a newly recognized species.

For the purposes of management and conservation, cetacean populations have usually been distinguished on the basis of geographic distance and barriers (Donovan 1991). Some of the examples given above indicate that geographic distance is not a good indicator of genetic differentiation in these species. In other cases it was evident that more than one reproductive stock was inhabiting a given geographic region. For example, killer whales (*Orcinus orca*) in the north-eastern Pacific near Vancouver Island, B.C., can be distinguished by behaviour. Some pods pursue mostly fish prey, and some pursue mostly marine mammal prey. These two 'populations', while inhabiting the same waterways, showed the same level of genetic distinction as killer whales from the North Pacific in comparison with killer whales from the South Atlantic (Hoelzel & Dover 1991c; Hoelzel 1991c). Another problem is associated with the fact that many species migrate between breeding and feeding grounds, but are typically hunted on the feeding grounds. Therefore management boundaries drawn on feeding grounds, no matter how finely partitioned, could be meaningless depending on the spatial and temporal pattern of mixing of whales from different breeding populations in a given feeding area (Hoelzel 1991d). Wada (1991) found evidence for mixing of Korean and Japanese coastal minke whale stocks on the feeding grounds north of Japan. The opposite pattern, whales from a given breeding ground feeding in several geographically isolated areas, has been shown for the humpback whale, based on mtDNA variation (Baker *et al.* 1990).

The genetic divisions between pinniped populations have been investigated less extensively, though a number of studies are under way. Tag recovery data in polygynous species often indicate considerable movement between local breeding rookeries from one year to the next (see Riedman 1990). However, this is not always the case. Southern elephant seal (*Mirounga leonina*) breeding rookeries are distributed on the Argentine mainland and on several islands in the sub-Antarctic. Hoelzel *et al.* (in press) investigated mtDNA variation, comparing elephant seal rookeries on South Georgia and the Argentine mainland. Variation was very high on South Georgia, and very low on the mainland. The pattern of variation suggested that the Argentine population was founded by seals from the sub-Antarctic about 100 000 years ago, and that there is little present-day dispersal between these two populations.

Conclusions

Over the last 20 years or so various technologies have been developed that have greatly facilitated our understanding of marine mammal behaviour and population biology. These have included the application of photography and radio telemetry to the identification and long-term tracking of individuals. Together with more traditional methods such as mark-recapture and the comparative analysis of morphology and life-history parameters, these techniques have begun to reveal the basic structure of some marine mammal populations. However, not all species are accessible for study, and at the best of times these approaches can produce equivocal results. Molecular genetic techniques address some of the same questions, and some otherwise intractable questions. Since the genetic structure of a population is a consequence of behaviour and demographic structure, a genetic analysis can reveal detailed information without requiring long-term access to a study subject. This facilitates the investigation of a much broader range of species, including those that are normally observed only very rarely. Moreover, questions such as the identification of paternity and the quantification of genetic distance between populations can only be addressed by these methods. Perhaps the most important strength of these analyses is that the effect of demographics and behaviour is being investigated, rather than one of several possible causes. For example, while immigration of an individual into a new population can be documented by mark-recapture, radio-telemetric or photo-identification methods, only genetic analysis can assess the extent to which immigrants are reproductively successful in a new population. At the same time, interpretation of data and the identification of the likely cause of observed patterns are greatly facilitated by combining genetic and non-genetic methods.

Several observations can be made about the genetic data reported for marine mammal species to date. First, apparent population structure based on geographic distance or boundaries is not well correlated with genetic distance. Distant populations separated by continental land masses (such as fin whales in the North Pacific and North Atlantic) can be highly similar, while sympatric (as with the killer whales in the north-eastern Pacific) or proximate populations (such as the southern elephant seals on South Georgia and the Argentine mainland) can show a high level of genetic differentiation. For some species (especially baleen whales), this is further complicated by migration patterns where breeding stocks form a mixed assemblage on feeding grounds. Second, in several studies, morphological variation was not well correlated with the degree of genetic differentiation between morphotypes. In the case of spinner dolphins, a dimorphism is maintained between two populations, though intermediate forms are common and mtDNA variation was as great within as between populations. This distinction could be maintained by selection, and may not be detectable by methods that measure general levels of variation. In this case variation should be conserved, even though no overall levels of genetic distance are evident. The small-form Bryde's and minke whale populations each showed relatively subtle morphological variation, but very high levels of genetic differentiation. Finally, it has been possible to use measures of present-day variation to interpret the effective population size and demographic history of some species. Many marine mammal species have been extensively exploited as part of a fishery, through incidental takes or through competition and destruction of habitat. Assessing the effective population size and the extent of inbreeding in these populations will be crucial to determining conservation measures that will preserve genetic diversity, necessary for the long-term survival of any species in a changing environment. In general, genetic data have improved our understanding of marine mammal populations, but most of all, results to date emphasize the need for further study.

References

Amos, W. (1991). Molecular techniques applied to problems in marine mammal social organization: evolution in progress! *Abstr. bienn. Conf. Biol. Mar. Mammals* 9. Society for Marine Mammalogy. Unpublished.

Amos, W., Barrett, J. & Dover, G. A. (1991). Breeding system and social structure in the Faroese pilot whale as revealed by DNA fingerprinting. *Rep. int. Whal. Commn spec. Issue* No. 13: 255–270.

Amos, W. & Hoelzel, A. R. (1991). Long-term preservation of whale skin for DNA analysis. *Rep. int. Whal. Commn spec. Issue* No. 13: 99–104.

Baker, C. S., Palumbi, S. R., Lambertsen, R. H., Weinrich, M. T., Calambokidis, J. & O'Brien, S. J. (1990). Influence of seasonal migration on geographic distribution of mitochondrial DNA haplotypes in humpback whales. *Nature, Lond.* **344**: 238–240.

Bigg, M. A., Olesiuk, P. F., Ellis, G. M., Ford, J. K. B. & Balcomb, K. C. (1990). Social organization and genealogy of resident killer whales (*Orcinus orca*) in the coastal waters of British Columbia and Washington State. *Rep. int. Whal. Commn spec. Issue* No. 12: 383–405.

Brown, M. W., Kraus, S. D. & Gaskin, D. E. (1991). Reaction of North Atlantic right whales (*Eubalaena glacialis*) to skin biopsy sampling for genetic and pollutant analysis. *Rep. int. Whal. Commn spec. Issue* No. 13: 81–90.

Bruford, M. W., Hanotte, O., Brookfield, J. F. Y. & Burke, T. (1992): Single-locus and multilocus DNA fingerprinting. In *Molecular genetic analysis of populations; a practical approach*: 225–270. (Ed. Hoelzel, A. R.). Oxford University Press, Oxford.

Cann, R. L., Stoneking, M. & Wilson, A. C. (1987). Mitochondrial DNA and human evolution. *Nature, Lond.* **325**: 31–36.

Danielsdottir, A. K., Duke, E. J., Joyce, P. & Arnason, A. I. (1991). Preliminary studies on genetic variation at enzyme loci in fin whales (*Balaenoptera physalus*) and sei whales (*B. borealis*) from the North Atlantic. *Rep. int. Whal. Commn spec. Issue* No. 13: 115–124.

Dizon, A. E., Southern, S. O. & Perrin, W. F. (1991). Molecular analysis of mtDNA types in exploited populations of spinner dolphins (*Stenella longirostris*). *Rep. int. Whal. Commn spec. Issue* No. 13: 183–202.

Donovan, G. P. (1991). A review of IWC stock boundaries. *Rep. int. Whal. Commn spec. Issue* No. 13: 39–70.

Dover, G. A. (1987). DNA turnover and the molecular clock. *J. molec. Evol.* **26**: 47–58.

Duffield, D. A. & Wells, R. S. (1991). The combined application of chromosome, protein and molecular data for the investigation of social unit structure and dynamics in *Tursiops truncatus*. *Rep. int. Whal. Commn spec. Issue* No. 13: 155–170.

Greenwood, P. J. (1980). Mating systems, philopatry and dispersal in birds and mammals. *Anim. Behav.* **28**: 1140–1162.

Hill, W. G. (1981). Estimation of effective population size from data on linkage disequilibrium. *Genet. Res.* **38**: 209–216.

Hoelzel, A. R. (1989). *Behavioural ecology and population genetics of the killer whale.* PhD thesis: University of Cambridge.

Hoelzel, A. R. (Ed.) (1991a). *Genetic ecology of whales and dolphins. Rep. int. Whal. Commn spec. Issue* No. 13: 1–311.

Hoelzel, A.R. (1991b). Killer whale predation on marine mammals at Punta Norte, Argentina; food sharing, provisioning and foraging strategy. *Behav. Ecol. Sociobiol.* **29**: 197–204.

Hoelzel, A.R. (1991c). Analysis of regional mtDNA variation in the killer whale; implications for cetacean conservation. *Rep. int. Whal. Commn spec. Issue* No. 13: 225–234.

Hoelzel, A. R. (1991d). Whaling in the dark. *Nature, Lond.* **352**: 481.

Hoelzel, A. R. & Dover, G. A. (1991a). Mitochondrial D-loop DNA variation within and between populations of the minke whale (*Balaenoptera acutorostrata*). *Rep. int. Whal. Commn spec. Issue* No. 13: 171–183.

Hoelzel, A. R. & Dover, G. A. (1991b). *Molecular genetic ecology*. Oxford University Press, Oxford.

Hoelzel, A. R. & Dover, G. A. (1991c). Genetic differentiation between sympatric killer whale populations. *Heredity, Lond.* **66**: 191–195.

Hoelzel, A. R. & Green, A. (1992). Analysis of population-level variation by sequencing PCR-amplified DNA. In *Molecular genetic analysis of populations; a practical approach*: 159–188. (Ed. Hoelzel, A. R.). Oxford University Press, Oxford.

Hoelzel, A. R., Halley, J., Campagna, C., Arnbom, T., Le Boeuf, B. J., O'Brien, S. J., Ralls, K. & Dover, G. A. (In press). Elephant seal genetic variation and the use of simulation models to investigate historical population bottlenecks. *J. Hered.*

Jeffreys, A. J., Wilson, V. & Thein, S. L. (1985). Hypervariable 'minisatellite' regions in human DNA. *Nature, Lond.* **314**: 67–73.

Lambertsen, R.H. (1987). A biopsy system for large whales and its use for cytogenetics. *J. Mammal.* **68**: 443–445.

Lewin, R. (1990). Molecular clocks run out of time. *New Scient.* **125**: 38–41.

Li, W.-H., Luo, C.-C. & Wu, C.-I. (1985). Evolution of DNA sequences. In *Molecular evolutionary genetics*: 1–94. (Ed. MacIntyre, R. M.). Plenum Press, New York.

Lynch, M. (1988). Estimation of relatedness by DNA fingerprinting. *Molec. Biol. Evol.* **5**: 584–599.

Rice, D. W. (1989). Sperm whale—*Physeter macrocephalus* Linnaeus 1758. In *Handbook of marine mammals* 4. *River dolphins and the larger toothed whales*: 177–233. (Eds Ridgway, S. H. & Harrison, R. J.). Academic Press, London, San Diego etc.

Riedman, M. (1990). *The pinnipeds: seals, sea lions and walruses*. University of California Press, Berkeley, Los Angeles, Oxford.

Southern, E. (1975). Detection of specific sequences among DNA fragments separated by gel electrophoresis. *J. molec. Biol.* **98**: 503.

Stewart, B. S. & Leatherwood, S. (1985). Minke whale, *Balaenoptera acutorostrata* Lacepède, 1804. In *Handbook of marine mammals 3. The sirenians and baleen whales*: 91–136. (Eds Ridgway, S. H. & Harrison, R. J.). Academic Press, London, Orlando etc.

Tautz, D. (1989). Hypervariability of simple sequences as a general source for polymorphic DNA markers. *Nucleic Acids Res.* **17**: 6463–6471.

Vawter, L. & Brown, W. M. (1986). Nuclear and mtDNA comparisons reveal extreme rate variation in the molecular clock. *Science* **234**: 194–196.

Wada, S. (1991). *Genetic distinction between two minke whale stocks in the Okhotsk Sea coast of Japan*. International Whaling Commission Working Paper SC/43/Mi32. Unpublished.

Wada, S., Kobayashi, T. & Numachi, K. I. (1991). Genetic variability and differentiation of mitochondrial DNA in minke whales. *Rep. int. Whal. Commn spec. Issue* No. 13: 203–216.

Wada, S. & Numachi, K. I. (1991). Allozyme analyses of genetic differentiation among the populations and species of the *Balaenoptera*. *Rep. int. Whal. Commn spec. Issue* No. 13: 125–154.

Waples, R. S. (1989). A generalized approach for estimating effective population size from temporal changes in allele frequency. *Genetics* **121**: 379–391.

Waples, R. S. (1991). Genetic methods for estimating the effective size of cetacean populations. *Rep. int. Whal. Commn spec. Issue* No. 13: 279–300.

Weinrich, M. T., Lambertsen, R. H., Baker, C. S., Schilling, M. R. & Belt, C. R. (1991). Behavioural responses of humpback whales (*Megaptera novaeangliae*) in the southern Gulf of Maine to biopsy sampling. *Rep. int. Whal. Commn spec. Issue* No. 13: 91–98.

Wilkinson, G. S. (1985). The social organization of the common vampire bat. II. Mating system, genetic structure and relatedness. *Behav. Ecol. Sociobiol.* **17**: 123–134.

Wilson, A. C., Cann, R. L., Carr, S. M., George, M., Gyllensten, U. B., Helm-Bychowski, K. M., Higuchi, R. G., Palumbi, S. R., Prager, E. M., Sage, R.D. & Stoneking, M. (1985). Mitochondrial DNA and two perspectives on evolutionary genetics. *Biol. J. Linn. Soc.* **26**: 375–400.

Winn, H. E., Bischoff, W. L. & Taruski, A. G. (1973). Cytological sexing of Cetacea. *Mar. Biol., Berl.* **23**: 343–346.

Wong, Z., Wilson, V., Jeffreys, A. J. & Thein, S. L. (1986). Cloning a selected fragment from a human DNA fingerprint: isolation of an extremely polymorphic minisatellite. *Nucleic Acids Res.* **14**: 4605–4617.

Symp. zool. Soc. Lond. (1993) No. 66: 33–48

Use of molecular probes to analyse pilot whale pod structure: two novel analytical approaches

Bill AMOS

*Department of Genetics
University of Cambridge
Downing Street
Cambridge CB2 3EH, UK*

Synopsis

DNA fingerprinting can be used to answer many questions in behavioural ecology. However, where the number of crosswise comparisons to be made is large, as in the study of genetic relationships within a group, single locus analysis is often preferable. Furthermore, the study of group structure is often complicated by the difficulty of defining a population mean. Two analytical approaches are presented which have been designed to circumvent this problem: one derives an estimate for the number of individuals accompanied by one or other parent, the other tests for strong philopatry. The two methods are investigated by means of computer simulations and, although applicable to other species, are illustrated with an analysis of pilot whale pod structure. Both appear to provide useful information concerning relatedness within the group.

Introduction

In 1985, the discovery of a class of hypervariable DNA sequences, known as minisatellites, led to the development of 'DNA fingerprinting', a technique allowing individual identification and positive parentage analysis in humans (Jeffreys, Wilson & Thein 1985). This was followed by the discovery that similar sequences could be detected in a wide range of animals and plants (Burke 1989). Coming after some 30 years of isozyme electrophoresis, a technique with significant limitations, and at around the same time that the study of mitochondrial DNA was beginning to be used widely, these events sparked a spectacular growth in the use of molecular probes to study natural populations.

In general, paternity testing with DNA fingerprinting has been straight-forward when the number of possible fathers is small, as in many avian

ZOOLOGICAL SYMPOSIUM No. 66
ISBN 0–19–854069–8

Copyright © 1993 The Zoological Society of London
All rights of reproduction in any form reserved

studies (for example, Burke *et al.* 1989). However, where many crosswise comparisons have to be made, for instance in the study of paternity in highly polygynous species or in the analysis of interrelationships within a group, interpretation can become prohibitively time-consuming.

DNA fingerprinting

The term DNA fingerprinting is often used generically, for any analysis depending on DNA sequences which show marginally higher levels of polymorphism than protein isozyme analysis. However, I wish to retain its original usage, referring to a molecular technique which uncovers sufficient genetic variability at many, independently segregating, highly polymophic loci to distinguish each individual uniquely. Occasionally this usage is emphasized by the term 'multilocus fingerprinting'. The bar-code-like bands which constitute a DNA fingerprint can be used in three primary ways:

1. Identity: in just the same way as traditional fingerprints are used, DNA fingerprints can be used to identify individuals uniquely. In theory, this could be used to study natural populations by mark-recapture. However, the logistic problems of creating a suitable database make this unlikely to be used in practice.

2. Parentage: this is the sphere in which DNA fingerprinting has, perhaps, the greatest practical and theoretical potential. In most species, the majority of bands are inherited as independently segregating characters. Those present in an offspring which are not assigned to one parent, say the mother, must derive from the other, in this case the father. By matching this subset of bands to candidate males, and assuming that sufficient variability is present, the unknown parent may be identified with considerable confidence.

3. Relatedness: in many species, the proportion of bands which are shared by chance between two unrelated individuals will be as low as 20–30%. This compares with more than 50% shared between a parent and its offspring, or between two full sibs. These observations open up the possibility of estimating the degree of relatedness between two individuals on the basis of their degree of band-sharing. For this purpose, an index of similarity, termed the band-sharing coefficient (defined as $2Nab/(Na + Nb)$, where Nab is the number of bands which are shared between two individuals, a and b, and $(Na + Nb)$ is the total number of bands scored in the two individuals combined), is commonly used. In practice, there tends to be little overlap between band-sharing values derived from unrelated individuals and first-degree relatives. These two classes may often, therefore, be told apart. It is also occasionally possible to discriminate

between second-degree relatives and either first-degree relatives or unrelated individuals, but beyond this level it is of little use (Lynch 1988, 1990).

Unfortunately, the very complexity of a fingerprint which gives it its power also limits its use; comparisons between samples run on different gels are unreliable. Computerized data bases can help (see, for example, Amos 1992), but all predicted paternities still have to be confirmed by rerunning samples in adjacent lanes. For this reason I wish to assume that between-gel comparisons are effectively unavailable as an option for large-scale screening, and can only be used as a confirmatory tool where other information can be applied to identify a small subset of candidates.

Single-locus systems

In view of the problems inherent in interpreting multilocus DNA fingerprints, workers are turning increasingly towards the use of individual highly variable loci. In such systems, genetic information is gathered from a single chromosomal site, resulting in either one (homozygous) or two (heterozygous) bands per individual. Subsequent analysis is proportionately simplified. Two classes of DNA sequences are currently used: single-locus minisatellites, and microsatellites.

Single-locus minisatellites

The basis of DNA fingerprint variability lies with a family of short, repeated DNA sequences known as minisatellites. Long strings of these repeat units occur at many chromosomal locations and vary in length because of frequent gain and loss of repeat units. By cloning individual minisatellite arrays, polymorphism at any one locus may be screened independently of the others (Wong et al. 1986). Minisatellites are the most variable DNA sequences yet discovered, and are thus the most informative for indicating relatedness. Unfortunately, since their variability originates from a functional instability which is recognized even in bacteria (Jeffreys et al. 1985), such sequences are technically difficult to clone.

Microsatellites

Microsatellites comprise short tracts of di- or trinucleotide repeats and occur abundantly throughout the genomes of higher organisms (Tautz & Renz 1984). Within these regions, molecular 'slippage', a process whereby repeat units become misaligned, probably during DNA replication, generates length variability (Tautz 1989). Microsatellites make good genetic markers on several counts. They are abundant, easy to clone and can be screened efficiently by using the polymerase chain reaction (PCR). Briefly, PCR is a method of enzymic DNA amplification by which a highly specific DNA region can be copied many times. If the amplification is designed so that the product includes a microsatellite region, changes in length of the array due to slippage will be reflected in the size of the resulting fragment. Since PCR will work on even a single starting molecule, low-grade samples may be used, small in quantity and/or poor in quality. A particular advantage of this method is its precision; the lengths of different alleles are determined to single nucleotide accuracy by means of polyacrylamide sequencing gels. The main drawback with microsatellite loci is that they usually show somewhat lower levels of variability than are found for many minisatellite loci. However, this shortfall may be overcome, to a great extent, by screening more loci.

Analysis of group structure

In many species, individuals come together to form groups, either transiently or for longer periods. There are several possible reasons for doing this, amongst which are to feed, to breed, and for mutual protection. In many cases, the formation, composition and interchange between groups will influence, if not determine, the dynamics of the population as a whole. The analysis of groups is particularly relevant to cetaceans. For the majority of species, behavioural observations are difficult, yet many odontocetes are highly social, and a number can be sampled opportunistically either from mass-stranding events or from incidental fishery mortalities.

One step towards understanding the behaviour of a group is to establish the genetic relationships between its members. Where group size is small, it may be feasible to achieve this by using DNA fingerprinting, for example in the analysis of lion prides (Packer et al. 1991). For larger groups this is not practical. An alternative is to look for patterns in the distribution of genetic markers within the group, using one or more highly variable loci. However, the analysis tends to be difficult because many approaches are based on tests for deviations of allele frequencies away from a population mean. Where a large number of groups can be sampled such methods can be informative, but for most species, few groups, perhaps only one, can be studied. Under these circumstances the best estimate of the population mean is poor, tending towards that of the group itself, and the analysis becomes circular.

A further problem with the analysis of groups is that sample sizes will be limited strictly to the number within the group or group sub-class. Increasing the sample size of groups will help, but this makes a further assumption that all groups are similar. In order to counter these problems, two novel tests have been constructed. Both avoid the need to define population allele frequencies and maximize sample sizes by drawing on information from as many individuals as possible, i.e. they are global tests which look for an overall pattern. One estimates the proportion of individuals which are present as part of a parent:offspring pair, and one examines the extent to which particular alleles remain with, and are propagated within, the group. These tests are illustrated in the analysis of genetic relationships between members of pilot whale pods.

Test 1: Estimating the proportion of parent–offspring pairs

For some individuals, behavioural or other observations will indicate likely maternal relationships, for example young mammals suckling and young birds in a nest. These cases can be confirmed by DNA fingerprinting. Amongst older animals, observational clues are unlikely to be available; consequently the number of pairwise combinations which need to be assessed is far greater. Analysis of one or more highly variable loci can be used to exclude a proportion of these. However, it is unlikely that sufficient variability will be present to allow true parent:offspring relationships to be identified with certainty. What is more likely is that each animal will be found to be accompanied by one or more genetically compatible 'parents', any or none of which could be the true one. The following analysis, illustrated by a search for maternal relationships in a pilot whale pod, provides an indication of whether the true parents have been sampled.

Consider a single animal of genotype ab at a multiallelic locus, in a group where x females are old enough to be its mother. The following numbers of each allelic class are scored:

n_a = number of a alleles amongst the x possible mothers
n_b = number of b alleles amongst the x possible mothers
$n_m = n_a + n_b$ = the total number of alleles of possible maternal origin present amongst the x possible mothers

In the absence of other information, any female carrying either an a or a b allele could be the mother. Let such individuals be indicated by inverted commas; viz. 'mothers'. Even if the true mother is not in the group, some 'mothers' could still be found. Assuming random assortment of alleles, the probability that any given member of our group is a 'mother' is given by:

$$p = 1 - (1 - n_m/2x)^2$$

It follows that, when the true mother has not been sampled, the number of 'mothers' we expect is simply

$$px \qquad\qquad (1)$$

However, if the true mother is present, these equations must be modified to exclude both her and the maternal allele she carries. The new best estimate for the frequency of maternal alleles thus becomes $(n_m - 1)/(2x - 1)$. Again assuming random assortment, each remaining female has a new probability, p_m, of being a 'mother':

$$p_m = 1 - \{1 - (n_m - 1)/(2x - 1)\}^2$$

In total, therefore, we expect to find the true mother plus $p_m(x - 1)$ non-mothers, i.e.

$$1 + p_m (x - 1) \tag{2}$$

Applied to the same group of females, equations (1) and (2) yield different values, the magnitude of the difference depending on both the size of the group and the number of maternal alleles. This provides the basis for a test of whether the true mother has been sampled. The expected number of 'mothers' (E), based on the true mother being absent, should be compared with the number of 'mothers' observed empirically (O). In cases where the true mother is absent, the statistic $\{O\text{-}E\}$ is distributed with a mean of 0. When she is present, however, $\{O\text{-}E\}$ will be distributed about a mean, V, calculated as the difference between equations (1) and (2):

$$V = 1 + p_m (x - 1) - px \tag{3}$$

In samples where the true mother is present in some cases but not in others, the statistic $\{O\text{-}E\}$ is expected to have a bimodal distribution with peaks at 0 and V. The mean value of $\{O\text{-}E\}$ in such a sample is, therefore, an estimator of the proportion of cases in which the true mother is present. However, if this approach is to be used on real data, both V and the confidence limits on such an estimate need to be determined.

Fortunately, although, in each instance, V can assume any value between 0 and 1, depending on the sizes of x and n_m, the majority of realistic conditions favour V lying close to unity (see Table 1a). It is only when x is small and n_m is not small that the magnitude of the difference between the last two terms in equation (3) becomes significantly greater than 0.

With respect to the confidence limits, a precise derivation is difficult because of the complexity of the distribution. One method is to take advantage of the fact that, whatever the nature of the distribution being sampled, estimates of its mean tend to be distributed approximately normally. Since it is the mean value of $\{O\text{-}E\}$ that is of interest, a simple standard error can thus be used to define the confidence limits.

In practice, all aspects of this analysis will benefit from V values which lie as close as possible to 1. First, this will minimize the variance in V itself. Second, the separation of the distribution's peaks, and hence the sensitivity

Table 1. Example solutions to equation (3) for a range of values of group size (x) and maternal allele number (n_m)

All parents share with their offspring at least one allele at each locus. If a young animal is compared with a group of candidate parents, the number which are likely to be genetically compatible, assuming random assortment of alleles, is readily calculated. However, such predicted values will differ, depending on whether the true parent is present or not. Equation (3) is a general expression for the difference, V, between these two states. Tables 1a, b and c below show how V varies over a range of group sizes and maternal allele frequencies, for one, two and four loci respectively.

(a) One locus

Number of maternal alleles, n_m	Number of individuals, x			
	10	20	40	60
1	1.00	1.00	1.00	1.00
3	0.80	0.90	0.95	0.97
5	0.62	0.81	0.90	0.93
15	0.07	0.41	0.68	0.78
25		0.15	0.48	0.64
40			0.24	0.43

(b) Two loci

Number of maternal alleles, n_m	Number of individuals, x			
	10	20	40	60
1	1.00	1.00	1.00	1.00
3	0.96	0.99	1.00	1.00
5	0.86	0.96	0.99	1.00
15	0.13	0.65	0.90	0.95
25		0.27	0.73	0.87
40			0.44	0.68

(c) Four loci

Number of maternal alleles, n_m	Number of individuals, x			
	10	20	40	60
1	1.00	1.00	1.00	1.00
3	1.00	1.00	1.00	1.00
5	0.98	1.00	1.00	1.00
15	0.25	0.88	0.99	1.00
25		0.47	0.93	0.98
40			0.70	0.91

of the test will be maximized. Since the size of the test group, x, will usually be fixed, it follows that n_m should be minimized. This can be achieved by including information from as many loci as possible. Loci can be combined by calculating p_m and p for each locus independently, then multiplying these values together to construct an overall value, indicated by an accessory subscript 't'; viz. p_t and p_{mt}. Thus, for j loci;

$$p_t = p_a\, p_b \dots p_j \text{ and } p_{mt} = p_{ma}p_{mb} \dots p_{mj}$$

where p_i and $p_{mi} = p$ and p_m at the ith locus respectively. p_t and p_{mt} are then substituted directly into equation (3). Tables 1b and 1c show how screening two and four loci respectively can increase the chance of V being equal to 1.

To investigate how this approach performs in practice, it was applied to data collected from entire pilot whale pods. The long-finned pilot whale, *Globicephala melas*, is a medium-sized whale which is relatively abundant in the North Atlantic. It is a highly social species, swimming in large, cohesive groups, or pods, numbering some 50–250 whales. Occasionally very large pods containing in excess of 1000 whales are observed. Man has exploited the relative ease with which pods may be herded to guide them into shallow water for slaughter. Today, this practice continues in the Faeroe Islands, where an average of some 1700 animals are taken annually. In most cases it is believed that no animals escape, either during the drive or the killing.

During the period 1986–1988 a collaborative research programme

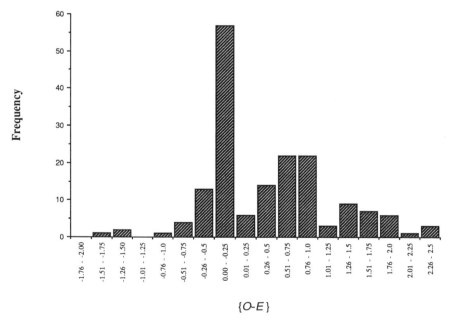

Fig. 1. Empirical distribution of the statistic {O-E}. {O-E} values (see text) were derived for all members of two pilot whale pods, Leynar (220787, $n = 89$) and Miðvágur (240787, $n = 98$), except those with fewer than five females old enough to be their mothers.

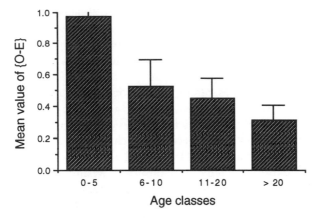

Fig. 2. Estimated proportion of pilot whales, in different age classes, accompanied by their mothers. {O-E} values (see text) are calculated for each whale in the pod Miðvágur (240787, $n = 98$). Whales are classified into four age classes and the mean value of {O-E} is plotted for each group. Assuming $V = 0.85$ (estimated from the data in Fig. 1)), these means provide an estimate, for each group, of the proportion of whales accompanied by their true mother. Error bars are standard errors of the mean.

sampled the vast majority of whales caught. A number of these pods have subsequently formed the basis of a genetic study, using DNA fingerprinting and single locus analysis to look into their breeding behaviour and social structure (Amos, Barrett & Dover 1991). Two pods were typed for six microsatellite loci and one minisatellite locus (Schloetterer, Amos & Tautz 1991; Amos, Schloetterer & Tautz 1993) and {O-E} values were calculated for each whale in turn (excluding those for which there were five or fewer possible mothers). The distribution obtained (see Fig. 1) is bimodal, as expected, and has a variance of approximately 1. It is not, however, very smooth, but this may be explained, in part, by the fact that, whereas E values are continuous, only whole numbers of mothers can be observed.

Figure 2 shows the data from one of these pilot whale pods, the Miðvágur drive (24–07–87), which comprises 98 animals. Individual whales were classified into four different age groups and the mean {O-E} value was plotted for each group. Using the approximation that $V = 0.83$, these data imply that some 95% of 0–5 year olds are accompanied by their mothers, declining to 40% amongst adults. Such proportions are biologically plausible, suggesting that the approach used is valid.

Test 2: Maintenance of alleles within a group: long-term stability

Test 1 can be used to estimate what proportion of animals are first-degree relatives of each other. Also of interest, however, are the relationships between older group members. How many are siblings themselves and how many are strangers from other groups? What is the long-term stability of the group? Such questions are not easy to answer. The following test was designed to analyse the overall degree of philopatry in a pilot whale pod.

Previous work has suggested that pilot whale pods are possibly matrilineal, and that males rarely, if ever, father offspring within a pod (Amos, Barret *et al.* 1991). If this is true, a pod may be thought of as comprising a maternal core into which novel alleles are fed annually from visiting males. Each pod should become enriched for those alleles which are characteristic of the matriline. This would readily explain the inter-pod genetic differentiation which has been observed. At any time, the commonest alleles will be those which have been transmitted to the most offspring, i.e. those belonging to the oldest females. As time passes, these animals will die and the pod will assume an allelic balance characteristic of the next oldest, and so on. The stochastic nature of lineage survival will ensure that, over time, each pod will continuously change its allelic balance, the degree of deviation from the population mean tending to be related to the rate of growth of the pod and the fidelity of the matriline (how often individuals or groups of animals move between pods).

From this basic model, a prediction can be made that older females should carry two alleles, both of which are common in the pod. In contrast, a young animal will carry one maternal allele, the frequency of which will depend on its matrilineal antiquity, and one paternal allele, which will be new to the pod. In a highly polymorphic system, this allele will usually be rare. Females of intermediate age will carry one matrilineal allele and one which, although paternal in origin, has begun to be woven into the genetic matrix of the pod through that animal's offspring. It is thus predicted that there should be a positive correlation between age and genotype probability based on random assortment of the pod's alleles. In the context of the pod as a whole, one might expect further that, as long as males also stay with their mothers, it should make little difference whether they are included or excluded from the analysis. Thus, although in most breeding systems it is unlikely that males will father any offspring within the pod, by merely carrying one maternal allele they will make a positive contribution to the relative frequency of matrilineal alleles.

Pod-specific genotype probabilities were derived for all members (males and females) of the two pilot whale pods mentioned above, using data collected at a hypervariable microsatellite locus (54 alleles in the two pods combined). In both cases, the genotype probabilities were found to be highly correlated with age (pod 1: correlation coefficient, $r = 0.41$, $n = 89$, $p < 0.001$; pod 2: $r = 0.33$, $n = 98$, $p < 0.001$: Amos, Schloetterer *et al.* 1993), fulfilling the prediction. To investigate how this degree of correlation relates to pod size and alternative allele distributions, a number of computer simulations were performed, using the individual ages and alleles present in pod 1.

Simulation 1

Although the correlation coefficients for the two pods were both highly significant according to statistical tables, it is possible that a skew in the age distribution could act to increase the variance of the estimates. This would lead to an overestimate of the level of significance achieved.

In the first simulation, therefore, all alleles in pod 1 were assigned randomly back to the original pod members and new correlation coefficients were calculated. The frequency of each allele and the starting age distribution were thus preserved. In 1000 runs the maximum correlation coefficient recorded was 0.367 (mean = 0.0009, S.D. = 0.112). This is less than the empirically derived value for this pod. It may be concluded that the distribution of alleles within the pod is not random, but is instead age-related, and therefore reflects some degree of pod substructure.

Simulation 2

In a pilot whale pod, test 1 showed that many individuals are accompanied by their mothers. This raises the question of whether a significant age: genotype correlation could be generated by a group of unrelated females, each accompanied by her own offspring. This possibility was examined by redistributing the alleles scored for pod 1 in such a way as to create mini 'families', each comprising a female and several offspring. Although alleles and 'mothers' were chosen at random, certain restrictions were imposed in order to keep within the bounds set by the original data. The procedure adopted was as follows. First a mature female was selected from amongst all those of sufficient age, assuming a maximum reproductive rate of one offspring born every three years after the age of five (Sergeant 1962). Second, random 'offspring' were selected, each at least five years younger than the 'mother', until the target family size was reached. Next each offspring was given one or other maternal allele, with equal probability. Maternal alleles were selected retrospectively, depending on the number of progeny carrying each. Thus, if six copies of an allele are required (one for each of five offspring plus one maternal), the choice of allele is restricted to only those with six unselected copies remaining as well as an initial frequency of six or more. Finally, random paternal alleles were assigned. Simulations were aborted whenever a random selection could not be fulfilled, i.e. when either no females of sufficient age or when no alleles of sufficient frequency remained. In practice, these restrictions limited the proportion of the pod which could be assigned to families to about 65%. Once all families had been established, remaining alleles were distributed randomly to the rest of the pod members.

Each set of simulation conditions was run 100 times and the maximum and mean correlations were recorded. Even under the most extreme conditions, with two thirds of the pod comprising nine-member families, the average correlation coefficient was only 0.082 (maximum value recorded 0.33). This suggests at least some relatedness between different putative family subgroups within a pod.

Simulation 3

Simulation 2 attempts to generate high correlations by using sets of rules to reassign alleles back to pod members. This approach, although interesting, is unrealistic for two reasons. First, each allele assignment is not an independent event; if a young animal receives a rare allele, this allele is no longer available to assign to an old whale. Second, the breeding system which operates acts to promote both an allelic bias away from the theoretical population mean and a correlation between allele frequency and age. By starting with the alleles from a real pod, the former element is established before the simulation begins. Since the allelic bias and any correlation between age and genotype are intimately connected, it is potentially misleading to consider one without the other.

The most plausible explanation for high levels of relatedness between members of a pilot whale pod is that the group is matrilineal, either females or both sexes remaining with their mother's pod. To investigate whether such a system is compatible with the pilot whale data, model computer pods were set up and allowed to reproduce, as matrilines, in a biologically realistic fashion. Pods were initiated by either a single female or a group of related females, and allowed to reproduce until a target pod size was achieved (for details, see Table 2 legend). The parameters used in each simulation are based on data from the Faeroese pilot whale fishery. Thus, the minimum age of conception is taken as five years old and females live to 45 (the oldest female in the Leynar drive, although older individuals have been recorded). Against a background of default values, three parameters were varied in turn: final pod size, allelic diversity and reproductive rate.

Summary results from these simulations are presented in Table 2. In many cases, the mean correlation coefficients were close to the empirical values and under no circumstances did the empirical values lie outside the 95% confidence limits of any simulation. Within the confines of the accuracy of the models, these results appear to confirm that it is the relationship between submatrilines which is important in generating age:genotype correlations, not merely the presence of groups of related individuals. In addition, several trends are apparent: higher correlations are found with higher female reproductive rates, lower pod size and where the number of alleles was approximately the same as the target pod size. Further, a single surviving founder tends to produce higher correlations than a group of

Table 2. The effect of pod structure on the degree of correlation between genotype frequency and age: summary results for a simulated pilot whale matriline

Pilot whale matrilines were simulated by means of a computer model. Matrilines were generated from single founding females according to simple rules. Reproduction cannot occur until the age of five years. Thereafter a single offspring is produced with probability p (the reproductive rate parameter) every three years. A birth sex ratio of 60 females to 40 males was used to mimic the real life bias (Sergeant 1962). Founding and paternal alleles at a single locus are drawn at random from a distribution in which all alleles are at a constant, equal frequency. A matriline is complete when the founding female reaches the age of 45 (the age of the oldest female in the Leynar drive; pilot whale females are thought to live to between 40 and 50 years old). Reproduction was halted when a predetermined target pod size was attained. Whenever the first matriline was smaller than this target, further, unrelated matrilines were initiated. Three parameters were investigated, each of which was varied with respect to a set of default values: target pod size (default value = 100); reproductive rate (default value = 0.7); number of alleles (default value = 60). For each parameter, the range over which it was varied is indicated, followed, in square brackets, by the stepwise increment used.

At the end of each run, the genotype frequency of every animal was calculated according to the formula: $\sqrt{(A1\ A2)}$, where $A1$ and $A2$ are the pod-specific frequencies of its two alleles. Every set of conditions was repeated 100 times and the mean and maximum correlation coefficients were recorded. For each parameter, the Table presents both the highest mean correlation and greatest single correlation coefficient obtained, followed, in brackets, by the corresponding parameter value of each. In addition, each set of conditions was applied both to a single founder female and to a group of related founders, simulated by allowing reproduction to continue up to the age of 60, then deleting all individuals older than 45.

Parameter	Single/Related founders (S/RF)	Highest mean	Highest value
Reproductive rate	S	0.38 (1.0)	0.60 (0.9)
Range 0.5–1.0 [0.1]	RF	0.26 (0.8)	0.56 (0.5)
Number in pod	S	0.44 (25)	0.80 (25)
Range 25–200 [25]	RF	0.25 (175)	0.62 (25)
Number of alleles	S	0.39 (100)	0.65 (100)
Range 25–200 [25]	RF	0.29 (100)	0.60 (100)

related founders. Although these findings are in line with intuitive expectations, further work is required in order to determine their significance.

Discussion

It is only recently that highly variable single-locus genetic markers have become readily available. One of the uses to which these may be put is the analysis of group structure in natural populations. Presented here are two examples of tests which are designed to investigate specific aspects of pilot whale pod structure.

The first test estimates the proportion of individuals in a group which are accompanied by either parent. Applied to different age classes in complete pilot whale pods, the method produced plausible results, suggesting that it is valid. Only two parameters need to be defined; the number of candidate females and the frequency of parental alleles within this group. In general terms, the method works best when group size is large and when parental alleles are rare. In this way it complements DNA fingerprinting, which works best in small groups, where the number of comparisons to be made is small.

It might be argued that the pilot whale data used here are better than will be available in most other studies. All pilot whales were aged accurately, by counting annual growth layers in the teeth, and the pods are thought to have been sampled completely. However, knowledge of an individual's age is only useful in as much as it restricts the focus of the analysis to a subgroup. For most purposes a simple mature/immature classification will perform adequately. Of two main circumstances which will undermine the validity of the test, neither is necessarily age-related. The first is where the group of females tested is not representative of the group from which the true mother originates. For example, in a highly polygynous, annually breeding species, different cohorts could be genetically dissimilar. By averaging over several cohorts such differences would be reduced or eliminated, leading to an underestimate of the number of 'mothers' one would expect to find by chance. The second situation is where there is non-random assortment of alleles. These two problems share several similarities and both, ostensibly, require retrospective knowledge of the breeding system in order to be detected. However, in practice, it may be sufficient to plot the empirical $\{O\text{-}E\}$ values as a histogram, as in Fig. 1. Wherever the resulting distribution is strongly bimodal, with peaks at 0 and around 1, the test is probably valid. Other distributions should be treated with caution.

The second test addresses the more general problem of the overall relatedness of members in a group. The method used here is to look for a correlation between age and genotype probability, on the assumption that the more alleles tend to be retained within the group, the greater will be the tendency for older animals to have group-characteristic genotypes.

Three sets of simulations were performed to investigate the sorts of conditions which can generate significant correlations. The first simply confirms the degree of significance indicated by statistical tables and hence that there is a genuine relationship between age and genotype. The second set of simulations investigated whether groups of unrelated families could explain the empirical correlation. Again, the correlation coefficients obtained fell short of the value found for the real pod. This implies that several unrelated family groups are alone insufficient to generate the empirical correlation. Finally, completely artificial, matrilineal pods were created by using a computer and simple rules for reproduction. Over a wide range of pod size, allelic diversity and reproductive rate, correlations were produced which were all compatible with the observed value for the true pod. Whilst none of these models can claim to be totally realistic, and many modifications could be made to improve them, it is interesting to note how close the artificial values are to the experimental ones. This is despite the fact that the models are perfect matrilines, they do not include any migration between pods. Combined with the high proportion of mother: offspring relationships identified by test 1, these results suggest that pilot whale pods mix little with each other and are stable over several generations.

Test 2 can be applied to investigate situations other than matrilines. If either sex shows strong natal group fidelity, each group will become enriched for those alleles carried by the stay-at-homes and their offspring. In just the same way as the special case of the pilot whale matriline, this will lead to an allelic bias. Using a highly polymorphic system such as those employed here, a significant age:genotype correlation should be found.

In conclusion, two new approaches for analysing single-locus data from animal groups are described. Both appear to lead to meaningful inferences about pilot whale pod structure, combining well to test for both close relatives (parent:offspring) and more distant relationships (matrilineal branches).

Acknowledgements

This work was supported by the British N.E.R.C. Microsatellite analysis was initiated by, and performed in collaboration with, Christian Schloetterer and Diethard Tautz. Helpful comments on the manuscript were made by Nick Galwey and David Bancroft. I am particularly thankful towards Ian Boyd for the thoroughness of his editing. I continue to express my gratitude to the members of the team which carried out the original pilot whale sampling, particularly Dorete Bloch, Genevieve Desportes and Rogvi Mouritsen.

References

Amos, W. (1992). Analysis of polygamous systems using DNA fingerprinting. *Symp. zool. Soc. Lond.* No. 64: 151–165.

Amos, W., Barrett, J. A. & Dover, G. A. (1991). Breeding behaviour of pilot whales revealed by DNA fingerprinting. *Heredity* 67: 49–55.

Amos, W., Schloetterer, C. & Tautz, D. (1993). Social structure of pilot whales revealed by analytical DNA profiling. *Science* 260: 670–672.

Burke, T. (1989). DNA fingerprinting and other methods for the study of mating success. *Trends Ecol. Evol.* 4: 139–144.

Burke, T., Davies, N. B., Bruford, M. W. & Hatchwell, B. J. (1989). Parental care and mating behaviour of polyandrous dunnocks, *Prunella modularis* related to paternity by DNA fingerprinting. *Nature, Lond.* **338**: 249–251.

Jeffreys, A. J., Wilson, V. & Thein, S. L. (1985). Hypervariable 'minisatellite' regions in human DNA. *Nature, Lond.* **314**: 67–73.

Lynch, M. (1988). Estimation of relatedness by DNA fingerprinting. *Molec. Biol. Evol.* **5**: 584–599.

Lynch, M. (1990). The similarity index and DNA fingerprinting. *Molec. Biol. Evol.* **7**: 478–484.

Packer, C., Gilbert, D. A., Pusey, A. E. & O'Brien, S. J. (1991). A molecular genetic analysis of kinship and cooperation in African lions. *Nature, Lond.* **351**: 562–565.

Schloetterer, C., Amos, W. & Tautz, D. (1991). Conservation of polymorphic simple sequence loci in cetacean species. *Nature, Lond.* **354**: 67–68.

Sergeant, D. E. (1962). The biology of the pilot or pothead whale *Globicephala melaena* (Traill) in Newfoundland waters. *Bull. Fish. Res. Bd Can.* No. 132: 1–84.

Tautz, D. (1989). Hypervariability of simple sequences as a general source of polymorphic DNA markers. *Nucleic Acids Res.* **17**: 6462–6471.

Tautz, D. & Renz, M. (1984). Simple sequences are ubiquitous components of eukaryotic genomes. *Nucleic Acids Res.* **12**: 4127–4138.

Wong, Z., Wilson, V., Jeffreys, A. J. & Thein, S. L. (1986). Cloning a selected fragment from a human DNA 'fingerprint': isolation of an extremely polymorphic minisatellite. *Nucleic Acids Res.* **14**: 4605–4616.

Symp. zool. Soc. Lond. (1993) No. 66: 49–60

Comparative levels of genetic variability in harbour seals and northern elephant seals as determined by genetic fingerprinting

Niles LEHMAN[1,2]

[1]Department of Molecular Biology
The Scripps Research Institute
10666 North Torrey Pines Road
La Jolla, CA 92037, USA

Robert K. WAYNE[2,3]

[2]Department of Biology
University of California
Los Angeles, CA 90024–1606, USA

[3]Institute of Zoology
The Zoological Society of London
Regent's Park, London NW1 4RY, UK

and Brent S. STEWART[4]

[4]Hubbs-Sea World Research
Institute
1700 South Shores Road
San Diego, CA 92109, USA

Synopsis

As a consequence of recent population bottlenecks, the amount of genetic variability in the northern elephant seal (*Mirounga angustirostris*) is thought to be extremely low. However, the species is apparently thriving throughout its breeding range in the eastern Pacific. By using analysis of hypervariable minisatellite loci in the northern elephant seal, we find that this species is indeed lacking in variation, as roughly 90% of alleles are shared among all the individuals tested. There is not enough variation to test hypotheses concerning island and beach philopatry. In contrast, harbour seals (*Phoca vitulina*) tested from the eastern Pacific possess much greater levels of variation at these loci and demonstrate a small degree of population-genetic structuring among the breeding sites studied.

Introduction

In recent years, many authors have discussed a possible causative relationship between a species' ability to withstand extinction in the face of

ZOOLOGICAL SYMPOSIUM No. 66
ISBN 0–19–854069–8
Copyright © 1993 The Zoological Society of London
All rights of reproduction in any form reserved

environmental perturbation and the amount of genetic variation possessed by the species. Several factors, including inbreeding depression avoidance (Ralls, Brugger & Ballou 1979), disease resistance (O'Brien & Evermann 1988), and phenotypic plasticity (Thompson 1991), have been cited as potential advantages of polymorphic species. In several cases, such as the cheetah (O'Brien, Wildt et al. 1983; O'Brien, Roelke et al. 1985), a strong correlation exists between declining census numbers (together with reduced fertility) and low variability at a suite of loci not thought to be subject to selective forces (for review, see O'Brien & Evermann 1988).

However, there are a number of species in which low variability has been reported, yet whose population size is either stable or increasing. The naked mole-rat (Reeve et al. 1990), the Channel Islands gray fox (Gilbert et al. 1990), and the polar bear (M. A. Ramsay pers. comm.) are all mammalian examples. The northern elephant seal (*Mirounga angustirostris*) also falls in this category and may in fact be the most extreme case. As a consequence of human hunting the species was driven near to extinction in the late 19th century (Townsend 1885, 1912). From as few as 100 individuals surviving at sea or on a single island refugium, the species has recovered to a current population of roughly 125 000, over 65% of which haul out to breed on the southern California Channel Islands (Bartholomew & Hubbs 1960; Stewart 1989; Stewart et al. in press). Furthermore, analysis of 23 blood protein loci revealed no polymorphisms by starch gel electrophoresis (Bonnell & Selander 1974), and mitochondrial DNA D-loop sequencing has detected only two genotypes in northern elephant seal populations (Hoelzel et al. 1991).

The explosive recovery of the northern elephant seal without any apparent concomitant genetic divergence presents some possible alternatives to the persistence/variability connection hypothesis. First, the repopulation of former habitat by the elephant seal may have occurred more rapidly than the generation of genetic variants (through either mutation or recombination events), and the species may currently still be at risk of extinction. Second, in this species, and perhaps others, variability may not be a requisite for ability of the species to survive. A corollary is that certain genomes may even be adapted to persist without variability. Third, the loci surveyed so far may not be representative of the true amount of variation present in the species, and northern elephant seal loci may actually be as polymorphic as those of other pinnipeds.

In this report, we have investigated the third of these alternatives by comparing levels of genetic variability at hypervariable 'minisatellite' nuclear loci in northern elephant seals and Pacific harbour seals (*Phoca vitulina richardsi*). Minisatellite loci are highly polymorphic in most vertebrate species and mutate at a rate of approximately 10^{-4} mutations to a new length allele per gamete per generation, at least in humans (Jeffreys,

Wilson & Thein 1985a); consequently in outbred populations no two non-sibs are expected to possess exactly the same set of alleles when assayed by the multilocus screening technique of DNA 'fingerprinting' (Jeffreys *et al.* 1985a, b). DNA fingerprinting can be used to estimate population-genetic structuring, both geographic and ecological, within a species (Burke 1989; Gilbert *et al.* 1990; Lehman *et al.* 1992), and has been used with success in marine mammals (Hoelzel & Dover 1991; Harris, Young & Wright 1991; van Pijlen, Amos & Dover 1991). We have chosen the harbour seal as a comparison to the northern elephant seal because the harbour seal has a wide geographic distribution and has persisted in large numbers without any known population bottlenecks. Some populations of harbour seals use the same southern California Channel Islands to breed as the northern elephant seal. We expect the harbour seal to exhibit 'standard' levels of genetic variability at minisatellite loci that can be contrasted to those found in elephant seals. Additionally, by surveying harbour seal populations along the North American west coast, we can determine whether minisatellite loci can be used to describe geographic partitioning in this species, and, by surveying elephant seal populations from several beach sites on two of the Channel Islands, we potentially can assay the amount of beach and island philopatry in this species.

Materials and methods

Harbour seal samples
Blood samples were obtained from individuals from the following localities: Prince William Sound, Alaska ($n = 3$), Puget Sound region, Washington State ($n = 34$), San Francisco Bay, California ($n = 11$), San Miguel Island, California ($n = 11$), San Nicolas Island, California ($n = 20$). San Miguel and San Nicolas Islands are two of the southern California Channel Islands; both harbour seals and northern elephant seals haul out in large numbers on these islands to breed.

Northern elephant seal samples
Blood samples were obtained from individuals from the following localities: San Miguel Island ($n = 69$, all from the same breeding site, Point Bennett), San Nicolas Island ($n = 21$ from west end beaches, $n = 4$ from south-central beaches, $n = 7$ from south-eastern beaches). A mixture of adult males, adult females and pups-of-the-year were sampled. Only for the pups was it known with certainty that the beach site of collection was the birth site.

DNA preparation and fingerprinting
DNA was extracted from white blood cells by standard methods (Maniatis,

Fritsch & Sambrook 1982). Restriction-enzyme digests and Southern transfers were performed as outlined in Gilbert *et al.* (1990) and Lehman *et al.* (1992). Hybridization of DNA bound to Amersham Hybond N+ nylon membranes was carried out using ^{32}P-radiolabelled minisatellite probes 33.6 and 33.15 (Jeffreys *et al.* 1985a). After two high-stringency washes of the membranes at 50 °C in 0.0015 mol/l sodium citrate/0.015 mol/l sodium chloride/0.5% SDS, autoradiography for 1–7 days was carried out to produce DNA fingerprinting bands of minisatellite loci (Lehman *et al.* 1992).

Analysis of minisatellite loci polymorphisms

Scoring of minisatellite bands was performed as described in Lehman *et al.* (1992). For each gel, the result was a presence-absence matrix of alleles that could be used to assess both overall levels of similarity and relative similarities between groups of individuals. Only individuals run on the same gel were compared, as scoring alleles from different gels is not reliable. Pairwise similarity between individuals x and y was calculated by the similarity index, S (Gill, Jeffreys & Werrett 1985):

$$S_{xy} = 2n_{xy}/n_x + n_y$$

where n_{xy} is the number of alleles in common between individuals x and y, and n_x and n_y are the total number of alleles possessed by individuals x and y, respectively. For a group of individuals, the average similarity, $\bar{S}$, was taken from all pairwise comparisons in the group. This statistic incorporates non-independent components, and its standard error must be calculated accordingly (Lynch 1990) as:

$$\text{S.E.} = \sqrt{2\bar{S}(1-\bar{S})(2-\bar{S})/\bar{n}(4-\bar{S})}$$

where $\bar{n}$ is the average number of alleles possessed per individual in the group. Note that the value of this standard error is not affected by the number of pairwise comparisons used in the generation of $\bar{S}$. For ease of presentation, all similarity values and their standard errors have been multiplied by 1000. To test for significances between average pairwise similarity values of various groups, computer permutation tests were performed to assess the probability that an average value difference as great as the one observed could have been generated by chance in 10 000 random samplings of the data (Lehman *et al.* 1992).

Results

In total, eight informative gels were produced; seven contained samples from only one of the two species and are summarized in Tables 1 and 2. On

Table 1. Sample pairwise similarity matrices. Similarity values were calculated from scoring pinniped genomic DNA digested with Hinf I, separated on a 1% agarose gel, transfered to a nylon membrane and hybridized with the minisatellite probe 33.15 (Jeffreys *et al.* 1985a). Both matrices shown here were derived from samples run on the same gel so that a direct comparison of genetic variability in northern elephant seals and harbour seals from the southern California Channel Islands can be made. For elephant seals, samples A, B, F, H, I, J and K are from individuals caught at San Nicolas Island, while samples C, D, E and G are from individuals caught at San Miguel Island. For harbour seals, all samples are from individuals caught at San Nicolas Island.

	A	B	C	D	E	F	G	H	I	J	K
A	—										
B	974	—									
C	875	833	—		Elephant seals $\bar{S}$ = 895, S.E. = 60.2						
D	842	821	914	—							
E	895	872	914	947	—						
F	914	889	938	857	857	—					
G	895	923	857	895	842	914	—				
H	842	872	914	947	895	857	947	—			
I	821	850	889	872	923	833	821	872	—		
J	919	947	882	811	865	941	919	865	895	—	
K	973	947	882	865	919	941	919	865	842	944	
A	—										
B	533	—									
C	698	474	—		Harbour seals $\bar{S}$ = 566, S.E. = 97.7						
D	542	512	634	—							

one gel, the relative amount of genetic variability between the two species was tested directly by running 11 northern elephant seal samples and four harbour seal samples (Table 1). When this last gel was hybridized with the 33.15 probe, the four harbour seal samples (all from San Nicolas Island) produced an average similarity of 566 (S.E. = 97.7), while the elephant seals produced an overall similarity of 895 (S.E. = 60.2). Because all of the pairwise similarity values between harbour seal samples fall below the range of values produced between elephant seals (Table 1), a permutation test for significance would conclude that, at least for these individuals, northern elephant seals have significantly lower genetic variability at minisatellite loci than harbour seals ($P = 0$). Within elephant seals analysed on this gel, there was no difference between average similarity of the four individuals sampled from San Miguel Island and the seven individuals sampled from San Nicolas Island, as each island produced an average similarity of 895 (S.E. = 60.2).

From the four gels devoted to northern elephant seals, absolute pairwise similarity values ranged from 687 to 1000, and average similarities within groups ranged from 821 to 968 (Table 2). Twenty-two pairwise comparisons

Table 2. Summary of DNA fingerprinting results for gels containing only northern elephant seal samples. Average number of bands scored per individual (*n*) for gels hybridized with the 33.15 probe was 29.6, while *n* for gels hybridized with the 33.6 probe was 12.8. SMI = San Miguel Island; SNI = San Nicolas Island.

Gel I.D. No. & Probe	Comparison	Average similarity	No. of pairwise comparisons	Standard error	Conclusions
Man 4 33.6	Same island, same beach	923	24	54.6	Cannot distinguish by these loci either islands or beaches (thus there is not enough variation, too much gene flow, or both)
	Same island, diff. beach	956	10	41.6	
	SMI v. SMI	968	6	35.6	
	SNI vs. SNI	925	28	53.9	
	SMI vs. SNI	948	32	45.1	
Man 5 33.15	Overall (SMI)	831	120	55.3	There are no sex-dependent trends on San Miguel Island
	Female vs. female	842	45	53.6	
	Male vs. male	835	15	54.7	
	Female vs. male	821	60	56.7	
Man 5 33.6	Overall (SMI)	867	120	96.8	Similarities higher with 33.6 than 33.16
Man 6 33.15	Overall (SNI)	879	55	48.1	Again, no beach-specific genetic partitioning . . . no evidence of philopatry
	Same beach	883	16	47.3	
	Different beach	878	39	48.2	
Man 7 33.15	Same island	886	34	57.0	More evidence of high inter-island similarities
	SMI vs. SMI	941	6	41.6	
	SNI vs. SNI	875	28	59.5	
	SMI vs. SNI	901	32	53.3	

gave similarity values of 1000 (identical genotypes). From these gels it can be seen that the similarity among individuals is consistently high, and there are no sex- or locality-dependent trends in average similarity (Table 2). Within single gels, no comparisons between groups generate a significant difference by a permutation test ($P > 0.5$ in all cases).

From the three gels devoted to harbour seals, absolute pairwise similarity values ranged from 75 to 1000, and average similarities within groups ranged from 279 to 928 (Table 3). Similarities with the 33.6 probe were higher than similarities with the 33.15 probe, a trend seen to a lesser extent in the elephant seal gels. Four pairwise similarity values were calculated as 1000; all of these were produced from the 33.6 probe. For each gel, average similarity values within a particular locality were generally higher than average similarity values between localities (Table 3). However, only for the Pvi 1 gel was the same vs. different area difference significant by a permutation test ($P < 0.05$). Nevertheless, the data do indicate that minisatellite loci can detect geographic partitioning in the harbour seal among the localities surveyed. A measure of the proportion of the total amount of genetic variation that is due to among-locality variation can be given by Wright's (1965) F_{ST} index. For fingerprinting data, F_{ST} can be approximated by the formula of Lynch (1991): $F_{ST} \approx (1 - S_b)/(2 - S_w - S_b)$, where S_b and S_w are average between- and within-similarities for the localities in consideration. Application of this formula to the data from the Pvi 3 gel probed with 33.6 gives an F_{ST} value of 0.52 when the Washington (WA), San Francisco (SF), San Miguel Island (SMI) and San Nicolas Island (SNI) localities are considered. In each gel, similarities within the Channel Island (CI) populations are higher than similarities within the Washington and San Francisco localities (as only one seal sample from Alaska was scored, variation in Alaskan populations could not be addressed). Although variability can always be ranked in the following order: SF > WA > CI, only the SF vs. CI comparisons generate significant differences ($P < 0.05$) by permutation tests.

Discussion

This chapter has documented three trends with respect to genetic variation in nuclear minisatellite loci in harbour seals and northern elephant seals. First, it is clear that variability at these loci is greater in the harbour seal than in the northern elephant seal. The average pairwise similarity value for the Channel Islands populations of northern elephant seals is roughly 900, ranking it as one of the highest reported to date. Thus we provide additional evidence that this species is depauperate in genetic diversity, leaving open the possibility that some species may not require large amounts of polymorphism to persist and even proliferate. However, we cannot reject

Table 3. Summary of DNA fingerprinting results for gels containing only eastern Pacific harbour seal samples. Average number of bands scored per individual (n) for gels hybridized with the 33.15 probe was 25.8, while n for gels hybridized with the 33.6 probe was 11.9. SMI = San Miguel Island; SNI = San Nicolas Island.

Gel I.D. No. & Probe	Comparison	Average similarity	No. of pairwise comparisons	Standard error	Conclusions
Pvi 1 33.6	Same area				There is some geographic partitioning and these loci can detect it. Gene flow follows a stepping-stone model along west coast. More variability in Washington than in the Channel Islands.
	Washington (WA) vs. WA	818	60	122.7	
	SMI vs. SMI	741	15	142.0	
	SNI vs SNI	839	28	116.2	
	Different area	800	1	127.8	
	WA vs. San Francisco (SF)	693	76	151.3	
	WA vs. Channel Islands (CI)	736	6	143.1	
	SF vs. CI	679	60	153.6	
		750	10	140.1	
Pvi 2 33.6	Same area				Same as above, with even a slight differentiation between SMI and SNI. Yet north-south trend not as strong: WA closer to CI than to SF. However, SF still closer to CI than to WA.
	WA vs. WA	815	70	75.9	
	SMI vs. SMI	760	15	84.6	
	SNI vs. SNI	835	21	72.1	
	Different area	840	6	71.1	
	WA vs. SF	731	83	88.5	
	WA vs. CI	711	6	90.9	
	SF vs. CI	723	66	89.5	
	SMI vs. SNI	791	11	79.9	
		824	28	74.2	

Pvi 3 33.6	Same area	828	30	109.6	Again, geographic partitioning seen. Channel Islands much more similar than any other comparison. WA and SF more heterogeneous, especially SF. AK is clearly distinct from other localities (sample size of one). As above, WA as close to CI as to SF, reflecting the high variability at SF or higher gene flow to and from localities outside the Channel Islands.
	Same area (all sample sizes adjusted to 3)	860	12	99.9	
	WA vs. WA	807	21	115.3	
	SF vs. SF	780	3	121.8	
	SMI vs. SMI	928	3	73.0	
	SNI vs. SNI	926	3	73.9	
	Different area	815	106	113.2	
	Alaska (AK) vs. WA	706	7	136.4	
	AK vs. SF	669	3	142.1	
	AK vs. SMI	653	3	144.3	
	AK vs. SNI	621	3	148.1	
	WA vs. SF	828	21	109.6	
	WA vs. SMI	821	21	111.6	
	WA vs. SNI	846	21	104.3	
	SF vs. SMI	826	9	110.2	
	SF vs. SNI	853	9	102.1	
	SMI vs. SNI	906	9	82.9	
Pvi 3 33.15	WA vs. WA	401	21	91.0	Same as above: within Channel Islands similarity highest, and Alaska and Washington populations distinct from Channel Islands.
	SF vs. SF	322	3	87.9	
	SNI vs. SNI	494	3	91.2	
	AK vs. WA	366	7	89.9	
	AK vs. SF	223	3	79.5	
	WA vs. SF	279	21	84.9	

the hypothesis that monomorphism would ultimately be detrimental to the species' survival.

Second, we cannot detect any island or beach philopatry in the northern elephant seal, as no trends in allelic similarity exist among localities. This situation may be a consequence of the low level of variation present in the elephant seal to use for comparison, or of high amounts of gene flow among breeding localities, or of both these factors. It would be of interest in the future to compare individuals that are born at Año Neuvo or Isla de Guadalupe, as these are the most geographically distant breeding localities of the northern elephant seal that support large populations (Stewart *et al.* in press).

Third, we suggest that there is geographic population-genetic partitioning in the harbour seal along the west coast of North America. The partitioning is not absolute, as only approximately half of the observed variation can be attributed to among-locality variation. Thus there is also evidence of gene flow along the coast. Because there is a slight trend for geographically closer localities to exhibit higher average similarities, the flow appears to follow a stepping-stone pattern latitudinally. Though harbour seals typically demonstrate site fidelity at haul-out grounds during successive years and forage in waters near their breeding sites, individuals have been observed migrating as much as 550 km along the west coast (Reeves, Stewart & Leatherwood 1992). Occasional movements of this distance could account for the high similarity values observed between individuals from San Francisco and the Channel Islands.

Acknowledgements

We thank K. Frost, L. Lowry, H. Huber, S. Jeffries and D. Kopec for providing harbour seal samples from Alaska, Puget Sound and San Francisco localities. We would also like to thank Pamela Yochem, Robert DeLong, Klaus-Peter Koepfli and Debra Decker for field and technical assistance. B.S.S. was supported by U.S Air Force Contract FO4701–81–C–0081. Research was conducted under authorization of Marine Mammal Permits Nos 71, 341, 367, 510 and 579.

References

Bartholomew, G. A. & Hubbs, C. L. (1960). Population growth and seasonal movements of the northern elephant seal, *Mirounga angustirostris. Mammalia* **24**: 313–324.

Bonnell, M. L. & Selander, R. K. (1974). Elephant seals: genetic variation and near extinction. *Science* **184**: 908–909.

Burke, T. (1989). DNA fingerprinting and other methods for the study of mating success. *Trends Ecol. Evol.* **4**: 139–144.

Gilbert, D. A., Lehman, N., O'Brien, S. J. & Wayne, R. K. (1990). Genetic fingerprinting reflects population differentiation in the California Channel Island fox. *Nature, Lond.* **344**: 764–767.

Gill, P., Jeffreys, A. J. & Werrett, D. J. (1985). Forensic applications of DNA 'fingerprints.' *Nature, Lond.* **318**: 577–579.

Harris, A. S., Young, J. S. F. & Wright, J. M. (1991). DNA fingerprinting of harbor seals (*Phoca vitulina concolor*): male mating behaviour may not be a reliable indicator of reproductive success. *Can. J. Zool.* **69**: 1862–1866.

Hoelzel, A. R. & Dover, G. A. (1991). Genetic differentiation between sympatric killer whale populations. *Heredity, Lond.* **66**: 191–195.

Hoelzel, A. R., Halley, J., Campagna, C., Arnbom, T. & Le Boeuf, B. J. (1991). Molecular genetic variation in elephant seals; the effect of population bottlenecks. *Abstr. bienn. Conf. Biol. mar. Mammals* **9**: 34 Society for Marine Mammalogy. Unpublished.

Jeffreys, A. J., Wilson, V. & Thein, S. L. (1985a). Hypervariable 'minisatellite' regions in human DNA. *Nature, Lond.* **314**: 67–73.

Jeffreys, A. J., Wilson. & Thein, S. L. (1985b). Individual-specific 'fingerprints' of human DNA. *Nature, Lond.* **316**: 76–79.

Lehman, N., Clarkson, P., Mech, L. D., Meier, T. J. & Wayne, R. K. (1992). A study of the genetic relationship within and among wolf packs using DNA fingerprinting and mitochondrial DNA. *Behav. Ecol. Sociobiol.* **30**: 83–94.

Lynch, M. (1990). The similarity index and DNA fingerprinting. *Molec. Biol. Evol.* **7**: 478–484.

Lynch, M. (1991). Analysis of population genetic structure by DNA fingerprinting. In *DNA fingerprinting: approaches and applications*: 113–126. (Eds Burke, T., Dolf, G., Jeffries, A. J. & Wolff, R.). Birkhauser Verlag, Basel, Switzerland.

Maniatis, T., Fritsch, E. F. & Sambrook, J. (1982). *Molecular cloning: a laboratory manual.* Cold Spring Harbor Laboratory, New York.

O'Brien, S. J. & Evermann, J. F. (1988). Interactive influence of infectious disease and genetic diversity in natural populations. *Trends Ecol. Evol.* **3**: 254–259.

O'Brien, S. J., Roelke, M. E., Marker, L., Newman, A., Winkler, C. A., Meltzer, D., Colly, L., Evermann, J. F. & Wildt, D. E. (1985). Genetic basis for species vulnerability in the cheetah. *Science* **277**: 1428–1434.

O'Brien, S. J., Wildt, D. E., Goldman, D., Merril, C. R. & Bush, M. (1983). The cheetah is depauperate in genetic variation. *Science* **221**: 459–462.

Ralls, K., Brugger, K. & Ballou, J. (1979). Inbreeding and juvenile mortality in small populations of ungulates. *Science* **206**: 1101–1103.

Reeve, H. K., Westneat, D. F., Noon, W. A., Sherman, P. W. & Aquadro, C. F. (1990). DNA 'fingerprinting' reveals high levels of inbreeding in colonies of the eusocial naked mole-rat. *Proc. natn. Acad. Sci. USA* **87**: 2496–2500.

Reeves, R. R., Stewart, B. S. & Leatherwood, S. (1992). *The Sierra Club handbook of seals and sirenians.* Sierra Club Books, San Francisco, California.

Stewart, B. S. (1989). *The ecology and population biology of the northern elephant seal, Mirounga angustirostris Gill 1866, on the Southern California Channel Islands.* PhD diss.: University of California, Los Angeles.

Stewart, B. S., Le Boeuf, B. J., Yochem, P. K., Huber, H. R., DeLong, R. L., Jameson, R., Sydeman, W. & Allen, S. G. (In press). History and present status of the northern elephant seal population. In *Elephant seals*. (Eds Le Boeuf, B. J. & Laws, R. M.). University of California Press, Berkeley, CA.

Thompson, J. D. (1991). Phenotypic plasticity as a component of evolutionary change. *Trends Ecol. Evol.* **6**: 246–249.

Townsend, C. H. (1885). An account of recent captures of the California sea-elephant, and statistics relating to the present abundance of the species. *Proc. U.S. natn. Mus.* **8**: 90–94.

Townsend, C. H. (1912). The northern elephant seal, *Macrorhinus angustirostris*, Gill. *Zoologica, N.Y.* **1**: 159–173.

van Pijlen, I. A., Amos, B. & Dover, G. A. (1991). Multilocus DNA fingerprinting applied to population studies of the minke whale *Balaenoptera acutorostrata*. *Rep. int. Whal. Commn spec. Issue* No. 13: 245–254.

Wright, S. (1965). The interpretation of population structure by F-statistics with special regard to systems of mating. *Evolution, Lancaster, Pa* **19**: 395–420.

Symp. zool. Soc. Lond. (1993) No. 66: 61–93

Implications of DNA fingerprinting for mating systems and reproductive strategies of pinnipeds

D. J. BONESS[1]
W. D. BOWEN[2]
and J. M. FRANCIS[1]

[1]*Department of Zoological Research
National Zoological Park
Smithsonian Institution
Washington, DC 20008, USA*

[2]*Marine Fish Division
Bedford Institute of Oceanography
Dartmouth
Nova Scotia B2Y 4A2 Canada*

Synopsis

Two areas of research on reproductive strategies where DNA fingerprinting is proving most useful include providing direct measures of parentage, to validate observational estimates of mating success, and of relatedness among individuals in a group, to assess the possible influence of kin selection on behaviour. Analysis of nuclear DNA with single- or multi-locus probes is the most widely used technique for investigating questions in these areas. A major gap in our knowledge of pinniped mating strategies is the lack of estimates of mating success among aquatically mating species, in which males are least able to monopolize females, and which thus represent one end of the continuum of potential monopoly. Determining paternity through fingerprinting will not only provide the sole estimates of variance in mating success in these species, but, given our limited ability to observe behaviour underwater, it may help us to discern the strategies by which males achieve success. Among terrestrially mating species, estimates of individual success exist for only a few species and these may be inaccurate. In many species female movements may occur for various reasons so that females associate with more than one male and we cannot infer paternity from observed associations. Females may also be promiscuous in some species, providing the potential for sperm competition. Improved knowledge of paternity among terrestrially mating species may help to assess the importance of mate choice where female movements occur, and to improve our understanding of the success of alternative strategies. Co-operative behaviour, which is associated with high relatedness in other species, generally is not apparent in pinnipeds, but some species exhibit frequent fostering. Fingerprinting could provide the evidence to implicate kin selection. Although DNA fingerprinting should help to advance our

ZOOLOGICAL SYMPOSIUM No. 66
ISBN 0–19–854069–8

Copyright © 1993 The Zoological Society of London
All rights of reproduction in any form reserved

knowledge of pinniped reproductive strategies, we must also improve data on behavioural and other phenotypic characteristics to accompany the better estimates of mating success and relatedness.

Introduction

Major advances in behavioural ecology and population biology are sometimes made possible by new technology such as isotope methods (reviewed in Huntley *et al.* 1987) or VHF and satellite telemetry (Amlaner & MacDonald 1980). Although mitochondrial DNA methods have been in use for nearly a decade to assess genetic differences among populations (Moritz, Dowling & Brown 1987), recent rapid advances in both nuclear and mitochondrial DNA methodologies are creating the opportunity for significant advances in behavioural ecology, particularly with respect to the evolution of reproductive strategies (Burke 1989). DNA fingerprinting techniques are now being used to assess parentage and thus mating success of individual animals (Quinn *et al.* 1987; Morton, Forman & Braun 1990; H. G. Smith *et al.* 1991; Hoelzel this volume) and the role of relatedness in explaining patterns of social behaviour (Reeve *et al.* 1990; Packer, Gilbert *et al.* 1991).

To date many applications of DNA techniques to marine mammals have addressed questions of phylogeny and stock structure and have involved cetaceans more than pinnipeds (e.g. Arnason & Widegren 1986; Hoelzel & Dover 1989; Hoelzel, Hancock & Dover 1991). The use of DNA fingerprinting to study cetacean mating systems and reproductive behaviour has also been most prevalent (Amos & Dover 1990; Amos & Hoelzel 1990). Two probable reasons for this difference are that: (1) for many pinnipeds reproductive behaviour can be observed directly, whereas in cetaceans direct observation of individuals is difficult, and (2) pinnipeds do not form small social groups or exhibit obvious co-operative behaviour which might suggest a role for kin selection, whereas co-operative behaviour is often found in cetacean social groups.

In this paper we discuss selected areas of pinniped behavioural ecology and population biology which may be advanced by DNA studies. These areas include mating systems, life-history patterns, gene flow between species resulting from interbreeding of expanding populations, and the importance of genetic relatedness (e.g. kinship) in behaviour such as fostering.

Genetic methods

A number of recently developed molecular genetic approaches using DNA sequence variation have revolutionized our ability to determine the genetic relationships and mating patterns of free-ranging animals. Several recent papers provide excellent introductory and comprehensive treatments of these methods and their applications in behavioural ecology (Burke 1989; Kirby 1990; Pemberton & Amos 1990; Hoelzel 1992, this volume). Thus we provide only a brief introduction to genetic methods here.

The most widely known of these techniques is termed DNA fingerprinting, which involves a family of minisatellite sequences in nuclear DNA that have in common a core sequence of about 12 nucleotides (Jeffreys, Wilson & Thein 1985a). Individual variation in the pattern of restriction-digested DNA fragments revealed by a specific radio-labelled probe is visually expressed as a series of bands on an autoradiograph, hence the analogy to a fingerprint or product code. Fingerprints can be used to determine parentage since approximately 50% of the bands in offspring will come from the mother while the remaining bands will come from the father. Thus we exclude individuals as parents if bands in offspring are not in either parent and determine a band-sharing coefficient with each individual that shares some bands to determine which male and/or female is the parent(s).

Some DNA probes will hybridize to only a single locus, others to many loci, depending on the probe specificity and the analysis conditions used by the investigator. Hypervariable single-locus probes produce fewer bands, making them easier to interpret than multilocus probes but, because they tend to be family-specific, new probes must be developed for each family studied.

The analysis of microsatellite DNA using PCR amplification (Morin & Woodruff 1992) appears to be gaining acceptance as the preferred fingerprinting technique because it permits resolving fragment lengths to the nearest two base pairs so that one can compare directly across samples on different gels. The method is also less time-consuming and less costly than minisatellite techniques, and the resulting banding patterns are more suitable for computerized cataloguing of individual animals.

Mitochondrial DNA (mtDNA) is an informative genetic marker for studies of matrilineal gene flow (Avise, Bowen & Lamb 1989). Because mtDNA is maternally inherited, these fingerprints might find special application in situations where maternity is in doubt, such as studies of fostering.

Mating systems and reproductive strategies

Mating systems are an expression of how individual animals behave to maximize their reproductive success (N. B. Davies 1991). Males should maximize fitness by competing for as many mates as possible, provided that females do not require their assistance to rear offspring, whereas females should maximize fitness by ensuring offspring survival and choosing 'good' mates (Bateman 1948; Trivers 1972; Clutton-Brock 1988). The extent to which the sexes exhibit the different strategies and the precise form of them in a given species or population should be a function of ecological and phylogenetic constraints (Bartholomew 1970; Emlen & Oring 1977; Bradbury & Vehrencamp 1977; Rubenstein & Wrangham 1986; Boness 1991).

Pinnipeds provide an opportunity to study the influence of ecological variation on differences in mating systems because there is both extreme variation in the reproductive strategies of each sex and considerable diversity in breeding habitats and climates used by these animals (Stirling 1983; Bonner 1984; Oftedal, Boness & Tedman 1987; Boness 1991; Le Boeuf 1991). Males play no role in rearing offspring and devote their entire effort during the reproductive period to the acquisition of mates. Females primarily devote their effort to rearing offspring although in some species they may also enhance fitness through mate choice (Cox & Le Boeuf 1977; Boness, Anderson & Cox 1982; Heath 1989; Kuroiwa & Majluf 1989).

Pinnipeds are most well known for their extreme levels of polygyny (Ralls 1976; Emlen & Oring 1977; Alexander et al. 1979; Borgia 1979; N. B. Davies 1991). However, it has become increasingly apparent that they also exhibit a wide range of levels of polygyny and mating strategies (Table 1). Although we have some understanding of the ecological, phylogenetic and social factors which may account for this variation, the empirical base is limited (Bartholomew 1970; Stirling 1983; Boness 1991; Le Boeuf 1991).

To advance our knowledge of the evolution of mating systems and variation in reproductive strategies we need better data relating fitness of different phenotypes and genotypes to ecological factors. To date such efforts have used imprecise estimates of mating success from behavioural observations. What have been assumed to be reasonable estimates of mating success in one species may not be reasonable in others. For example, using DNA fingerprinting, recent studies have found that extra-pair copulations in avian species were successful so that species considered to be monogamous are now viewed as polygynous (Rabenold, Rabenold, Piper, Haydock & Zack 1990; Westneat 1990; Birkhead et al. 1990). However, in other socially monogamous birds and mammals, DNA analysis showed that extra-pair copulations were not successful (Gyllensten, Jakobsson & Temrin 1990; Ribble 1991). Fingerprinting analyses of polygynous red deer, *Cervus elaphus*, (Pemberton et al. 1992) and red-winged blackbirds, *Agelaius phoeiceus*, (Gibbs et al. 1990) have markedly changed estimates of the mating success of individuals and of the population variance in reproductive success. In the red deer study, observational measures of mating success underestimated the success of the most successful males and overestimated that of the less successful males (Pemberton et al. 1992).

The potential for similar errors in assessment of mating success exists

Table 1. Summary of current knowledge about pinniped mating systems

Species	Mating system	Principal and alternative male strategies[a]	Estimate of mating success					References
			Max.	% FF[b]	Mean	Variance (n)[c]	Basis for estimate	
Terrestrially mating species								
Arctocephalus australis	Resource defence polygyny?	a. Defend large birth/thermoregulatory sites	?	?	16	?	Number of FF/MM	Majluf (1987)
		b. Defend small thermoregulatory sites	?	?	10	?		
		c. Roam/defend peripheral territories	?	?	?	?		
	Lek polygyny?	a. Defend thermoregulatory sites to which females come to mate	?	?	?	?	—	Kuroiwa & Majluf (1989)
		b. Defend birth sites	?	?	?	?		
A. forsteri	Resource defence polygyny	a. Defend birth sites—shoreline & inland territories	30	35	5.3	10 (32)	Copulations (uncorr.) & number of FF-MM	Miller (1975)
		b. Roam/defend peripheral territories						Stirling (1971)
A. gazella	Resource defence polygyny	a. Defend birth sites—shoreline land-locked	?	?	11.2	?	Copulations (corr.)	McCann (1980)
			?	?	8.0	?		
			19	17	9.2	3 (12)	Number of FF/MM (est.)	Boyd (1989)
A. galapagoensis	Resource defence polygyny?	a. Defend large birth sites—continuous & discontinuous territory holders	14	24	5.9	3 (10)	Copulations (uncorr.)	Trillmich (1984, 1987)
		b. Roam/defend peripheral territories	?	?	?			
A. philippii	Resource defence polygyny	a. Defend thermoregulatory sites (aquatic)	8.0	24	2.3	2 (33)	Copulations (uncorr.)	Francis & Boness (1991)

Table 1. (*cont.*)

Species	Mating system	Principal and alternative male strategies[a]	Estimate of mating success				Basis for estimate	References
			Max.	% FF[b]	Mean	Variance (n)[c]		
		b. Defend birth/thermo-regulatory sites	6.5	20	1.3			
		c. Defend birth sites	5.0	12	0.7			
		d. Roam/defend peripheral territories	?	?	?			
A. pusillus	Resource defence polygyny?	a. Defend birth sites	66	40	24.7	20 (7)	Number of birth/MM	Rand (1967)
		b. Roam/defend peripheral territories	?	?	?			
		c. Roam aquatic periphery of breeding area	?	?	?			
A. p. doriferus	Resource defence polygyny?	a. Defend birth sites	63	?	53	?	Number of births/MM	David (1987)
		b. Roam/defend peripheral territories	?	?	9	?	Number of births/MM	Warneke & Shaughnessy (1985)
A. townsendi	Resource defence polygyny	a. Defend birth/thermo-regulatory site	13	?	3	?	Number of FF/MM	Pierson (1987)
		b. ?						
A. tropicalis	Resource defence polygyny?	a. Defend birth sites	?	?	6.6	?	Number of births/MM	Bester (1982)
		b. Roam/defend peripheral territories	?	?	?			
Callorhinus ursinus	Resource defence polygyny	a. Defend large birth sites	50	28	20.1	14 (8)	Number of FF/MM	Peterson (1965, 1968)
		b. Roam/defend peripheral territories						
		c. Group raiding to acquire territories						J. M. Francis (pers. comm.)
	Resource defence polygyny	a. Defend birth sites	161 153	16 21	56.3 42.4	41 (19) 36 (19)	Copulations (corr.) Number of FF/MM (est.)	Bartholomew & Hoel (1953)
		b. Roam/defend peripheral territories	?	?	?			

Species	Mating system	Behaviour					Measurement	Reference
	Resource defence polygyny	a. Defend birth sites	?	?	?	—	—	DeLong (1982: Castle Rock)
		b. Roam/defend peripheral territories	?	?	?			
	Female defence polygyny?	a. Shift areas as females move	?	?	?	—	—	DeLong (1982: Adams Cove)
		b. Defend fixed (birth?) site	?	?	?			
Eumetopias jubatus	Resource defence polygyny	a. Defend birth sites	32	21	11.9	10 (17)	Copulations (uncorr.)	Gentry (1970)
		b. Defend resting sites	?	?	?			
		c. Defend thermoregulatory sites						
		d. Roam terrestrially & aquatically						
	Resource defence polygyny	a. Defend birth sites	?	?	3.1	?	Number of FF/MM	Sandegren (1970)
		b. Defend birth/thermoregulatory sites	?	?	14.7			
		c. Defend thermoregulatory sites			2.3			
		d. Roam/defend peripheral territories	?	?	?			
Neophoca cinerea	Resource defence/female defence polygyny?	a. Defend very large birth sites? or defend females sequentially?	7	7	2.0	1 (90)	Copulations & number of FF/MM	Higgins (1990)
		b. Roam periphery of main breeding area	?	?	?			
		c. Group raiding to acquire females/ positions						
Otaria byronia	Female defence polygyny	a. Defend females in central breeding area	18	3	4.5	3 (23)	Copulations (corr.)	Campagna & Le Boeuf (1988a: Punta Norte)
		b. Defend females in peripheral area	?	?	?			
		c. Roam periphery of main breeding area	?	?	?			
		d. Group raiding to acquire females/ positions	?	?	?			

Table 1. (*cont.*)

Species	Mating system	Principal and alternative male strategies[a]	Estimate of mating success					References
			Max.	% FF[b]	Mean	Variance (n)[c]	Basis for estimate	
	Resource defence polygyny	a. Defend birth/thermo-regulatory sites	20	?	11.6	5 (11)	Copulations (?)	Campagna & Le Boeuf (1988b: Puerto Piramide)
		b. Defend birth sites	?	?	1.4			
		c. Roam/defend peripheral territories	?	?	?			
		d. Group raiding to acquire females/positions	?	?	?			
Phocarctos hookeri	Resource/female defence/lek polygyny?	a. Defend females and/or fixed site	19	12	5.2	3 (45)	Copulations (uncorr.)	Genty & Robert in Boness (1991)
		b. Roam/defend peripheral territories?	?	?	?			
			?	?	?			
	Resource defence	a. Defend birth sites	50	?	27.5	?	Number of FF/MM	Marlow (1975)
		b. Roam/defend peripheral territories	?	?	?			
Zalophus californianus	Lek polygyny?	a. Defend thermo-regulatory sites which females visit and mill (solicit mating)	30	30	4.7	8 (50)	Copulations (uncorr.)	Heath (1989: Reef)
			15	20	3.2	3 (28)	Copulations (uncorr.)	Heath (1989: White Scar)
		b. Defend birth/resting sites	?	?	?			
		c. Roam/mimic females	?	?	?			J. M. Francis (pers. comm.)
	Lek polygyny?	a. Defend thermo-regulatory sites which females visit and mill (solicit mating)	?	?	?	?	—	Heath (1989: Santa Barbara Is.)
		b. Defend birth/resting sites						

Species	Resource defence/lek polygyny?	Behaviour					Method	Reference
	Resource defence/lek polygyny?	a. Defend aquatic thermo-regulatory sites in which females raft in groups	18	95	4.8	17 (4)	Copulations (uncorr.)	Heath (1989: Isla de la Guarda)
		b. Defend birth/resting sites	0	0	0			
Mirounga angustirostris	Female defence polygyny	a. Defend females by establishing dominance hierarchy before female arrival	90	67	17.5	41 (8)	Copulations (corr.)	Le Boeuf (1974)
		b. Roam periphery/try to sneak copulations	?	?	?			
Mirounga leonina	Female defence polygyny	a. Defend females by establishing dominance before female arrival	126	38	28	52 (10)	Copulations (uncorr.)	McCann (1981)
		b. Roam periphery/try to sneak copulations						
	Female defence polygyny	a. Defend large harem of females	?	?	30	?	Number of FF/MM	van Aarde (1980)
		b. Roam periphery of harems	?	?	?			
Halichoerus grypus	Female defence polygyny	a. Co-dominant males defend small clusters of females	6	10	3.0	1 (34)	Copulations (uncorr.)	Boness & James (1979)
		b. Roam periphery/try to sneak copulations	?	?	?			
	Female defence polygyny	a. Defend small clusters of females through poorly defined hierarchy	11	10	4.1	3 (32)	Copulations (uncorr.)	Anderson et al. (1975)
		b. Roam periphery of female herd	?	?	?			
Lobodon carcinophagus	Female defence polygyny?	a. Defend females sequentially on ice?	?	?	?	?	—	Siniff et al. (1979)
		b. ?						Shaughnessy & Kerry (1989)

Table 1. (*cont.*)

Species	Mating system	Principal and alternative male strategies[a]	Estimate of mating success					References
			Max.	% FF[b]	Mean	Variance (n)[c]	Basis for estimate	
Aquatically mating species								
Odobenus rosmarus	Lek?	a. Defend territory off ice floes & whistle display b. Roam periphery of displaying males	?	?	?	?	—	Fay *et al.* (1984)
Halichoerus grypus	Resource defence polygyny?	a. Defend aquatic resting sites b. ?	? ?	? ?	2.3 ?	? ?	Number of FF/MM	Hewer (1957)
Leptonychotes weddelli	Resource defence/lek polygyny?	a. Defend territories at traditional ice leads b. Associate with females on ice & roam	8 ?	52 ?	2.5 ?	3 (20) ?	Number of females with grease marks	Hill (1987) Kaufman *et al.* (1975)
Cystophora cristata	Female defence?	a. Sequentially defend & mate females b. Roam among herd and challenge males?	? ?	? ?	? ?	? ?	—	Boness, Bowen & Oftedal (1988) Kovacs (1990)
Erignathus barbatus	Lek polygyny?	a. Sing under ice (displaying to females)? b. ?	?	?	?	?	—	Ray *et al.* (1969)
Phoca vitulina	Resource defence/lek polygyny?	a. Defend aquatic resting area/beach access? c. Roam?	? ?	? ?	? ?	? ?	—	Sullivan (1981) Perry (1993)
Monachus schauinslandi	Scramble competition polygyny?	a. Patrol along rookery for receptive females b. ?	?	?	?	?	—	Deutsch (1985)
Phoca groenlandica	Polygyny?	?	?	?	?	?	—	Merdsoy, Curtsinger & Renouf (1978)

Species	Strategy[a]							Reference
Phoca hispida	Monogamy?	?	?	?	?	?	—	Stirling (1977)
	Resource defence polygyny?	a. Defend territory with multiple birth lairs b. ?	?	?	?	?	—	T. G. Smith & Hammill (1981)
Phoca largha	Monogamy?	?	?	?	?	?	—	Fay (1974)
Phoca sibirica	Polygyny?	?	?	?	?	?	—	Thomas *et al.* (1982)
Hydrurga leptonyx	?	?	?	?	?	?	—	
Phoca caspica	?	?	?	?	?	?	—	
Phoca fasciata	?	?	?	?	?	?	—	

[a] Male strategies for acquiring mates are listed in the probable order of success. The strategies are our categorization based on descriptions by the authors.

[b] Percentage of total number of females in the study or the number of observed copulations, depending on the basis for the estimate.

[c] Variance among individual males divided by the mean. Calculation of mean and variance included only one male with no matings to standardize this across species. Data on males with no copulations are lacking for many species. *n* = the number of males observed.

among pinnipeds. In the following paragraphs we review estimates of individual mating success, discuss how these estimates might be biased, discuss how terrestrial and aquatic mating affect our ability to measure mating success, and briefly describe what additional information might improve our understanding of pinniped mating systems.

Estimates of mating success

Terrestrially mating species

We make a distinction between terrestrially and aquatically mating species because differences in the mobility of animals in these two media and in our ability to observe them result in somewhat different problems in estimating mating success. For the same reason, we consider on-ice mating as equivalent to terrestrial mating. In 16 of the 34 pinniped species all mating is predominantly or exclusively terrestrial. This is true for 13 of the 15 otariids and three of the 18 phocids. The California sea lion, *Zalophus californianus*, and the Juan Fernández fur seal, *Arctocephalus philippii*, appear to be exceptions among the otariids. The grey seal, *Halichoerus grypus*, one of the three terrestrial phocids, exhibits aquatic mating at some colonies (see below). Additionally, there are two species that give birth on ice which may also mate there. In the crabeater seal, *Lobodon carcinophagus*, although direct observation of mating is lacking, observation of sexual behaviour (Shaughnessy & Kerry 1989) and speculation on the effect of leopard seal, *Hydrurga leptonyx*, predation (Siniff *et al.* 1979) suggest that on-ice mating should be typical. Harp seals, *Phoca groenlandica*, may also occasionally mate on the ice in years of particularly heavy ice coverage (Popov 1966).

Among the terrestrially mating pinnipeds, estimates of mating success are basically of two types, an overall or average estimate for the population (the ratio of the number of reproductive females, or pups born, to the number of reproductive males; see Emlen & Oring 1977) and individual records for a representative sample of individuals in a given reproductive season. Average estimates of mating success for the population are available from most studies and suggest that the terrestrially mating species are polygynous (Table 1). Although such average estimates may be useful for making broad comparisons of the influence of ecological factors on sexual selection and mating systems, they are insufficient for estimating the intensity or degree of sexual selection because they do not directly measure mating success. A measure of standardized variance in reproductive success among males would be more informative (Wade & Arnold 1980). A variance measure requires records of mating success for individual males in the population and thus permits questions regarding strategies of individuals to be addressed. For example, one can assess: (1) the success of individuals exhibiting different behavioural or other phenotypic (e.g. age, size, adornments for display or competition) patterns within or between colonies; (2) variation in individual success from one reproductive period to another; or (3) differences in lifetime reproductive success of individual animals that may follow different behavioural courses throughout life. Estimates of variance in mating success among males are available for 14 terrestrially mating species (Table 1). For most species estimates are available from only a single colony and a single reproductive season. Few studies have tried to assess the mating success of males exhibiting different phenotypes in the form of either behavioural strategies or physical characteristics (cf. Le Boeuf 1974; McCann 1981; Deutsch, Haley & Le Boeuf 1990; Francis & Boness 1991; Godsell 1991). The northern elephant seal, *Mirounga angustirostris*, and the Steller's sea lion, *Eumetopias jubatus*, are the only species for which longitudinal records of individual male mating success are available to estimate lifetime reproductive success (Gisiner 1985; Le Boeuf & Reiter 1988).

Aquatically mating species

Fifteen of the 18 species of phocid seals are thought to mate aquatically. Aquatic mating is generally assumed because mating has rarely or never been observed in many of these species (Le Boeuf 1991). Also our knowledge of when mating occurs is based generally on the morphological or histological examination of ovaries, not on observations of mating behaviour. In 12 of the 15 species, aquatic mating is associated with the use of ice as a rearing habitat.

Aquatic mating has also been reported at some colonies of several pinniped species which typically mate on land. Grey seal males appear to maintain aquatic territories or at least mate aquatically at a number of colonies in Scotland and Wales (J. L. Davies 1949; Hewer & Backhouse 1960). Aquatic mating is suspected at ice-breeding concentrations in the southern Gulf of St Lawrence, Canada (M. Hammill pers. comm.) and the Baltic Sea (Curry-Lindahl 1970). In the California sea lion, most mating occurs in the intertidal area of the rookery at San Nicolas and Santa Barbara Islands in California and all mating in a colony in Mexico was aquatic (Heath 1989). In the Juan Fernández fur seal about half the observed matings occurred in entirely aquatic territories in relatively calm waters adjacent to the rookery (Francis & Boness 1991). Females of several otariid species begin foraging trips before becoming receptive and therefore some females may become receptive away from the rookery (California sea lions—Heath 1989; African fur seal, *Arctocephalus pusillus*—David & Rand 1986; Steller's sea lion—Gentry 1970; Higgins *et al.* 1988), making it possible for some males to adopt an alternative strategy and follow females to sea.

Direct estimates of mating success in aquatically mating species do not exist, with the limited exception of recent data on the Weddell seal, *Leptonychotes weddelli*, the California sea lion and the Juan Fernández fur seal (Hill 1987; Heath 1989; Francis & Boness 1991). In the Weddell seal where aquatic mating occurs under the ice, Hill applied coloured grease to males and showed that males came into physical contact with as many as eight females, suggesting a moderate degree of polygyny. Although mating can be reasonably inferred from this ingenious method, a direct measure of mating success is still lacking in this and most other aquatically mating species. The use of molecular genetic techniques could significantly improve this lack of information.

Possible sources of error in estimates of mating success

The measures used to estimate individual mating success in pinnipeds are similar to those used for other animals and include: (1) the number of females in association with a given male, (2) the number of young born in close proximity to him, or (3) the number of copulations that he is observed to achieve. Copulations might be expected to be a better measure than those based on associations, but the error in various measures may not be constant across species or even across populations within species.

Associations of either females or pups with a male may provide reasonable estimates of mating success in only a small number of species because females in many species move about the rookery during the time when they become receptive and these movements tend to place females in different males' areas of influence (Table 2). This is probably true of most, if not all, aquatically mating species and is particularly true among the terrestrial otariids, where the tendency seems to be for males to defend a territory rather than to move with females as the females move. Strong evidence to indicate that most females are sedentary prior to sexual receptivity is available for only two otariids, the northern fur seal, *Callorhinus ursinus* (Bartholomew & Hoel 1953; Peterson 1968) and the Antarctic fur seal, *Arctocephalus gazella* (McCann 1980; Boyd 1989). There is also some evidence which suggests most females are sedentary in the New Zealand fur seal, *Arctocephalus forsteri* (Miller 1975) and the Galapagos fur seal, *Arctocephalus galapagoensis* (Trillmich 1984). However, in these species there may be differences among populations and/or conflicting evidence. DeLong (1982) describes daily thermoregulatory movements for northern fur seal females at two colonies in the southernmost part of their range. In one of these colonies males remain on their territories while females move. In the other, males may abandon territoriality for defence of females, although DeLong interprets the males' behaviour as shifting territoriality. Furthermore, Bartholomew & Hoel (1953) refer to unexpectedly high rates of copulations for some territorial northern fur seal males and suggest that they are the result of previously mated females copulating as they depart from the rookery for their first foraging trip.

The problem of movement of females and uncertain paternity is not restricted to those species or populations in which males defend territories. In grey seals, where males defend females directly, the nature of female movement and male defence is such that individual females shift from one male to another over time and may be associated with more than one male during the time they are receptive (Boness & James 1979).

Another source of error in estimating mating success from the number of females (or pups born) associated with a male is that, in some species, there are males which roam the rookery and 'sneak' or 'steal' mates. In the Galapagos fur seal, the rugged-boulder topography of the rookery sometimes prevents a male from seeing an intruding male mate with a female he had been defending (Trillmich 1984). In grey seals and elephant seals subordinate males or neighbouring dominant males may mate with a female left undefended while her consort male mates with another female or is occupied in a fight (Carrick, Csordas & Ingham 1962; Le Boeuf 1974; Boness & James 1979). Subordinate male southern sea lions, *Otaria byronia*, and Australian sea lions, *Neophoca cinerea*, engage in group raids to steal females (Marlow 1975; Campagna, Le Boeuf & Cappozzo 1988).

Estimating mating success from observed copulations should produce fewer errors in assigning paternity than using the number of females or births as a measure. However, direct observations in any study at best cover 12–18 hours of the day so that not all matings are observed. In some studies where copulations are used to estimate mating success, adjustments are made assuming that behaviour outside the period of observation is comparable to that during observation (e.g. Bartholomew & Hoel 1953; Le Boeuf 1974; McCann 1980); this is probably not true for species that show daily thermoregulatory movements. Attempts to compensate for unequal duration of observation of different males have also been made (Gisiner 1985; Campagna & Le Boeuf 1988a).

A second source of error in inferring paternity and mating success from observed copulations is the prevalence of female promiscuity in many species. Female promiscuity may result in sperm competition (Ginsberg & Huck 1989), making it impossible to assess paternity without DNA techniques. Observed multiple matings are common among the three terrestrially mating phocids (Table 2). The average number of copulations per female in two studies of grey seals was 2.5 and 2.9 respectively (Anderson, Burton & Summers 1975; Boness & James 1979). In the northern elephant seal a similar value was obtained by Le Boeuf (1972). In both species a high proportion of females copulating more than once also mated with more than one male. The assumption that the first copulation is the one which is successful (Le Boeuf & Peterson 1969; Boness & James

Table 2. Potential sources of error in inferring paternity. Superscripts refer to the references in the final column.

Species	Female movement	Multiple copulations	Female promiscuity & others	References
Arctocephalus australis	Daily thermoregulatory[a,b,c]	8/31 mated twice[a]	Mating observed outside normal range[c] Females observed mating multiple males[a]	[a]Harcourt (1990); [b]Majluf (1987); [c]Kuroiwa & Majluf (1989)
A. forsteri	Some thermoregulatory[d] Some storm-related[e]	1/10 mated twice[d]	Observed mating away from birth site[e]	[d]Miller (1974, 1975) [e]Stirling (1971)
A. gazella	None regularly	1/30 mated twice	Observed mating away from birth site Observed mating by peripheral males	McCann (1980)
A. galapagoensis	May begin foraging before oestrus Some thermoregulatory	Some mated twice	Observed mating by peripheral males	Trillmich (1984, 1987)
A. philippii	Daily thermoregulatory	2/14 mated twice	Females observed mating multiple males	Francis & Boness (1991)
A. pusillus	Foraging trips before oestrus	?	?	Warneke & Shaughnessy (1985)
A. p. doriferus	Some thermoregulatory	?	?	Warneke & Shaughnessy (1985)
A. tropicalis	Some thermoregulatory	?	?	Bester (1982)
A. townsendi	Daily thermoregulatory	?	?	Pierson (1987)
Eumetopias jubatus	Foraging trips before oestrus[f,g] Shifting of resting places[g]	4/26 mated twice[g]	?	[f]Higgins et al. (1988) [g]Gentry (1970)

Species	Movement/behaviour	Mating frequency	Mating observations	References
Callorhinus ursinus	Daily thermoregulatory[h]	7% mated twice[h] 1.2 matings/female[j] ~30% mated twice[j]	Females observed mating multiple males[i,j]	[h]DeLong (1982) [i]Peterson (1968) [j]Bartholomew & Hoel (1953)
Neophoca cinerea	Some thermoregulatory	Some mated twice[k] 4/54 mated twice[l]	?	[k]Marlow (1975) [l]Higgins (1990)
Otaria byronia	Daily thermoregulatory	13/34 mated twice	Females observed mating multiple males 15% of matings in peripheral areas	Campagna & Le Boeuf (1988a, b)
Phocarctos hookeri	Gradual shift of herd[m,n] Individuals roam rockery at oestrus[n]	?	?	[m]Marlow (1975) [n]Gentry & Roberts in Boness (1991)
Zalophus californianus	Daily thermoregulatory[o,p,q] Begin foraging before oestrus[o] Shift resting places[q]	?	Observed mating by peripheral males[a,p]	[o]Heath (1989) [p]Odell (1972) [q]Peterson & Bartholomew (1967)
Halichoerus grypus	Shift resting places[s]	2.5 matings/female[r] 2.9 matings/females[s]	Females observed mating multiple males[r,s]	[r]Anderson et al. (1975) [s]Boness & James (1979)
Mirounga angustirostris	None regularly	Most mated several times	Females observed mating multiple males Observed mating by peripheral males	Le Boeuf (1972, 1974)
Mirounga leonina	None regularly	Some presumed to mate twice	Observed mating in water at periphery	McCann (1981)

1979) may be incorrect, as studies of order effects on mating success in other promiscuous species are equivocal (Ginsberg & Huck 1989). Clearly, DNA studies are needed in such species.

It is commonly accepted that otariid females copulate once. Yet the estimated proportion of females copulating more than once ranges from 0% to 38% (Table 2). The additional copulations are often with males different from those with which the female first mated (Table 2). Thus, the potential error in estimating individual mating success from copulations or associations may indeed be higher than previously thought.

What might we learn from improved estimates?

Aquatically mating species

Our greatest gains in understanding pinniped mating systems as a result of DNA analyses of paternity should come from studies of aquatically mating species because of the difficulty of observing reproductive behaviour in these species. A knowledge of reproductive strategies and mating success in these species is particularly important to our understanding of the interplay between ecological, social and phylogenetic constraints because the aquatically mating species, in which males are least able to monopolize mates, represent one end of the continuum of potential monopoly. They also represent almost an entire taxonomic family, the Phocidae.

Based on limited information from observations of spatial distribution of adults, the presence of wounds on males, male aquatic displays, and body size differences between the sexes, current characterizations of the mating systems of aquatically mating species include: (1) monogamy—hooded and Largha seals (Olds 1950; Fay 1974), (2) male territorial defence polygyny—Weddell and harbour seal (Kaufman, Siniff & Reichle 1975; Sullivan 1981; Perry 1993), (3) sequential polygyny—hooded and crabeater seals (Siniff *et al.* 1979; Boness, Bowen & Oftedal 1988; Kovacs 1990), (4) scramble competition—Hawaiian monk seal, *Monachus schauinslandi* (Deutsch 1985), and (5) lekking—walrus, *Odobenus rosmarus*, and bearded seal, *Erignathus barbatus* (Ray, Watkins & Burns 1969; Fay, Ray & Kibal'chich 1984; Stirling, Calvert & Spencer 1987; Cleator, Stirling & Smith 1989).

The validity of these descriptions is difficult to evaluate in the absence of data on matings of individual males and females or evidence of paternity from DNA fingerprinting. For example, it would be surprising for aquatic territorial defence to occur as a resource defence strategy in most aquatically mating species. Females are not confined to a particular location because they have probably weaned their pups and females are generally very dispersed compared to females of terrestrially mating otariids that exhibit territorial resource defence. Thus in Weddell seals and harbour seals, where males hold positions off the land or ice rookery, it may be that males are simply waiting to intercept departing females that they will then 'escort' until they mate (i.e. a system to defend females rather than territory). Alternatively, these males may be holding a position on a lek to which females leaving the rookery come to choose among the males. Interestingly, evidence based on the transfer of coloured grease from males to females in Weddell seals suggests that females mate with males other than the ones off the ice where the females reside (W. Testa pers. comm.).

Similarly, until we have evidence of mating success and paternity in the hooded seal it is difficult to know whether a male defending a female on the ice is employing a successful strategy. Females may be choosing among competing or displaying males under the ice after they depart (i.e. a lekking system). Kovacs (1990) reported that females often do not have males attending them on the ice. Interestingly, the conditions for many aquatically mating species seem to be ideal for classical lekking to have evolved (Emlen & Oring 1977; Bradbury & Gibson 1983). There are no resources clearly needed by females that can be defended nor does it appear possible directly to defend several females. At the same time, males and females are relatively clustered compared to their normal distribution outside the period of pupping.

However, it should be noted that territorial defence of aquatic resources can occur if conditions are suitable. A recent study of the Juan Fernández fur seal (Francis & Boness 1991), which exhibits both aquatic and terrestrial mating in the same colony, revealed successful aquatic territoriality. In this case, females clustered daily in predictable aquatic locations adjacent to the beach to obtain reprieve from the high midday temperatures.

Paternity assessments from DNA studies are critical to distinguish among these potential mating strategies of the aquatically mating species. If sequential female defence occurs in species such as hooded and crabeater seals, then males guarding females on the ice should be fertilizing them, and we would expect a low variance in mating success. If males are defending territories off the rookery because it gives them access to females on the rookery adjacent to their territories, as suggested for Weddell seals and harbour seals, then males should be fertilizing primarily those females that depart through their territory, and given the widespread spacing of females, variance in mating success should be moderate. If the mating system of the walrus and/or some of the above species is lek polygyny, then DNA studies should show that males sire the offspring of females from any part of the rookery. Also, from the variance in mating success of other mammalian and avian species that exhibit lek systems, we would expect a high variance in mating success (Wylie 1973; Clutton-Brock et al. 1988; Alatalo et al. 1992).

Two studies using DNA fingerprinting provide some evidence in the harbour seal to suggest that females may be choosing mates and that males positioned off the beach do not simply have exclusive mating rights to females that depart through their territories. The first study was on captive animals and showed that the male displaying the most sexual attention towards the females did not sire their offspring (Harris, Young & Wright 1991). The second, a field study, revealed that males holding territory off the beach accounted for only a small proportion of the offspring born to the females hauled out on the beach (Perry 1993).

Terrestrially mating species

Among terrestrially mating pinnipeds the immediate value of DNA fingerprinting studies will be to validate and supplement observational estimates of mating success. There are two areas in particular in which improved estimates of paternity and mating success will be most useful in advancing our understanding. First, inferences of lekking systems and female choice of males in two otariids (Heath 1989; Kuroiwa & Majluf 1989) are based primarily on observations of female movements prior to and at the time of receptivity. However, because the movements of females are associated with thermoregulation, without precise knowledge of paternity we cannot determine whether a female is choosing a male or simply being mated by the male in whose territory she spends the most time, since thermoregulation might be an alternative explanation of her movements in the former case and care of young of her usual location in the latter. The former case would suggest a form of lekking, and the latter a resource defence system. To distinguish between these alternatives, it is necessary to demonstrate, as Heath's (1989) behavioural observations suggest, that females are fertilized by a particular male or at a particular location disproportionately more often than expected given the amount of time spent with the male or in the territory.

The second area in which DNA studies of terrestrially mating pinnipeds will improve our knowledge is in assessing the success and nature of alternative mating strategies. Within a population, besides what is usually the principal mating strategy of males there may be floaters or satellite males that roam throughout or hang on the periphery of territorial males or the edge of leks (Alcock & Houston 1987; Gosling 1991; Shutler & Weatherhead 1991), or males that mimic females and sneak their way in among females (Arak 1984; Forsyth & Alcock 1990). It has often been assumed that males exhibiting an alternative strategy are doomed to failure in reproduction (J. N. M. Smith & Arcese 1989). However, recent evidence indicates that secondary strategies may often be associated with poor phenotypes, but that there frequently is a payoff, suggesting that these males are 'making the best of a bad situation' (J. N. M. Smith & Arcese 1989; Verrell 1991). In some species the alternative strategies may be as successful as the principal one (Gosling & Petrie 1990; Francis & Boness 1991).

Alternative strategies are evident in terrestrially mating pinnipeds, though not well documented. Descriptions of strategies exhibited by males within a population include 'sneaky' and 'satellite' or 'floater' behaviour patterns, as well as 'gang' behaviour in which males form groups to compete with dominant males for females (Table 1). We currently know little about the mating success of males exhibiting alternative strategies within a population (but see McCann 1981; Campagna & Le Boeuf 1988b; Deutsch et al. 1990; Godsell 1991; Francis & Boness 1991). In some cases this is because we simply have not focused observations on them. In other cases, the greater mobility of peripheral males makes it difficult to observe their mating behaviour directly. In both cases, DNA fingerprinting will enhance our understanding of the interrelationship between phenotypes (e.g. age, size), mating strategy and mating success. The need for accurate assessment of the success of different strategies is particularly important in that males may achieve maximum fitness by shifting strategies appropriately over the course of their reproductive life (Arak 1984; Apollonio et al. 1992). Once we have a better knowledge of the mating success of different phenotypes, we can begin to understand how different ecological conditions may favour different reproductive strategies.

Hybridization

The discussion of reproductive behaviour and applications of DNA fingerprinting can extend to the effects of mate choice expressed at the level of species recognition. Studies over the last decade document hybridization between several pinniped species. These include hybrids between Antarctic and subantarctic fur seals (Condy 1978; Kerley 1983; Kerley & Robinson 1987; Shaughnessey & Fletcher 1987), between the northern fur seal and either the California sea lion or the Guadalupe fur seal (Jameyson et al. 1981) and presumed hybrids between subantarctic and Juan Fernández fur seals (J. Francis pers. obs.). Mating has been observed between Guadalupe fur seal males and California sea lion females (Stewart et al. 1987) but to date no hybrids have been identified. Steller's sea lions have also been reported breeding with California sea lions (DeLong 1982). Among phocids, grey seal males have been reported attempting to copulate (Wilson 1975) and successfully copulating (D. Boness & Z. Lucas pers. obs.) with harbour seal females.

Identification of hybrids and their parentage by morphological comparisons is challenging. Even pups of different pure-bred species are sometimes difficult to distinguish on the basis of physical characteristics alone (Bester & Wilkinson 1989). Further, those adults with unusual combinations of characters that clearly appear to be hybrid (see photographs in Condy 1978 for example) present problems with identification of the parent species involved and their respective sexes.

DNA fingerprinting can be used to great advantage in these situations. Not only can mtDNA analysis help to identify the maternal origin of an individual's genotype but, in conjunction with allozyme data at the population level, mtDNA analysis can show measures of asymmetries in mate choice or reproductive isolation between hybridizing taxa (Harrison 1989).

Relatedness and kinship

DNA studies are also proving useful in testing theoretical predictions regarding kin selection. Hamilton (1964) proposed the concept of inclusive fitness (the basis for kinship theory) to help to explain altruistic behaviours, i.e. behaviours which appear to benefit recipients at the expense of the donor. Kin selection is generally thought to be more important in shaping social behaviour in species which live in small social groups formed by the retention of offspring born into the group (e.g. Eberhard 1975; Trivers 1985; Armitage 1987). It is also suspected to operate in species where dispersal patterns result in a high probability of neighbouring animals being related (Emlen 1991).

Recent efforts to address questions of kinship and relatedness by means of DNA fingerprinting appear to be promising. For example, Packer, Gilbert *et al.* (1991) used multilocus probes to assess degree of band sharing among individuals in various lion coalitions, and found that non-reproductive male helpers occur only in groups composed of close relatives. Using similar techniques, Reeve *et al.* (1990) discovered that relatedness among members of naked mole-rat colonies was relatively high, and Everitt *et al.* (1991) found a negative correlation between aggressiveness in wild-caught house mice and the degree of band sharing.

Pinnipeds generally do not exhibit co-operative behaviour or social conditions which would suggest the operation of kin selection. However, foster nursing is one behaviour for which kin selection might be important (Fogden 1971; Riedman & Le Boeuf 1982; Boness 1990; Boness, Bowen, Iverson & Oftedal 1992). The apparently altruistic nature of fostering and adoption has led researchers of a variety of birds and mammals to suggest kin selection as a possible explanation (Riedman 1982; Packer, Lewis & Pusey 1992). Both the Hawaiian monk seal and the grey seal are species which exhibit a high frequency (more than 60% of the females in a colony) of fostering and this occurs at small rookeries (Fogden 1971; Boness 1990). Evidence of breeding site-fidelity of adults has been reported in these species (Boness 1979; Johnson & Kridler 1983), and insofar as young return to their birth colony, there could be a high degree of relatedness among individuals in these populations. DNA studies could be instrumental in assessing both the general level of relatedness among colony members and the relatedness of mothers to the young they foster. Furthermore, to address questions of fostering in general requires identifying non-filial relationships between females and pups. Both genomic and mtDNA can be used to assess maternity (Avise *et al.* 1989; Rabenold, Rabenold, Piper & Minchella 1991).

Practical considerations and limitations of DNA analysis

Sample collection

Nuclear DNA from blood collected in EDTA tubes can be recovered by procedures outlined by Kirby (1990) and Hoelzel & Dover (1989). Amos (1989) reported good DNA recovery from skin samples stored for up to one month in a saturated saline solution. Skin samples may be stored for at least one year in a 20% dimethyl sulphoxide (DMSO) solution saturated with salt. Polymerase chain reaction (PCR) can be used to amplify DNA from very small quantities of tissue (Arnheim, White & Rainey 1990). Tissue stored in formaldehyde generally cannot be used for DNA analysis. Kirby (1990) and Hoelzel & Dover (1989) review the recovery of both mtDNA and nuclear DNA from different tissues. Given the breakneck speed with which new methods are being developed, close collaboration with a molecular biologist is advisable.

Limitations

Many of the statistical methods used to determine exclusion probabilities with respect to paternity analysis and band-sharing coefficients in assessing pedigrees are based on the assumption that alleles at different loci are not linked (Brock & White 1991). Many of these methods also require the estimation of allele frequencies. Cohen (1990) has critically reviewed the statistical methods used to calculate the probability of band sharing from DNA fingerprinting. His analysis points out the need to carefully consider heterogeneity in gene frequencies between subpopulations when estimating the probability of obtaining matching fingerprints by chance. The probability of determining a match between bands can also be affected by fluctuation in band position and band shifting (Gill, Sullivan & Werrett 1990). Although the human minisatellite probes (Jeffreys *et al.* 1985a, b) cross-hybridize with many other vertebrate taxa (Jeffreys 1987; Burke & Bruford 1987), including harbour seals (Harris *et al.* 1991), probes may behave differently in different species (Cummings & Hallet 1991). Thus a number of probes may need to be tested to obtain the best results in a new species. This will be particularly true in the case of single locus probes which tend to be taxa-specific (Hanotte *et al.* 1991).

Another problem of using fingerprinting techniques is that the similar-sized fragments of DNA may represent homologous DNA from different individuals which could be interpreted as indicating relatedness when it does not. In fact, studies have shown that unrelated individuals will share 20–40% of their bands simply by chance (Lynch 1988). This 'noise' in the system limits the extent to which multilocus fingerprints can be used to determine relatedness with high confidence much beyond parent-offspring relationships. Because estimates of band sharing may differ between species and within species when different probes are used, it is essential that baseline estimates of band sharing between unrelated individuals be determined before attempts are made to determine relatedness among individuals in any population (Lynch 1988; Cummings & Hallet 1991).

Conclusions

We have suggested three areas of behavioural ecology in which DNA fingerprinting can offer significant advances to our current knowledge of pinnipeds: mating systems, interbreeding between species and fostering behaviour. Interbreeding of species and fostering behaviour received only brief attention because of the limitation of space, but recent investigations indicate that both phenomena are more prevalent than initially thought. Our focus on mating systems reflects past emphasis on this taxonomic group. For many pinniped species, DNA fingerprinting may be the only means by which we can obtain information on mating success. In others DNA studies will provide a means to validate observational estimates of paternity. Assessing paternity will not be accomplished easily. Females need to be relocated a year after mating so that a blood sample can be taken from their pups. In the aquatically mating species, sampling of males may be a challenge because males in many species do not come onto the ice often during the mating period. Therefore, it is clear that we must also improve our behavioural data through innovative studies (e.g. Hill 1987; Kelly & Wartzok 1991; Walker & Bowen in press) to match the wealth of information expected from DNA fingerprinting.

Acknowledgements

We would like to thank Drs Kim Derrickson and Rob Fleischer for their helpful comments on the manuscript. Discussions with Rob Fleischer and Johnathon Wright on DNA fingerprinting methodology were also very helpful. During writing of this paper, John Francis was supported as a Visiting Scientist to the National Zoological Park by grants from the Smithsonian Scholarly Studies Fund and the Friends of the National Zoo.

References

Alatalo, R. V., Höglund, J., Lundberg, A. & Sutherland, W. J. (1992). Evolution of black grouse leks: female preferences benefit males in larger leks. *Behav. Ecol.* **3**: 60–65.

Alcock, J. & Houston, T. F. (1987). Resource defense and alternative mating tactics in the Banksia bee, *Hylaeus alcyoneus* (Erichson). *Ethology* **76**: 177–188.

Alexander, R. D., Hoogland, J. L., Howard, R. D., Noonan, K. M. & Sherman, P. W. (1979). Sexual dimorphism and breeding systems in pinnipeds, ungulates, primates, and humans. In *Evolutionary biology and human social behavior*: 432–435. (Eds Chagnon, N. A. & Irons, W.). Duxbury Press, North Situate.

Amlamer, C. J. & Macdonald, D. W. (Eds) (1980). *A handbook on biotelemetry and radio tracking*. Pergamon Press, Oxford.

Amos, B. (1989). Preserving tissue without refrigeration. *Fingerprint News* **3**: 20.

Amos, W. & Dover, G. A. (1990). DNA fingerprinting and the uniqueness of whales. *Mammal. Rev.* **20**: 23–30.

Amos, W. & Hoelzel, A. R. (1990). DNA fingerprinting cetacean biopsy samples for individual identification. *Rep. int. Whal. Commn spec. Issue* No. 12: 79–85.

Anderson, S. S., Burton, R. W. & Summers, C. F. (1975). Behaviour of grey seals (*Halichoerus grypus*) during a breeding season at North Rona. *J. Zool., Lond.* **177**: 179–195.

Apollonio, M., Festa-Bianchet, M., Mari, F., Mattioli, S. & Sarno, B. (1992). To lek or not to lek: mating strategies of male fallow deer. *Behav. Ecol.* **3**: 25–31.

Arak, A. (1984). Sneaky breeders. In *Producers and scroungers: strategies of exploitation and parasitism*: 154–194. (Ed. Barnard, C. J.). Croom Helm, London.

Armitage, K. B. (1987). Social dynamics of mammals: reproductive success, kinship and individual fitness. *Trends Ecol. Evol.* **2**: 279–284.

Arnason, U. & Widegren, B. (1986). Pinniped phylogeny enlightened by molecular hybridizations using highly repetitive DNA. *Molec. Biol. Evol.* **3**: 356–365.

Arnheim, N., White, T. & Rainey, W. E. (1990). Application of PCR: organismal and population biology. *Bioscience* **40**: 174–182.

Avise, J. C., Bowen, B. W. & Lamb, T. (1989). DNA fingerprints from hypervariable mitochondrial genotypes. *Molec. Biol. Evol.* **6**: 258–269.

Bartholomew, G. A. (1970). A model for the evolution of pinniped polygyny. *Evolution, Lancaster, Pa* **24**: 546–559.

Bartholomew, G. A. & Hoel, P. G. (1953). Reproductive behavior of the Alaska fur seal, *Callorhinus ursinus*. *J. Mammal.* **34**: 417–436.

Bateman, A. J. (1948). Intrasexual selection in *Drosophila*. *Heredity, Lond.* **2**: 349–368.

Bester, M. N. (1982). Distribution, habitat selection and colony types of the Amsterdam Island fur seal *Arctocephalus tropicalis* at Gough Island. *J. Zool., Lond.* **196**: 217–231.

Bester, M. N. & Wilkinson, I. S. (1989). Field identification of Antarctic and subantarctic fur seal pups. *S. Afr. J. Wildl. Res.* **19**: 140–144.

Birkhead, T. R., Burke, T., Zann, R., Hunter, F. M. & Krupa, A. P. (1990). Extra-pair paternity and intraspecific brood parasitism in wild zebra finches *Taeniopygia guttata*, revealed by DNA fingerprinting. *Behav. Ecol. Sociobiol.* **27**: 315–324.

Boness, D. J. (1979). *The social system of the grey seal*, Halichoerus grypus *(Fab.), on Sable Island, Nova Scotia*. PhD thesis: Dalhousie University.

Boness, D. J. (1990). Fostering behavior in Hawaiian monk seals: is there a reproductive cost? *Behav. Ecol. Sociobiol.* **27**: 113–122.

Boness, D. J. (1991). Determinants of mating systems in the Otariidae (Pinnipedia). In *The behaviour of pinnipeds*: 1–44. (Ed. Renouf, D.). Chapman & Hall, London etc.

Boness, D. J., Anderson, S. S. & Cox, C. R. (1982). Functions of female aggression during the pupping and mating season of grey seals, *Halichoerus grypus* (Fabricius). *Can. J. Zool.* **60**: 2270–2278.

Boness, D. J., Bowen, W. D., Iverson, S. J. & Oftedal, O. T. (1992). Influence of storms and maternal size on mother–pup separations and fostering in the harbor seal, *Phoca vitulina*. *Can. J. Zool.* **70**: 1640–1644.

Boness, D. J., Bowen, W. D. & Oftedal, O. T. (1988). Evidence of polygyny from spatial patterns of hooded seals (*Cystophora cristata*). *Can. J. Zool.* **66**: 703–706.

Boness, D. J. & James, H. (1979). Reproductive behaviour of the grey seal (*Halichoerus grypus*) on Sable Island, Nova Scotia. *J. Zool., Lond.* **188**: 477–500.

Bonner, W. N. (1984). Lactation strategies in pinnipeds: problems for a marine mammalian group. *Symp. zool. Soc. Lond.* No. 51: 253–272.

Borgia, G. (1979). Sexual selection and the evolution of mating systems. In *Sexual selection and reproductive competition in insects*: 19–79. (Eds Blum, M. S. & Blum, N. A.). Academic Press, New York.

Boyd, I. L. (1989). Spatial and temporal distribution of Antarctic fur seals (*Arctocephalus gazella*) on the breeding grounds at Bird Island, South Georgia. *Polar Biol.* **10**: 179–185.

Bradbury, J. W. & Gibson, R. M. (1983). Leks and mate choice. In *Mate choice*: 109–138. (Ed. Bateson, P.). Cambridge University Press, Cambridge.

Bradbury, J. W. & Vehrencamp, S. L. (1977). Social organization and foraging in emballonurid bats. III. Mating systems. *Behav. Ecol. Sociobiol.* **2**: 1–17.

Brock, M. K. & White, B. N. (1991). Multifragment alleles in DNA fingerprints of the parrot, *Amazona ventralis*. *J. Hered.* **82**: 209–212.

Burke, T. (1989). DNA fingerprinting and other methods for the study of mating success. *Trends Ecol. Evol.* **4**: 139–144.

Burke, T. & Bruford, M. W. (1987). DNA fingerprinting in birds. *Nature, Lond.* **327**: 149–152.

Campagna, C. & Le Boeuf, B. J. (1988a). Reproductive behaviour of southern sea lions. *Behaviour* **104**: 233–261.

Campagna, C. & Le Boeuf, B. J. (1988b). Thermoregulatory behaviour of southern sea lions and its effect on mating strategies. *Behaviour* **107**: 72–90.

Campagna, C., Le Boeuf, B. J. & Cappozzo, H. L. (1988). Group raids: a mating strategy of male southern sea lions. *Behaviour* **105**: 224–249.

Carrick, R., Csordas, S. E. & Ingham, S. E. (1962). Studies on the southern elephant seal, *Mirounga leonina* (L.) IV. Breeding and development. *CSIRO Wildl. Res.* **7**: 161–197.

Cleator, H. J., Stirling, I. & Smith, T. G. (1989). Underwater vocalizations of the bearded seal (*Erignathus barbatus*). *Can. J. Zool.* **67**: 1900–1910.

Clutton-Brock, T. H. (1988). *Reproductive success. Studies of individual variation in contrasting breeding systems.* University of Chicago Press, Chicago & London.

Clutton-Brock, T. H., Green, D., Hiraiwa-Hasegawa, M. & Albon, S. D. (1988). Passing the buck: resource defence, lek breeding and mate choice in fallow deer. *Behav. Ecol. Sociobiol.* **23**: 281–296.

Cohen, J. E. (1990). DNA fingerprinting for forensic identification: potential effects on data interpretation of subpopulation heterogeneity and band number variability. *Am. J. hum. Genet.* **46**: 358–368.

Condy, P. R. (1978). Distribution, abundance, and annual cycle of fur seals (*Arctocephalus* spp.) on the Prince Edward Islands. *S. Afr. J. Wildl. Res.* **8**: 159–168.

Cox, C. R. & Le Boeuf, B. J. (1977). Female incitation of male competition: a mechanism in sexual selection. *Am. Nat.* **111**: 317–335.

Cummings, S. A. & Hallett, J. G. (1991). Assessment of DNA fingerprinting for studies of small-mammal populations. *Can. J. Zool.* **69**: 2819–2825.

Curry-Lindahl, K. (1970). Breeding biology of the Baltic grey seal (*Halichoerus grypus*). *Zool. Gart., Lpz.* **38**: 16–29.

David, J. H. M. (1987). South African fur seal, *Arctocephalus pusillus pusillus*. In *Status, biology, and ecology of fur seals*: 65–71. (Eds Croxall, J. P. & Gentry, R. L.). US Dept. of Commerce, Washington, D.C. (*NOAA tech. Rep. NMFS* **51**.)

David, J. H. M. & Rand, R. W. (1986). Attendance behavior of South African fur seals. In *Fur seals: maternal strategies on land and at sea*: 126–141. (Eds Gentry, R. L. & Kooyman, G. L.). Princeton University Press, Princeton.

Davies, J. L. (1949). Observations on the grey seal (*Halichoerus grypus*) at Ramsey Island, Pembrokeshire. *Proc. zool. Soc. Lond.* **119**: 673–692.

Davies, N. B. (1991). Mating systems. In *Behavioural ecology* (3rd edn): 263–299. (Eds Krebs, J. R. & Davies, N. B.). Blackwell Scientific, London.

DeLong, R. L. (1982). *Population biology of northern fur seals at San Miguel Island, California.* PhD thesis: University of California, Berkeley.

Deutsch, C. J. (1985). Male-male competition in the Hawaiian monk seal. *Abstr. bienn. Conf. Biol. mar. Mammals* **6**: 25. Society for Marine Mammalogy. Unpublished.

Deutsch, C. J., Haley, M. P. & Le Boeuf, B. J. (1990). Reproductive effort of male northern elephant seals: estimates from mass loss. *Can. J. Zool.* **68**: 2580–2593.

Eberhard, M. J. W. (1975). The evolution of social behavior by kin selection. *Q. Rev. Biol.* **50**: 1–33.

Emlen, S. T. (1991). Evolution of cooperative breeding in birds and mammals. In *Behavioural ecology* (3rd edn): 301–337. (Eds Krebs, J. R. & Davies, N. B.). Blackwell Scientific, London.

Emlen, S. T. & Oring, L. W. (1977). Ecology, sexual selection and the evolution of mating systems. *Science* **197**: 215–223.

Everitt, J., Hurst, J. L., Ashworth, D. & Barnard, C. J. (1991). Aggressive behaviour among wild-caught house mice, *Mus domesticus* Rutty, correlates with a measure of genetic similarity using DNA fingerprinting. *Anim. Behav.* **42**: 313–316.

Fay, F. H. (1974). The role of ice in the ecology of marine mammals of the Bering

Sea. In *Oceanography of the Bering Sea*: 383–389. (Eds Hood, D. W. & Kelley, E. J.). University of Alaska, Fairbanks.

Fay, F. H., Ray, G. C. & Kibal'chich, A. A. (1984). Time and location of mating and associated behavior of the Pacific walrus, *Odobenus rosmarus divergens* Illiger. *NOAA tech. Rep. NMFS* **12**: 89–100.

Fogden, S. C. L. (1971). Mother–young behaviour at grey seal breeding beaches. *J. Zool., Lond.* **164**: 61–92.

Forsyth, A. & Alcock, J. (1990). Female mimicry and resource defense polygyny by males of a tropical rove beetle, *Leistotrophus versicolor* (Coleoptera: Staphylinidae). *Behav. Ecol. Sociobiol.* **26**: 325–330.

Francis, J. M. & Boness, D. J. (1991). The effect of thermoregulatory behavior on the mating system of the Juan Fernández fur seal, *Arctocephalus philippii*. *Behaviour* **119**: 104–127.

Gentry, R. L. (1970). *Social behavior of the Steller sea lion*. PhD thesis: University of California, Santa Cruz.

Gibbs, H. L., Weatherhead, P. J., Boag, P. T., White, B. N., Tabak, L. M. & Hoysak, D. J. (1990). Realized reproductive success of polygynous red-winged blackbirds revealed by DNA markers. *Science* **250**: 1394–1397.

Gill, P., Sullivan, K. & Werrett, D. J. (1990). The analysis of hypervariable DNA profiles: problems associated with the objective determination of the probability of a match. *Hum. Genet.* **85**: 75–79.

Ginsberg, J. R. & Huck, U. W. (1989). Sperm competition in mammals. *Trends Ecol. Evol.* **4**: 74–79.

Gisiner, R. C. (1985). *Male territorial and reproductive behavior in the Steller sea lion* Eumetopias jubatus. PhD thesis: University of California.

Godsell, J. (1991). The relative influence of age and weight on the reproductive behaviour of male grey seals *Halichoerus grypus*. *J. Zool., Lond.* **224**: 537–551.

Gosling, L. M. (1991). The alternative mating strategies of male topi, *Damaliscus lunatus*. *Appl. Anim. Behav. Sci.* **29**: 107–119.

Gosling, L. M. & Petrie, M. (1990). Lekking in topi: a consequence of satellite behaviour by small males at hotspots. *Anim. Behav.* **40**: 272–287.

Gyllensten, U. B., Jakobsson, S. & Temrin, H. (1990). No evidence for illegitimate young in monogamous and polygynous warblers. *Nature, Lond.* **343**: 168–170.

Hamilton, W. D. (1964). The genetical theory of social behavior I. *J. theor. Biol.* **7**: 1–16.

Hanotte, O., Burke, T., Armour, J. A. L. & Jeffreys, A. J. (1991). Hypervariable minisatellite DNA sequences in the Indian Peafowl *Pavo cristatus*. *Genomics* **9**: 587–597.

Harcourt, R. G. (1990). *Maternal influences on pup survival and development in South American fur seal*. PhD thesis: University of Cambridge.

Harris, A. S., Young, J. F. S. & Wright, J. M. (1991). DNA fingerprinting of harbour seals (*Phoca vitulina concolor*): male mating behaviour may not be a reliable indicator of reproductive success. *Can. J. Zool.* **69**: 1862–1866.

Harrison, R. G. (1989). Animal mitochondrial DNA as a genetic marker in population and evolutionary biology. *Trends Ecol. Evol.* **4**: 6–11.

Heath, C. B. (1989). *The behavioral ecology of the California sea lion*. PhD thesis: University of California, Santa Cruz.

Hewer, H. R. (1957). A Hebridean breeding colony of grey seals, *Halichoerus grypus* (Fab.), with comparative notes on the grey seals of Ramsey Island, Pembrokeshire. *Proc. zool. Soc. Lond.* **128**: 23–66.

Hewer, H. R. & Backhouse, K. M. (1960). A preliminary account of a colony of grey seals *Halichoerus grypus* (Fab.) in the southern Inner Hebrides. *Proc. zool. Soc. Lond.* **134**: 157–195.

Higgins, L. V. (1990). *Reproductive behavior and maternal investment of Australian sea lions*. PhD thesis: University of California, Santa Cruz.

Higgins, L. V., Costa, D. P., Huntley, A. C. & Le Boeuf, B. J. (1988). Behavioural and physiological measurements of maternal investment in the Steller sea lion, *Eumetopias jubatus*. *Mar. Mamm. Sci.* **4**: 44–58.

Hill, S. E. B. (1987). *Reproductive ecology of Weddell seals (Leptonychotes weddelli) in McMurdo Sound, Antarctica*. PhD thesis: University of Minnesota.

Hoelzel, A. R. (1992). *Molecular genetics of populations, a practical approach*. Oxford University Press, Oxford.

Hoelzel, A. R. & Dover, G. A. (1989). Molecular techniques for examining genetic variation and stock identity in cetacean species. *Rep. int. Whal. Commn spec. Issue* No. 11: 81–120.

Hoelzel, A. R., Hancock, J. M. & Dover, G. A. (1991). Evolution of cetacean mitochondrial D-loop region. *Molec. Biol. Evol.* **8**: 475–493.

Huntley, A. C., Costa, D. P., Worthy, G. A. J. & Castellini, M. A. (Eds) (1987). *Marine mammal energetics*. Society for Marine Mammalogy, Lawrence.

Jameyson, E. C., Duffield, D., Cornell, L., Antonelis, G. & DeLong, R. L. (1981). Discovery of a *Callorhinus ursinus* × ? hybrid within the San Miguel Island population. *Abstr. bienn. Conf. Biol. mar. Mammals* **4**: 63. Society for Marine Mammalogy. Unpublished.

Jeffreys, A. J. (1987). Highly variable minisatellites and DNA fingerprints. *Biochem. Soc. Trans.* **15**: 309–317.

Jeffreys, A. J., Wilson, V. & Thein, S. L. (1985a). Hypervariable 'minisatellite' regions in human DNA. *Nature, Lond.* **314**: 67–73.

Jeffreys, A. J., Wilson, V. & Thein, S. L. (1985b). Individual-specific 'fingerprints' of human DNA. *Nature, Lond.* **316**: 76–79.

Johnson, A. M. & Kridler, E. (1983). Interisland movement by Hawaiian monk seals. *Elepaio* **44**: 43–45.

Kautman, G. W., Siniff, D. B. & Reichle, R. (1975). Colony behavior of Weddell seals, *Leptonychotes weddelli*, at Hutton Cliffs, Antarctica. *Rapp. P.-v. Réun. Cons. perm. int. Explor. Mer* **169**: 228–246.

Kelly, B. P. & Wartzok, D. (1991). Three-dimensional movements of ringed seals under fast ice. *Abstr. bienn. Conf. Biol. mar. Mammals* **9**: 39. Society for Marine Mammalogy. Unpublished.

Kerley, G. I. H. (1983). Relative population sizes and trends, and hybridization of fur seals *Arctocephalus tropicalis* and *A. gazella* at the Prince Edward Islands, Southern Ocean. *S. Afr. J. Zool.* **18**: 388–392.

Kerley, G. I. H. & Robinson, T. J. (1987). Skull morphometrics of male Antarctic and subantarctic fur seals, *Arctocephalus gazella* and *A. tropicalis*, and their interspecific hybrids. In *Status, biology, and ecology of fur seals*: 121–131. (Eds Croxall, J. P. & Gentry, R. L.). US Dept. of Commerce, Washington, D.C. (*NOAA tech. Rep. NMFS* **51.**)

Kirby, L. T. (1990). *DNA fingerprinting*. Stockton Press, New York.

Kovacs, K. M. (1990). Mating strategies in male hooded seals (*Cystophora cristata*)? *Can. J. Zool.* **68**: 2499–2502.

Kuroiwa, M. I. & Majluf, P. (1989). Do South American fur seals in Peru lek? *Abstr. bienn. Conf. Biol. mar. Mammals* **8**: 36. Society for Marine Mammalogy. Unpublished.

Le Boeuf, B. J. (1972). Sexual behavior in the northern elephant seal *Mirounga angustirostris*. *Behaviour* **41**: 1–25.

Le Boeuf, B. J. (1974). Male-male competition and reproductive success in elephant seals. *Am. Zool.* **14**: 163–176.

Le Boeuf, B. J. (1991). Pinniped mating systems on land, ice, and in the water: emphasis on the Phocidae. In *The behaviour of pinnipeds*: 45–65. (Ed. Renouf, D.). Chapman & Hall, London etc.

Le Boeuf, B. J. & Peterson, R. S. (1969). Social status and mating activity in elephant seals. *Science* **163**: 91–93.

Le Boeuf, B. J. & Reiter, J. (1988). Lifetime reproductive success in northern elephant seals. In *Reproductive success: studies of individual variation in contrasting breeding systems*: 344–362. (Ed. Clutton-Brock, T. H.). University of Chicago Press, Chicago & London.

Lynch, M. (1988). Estimation of relatedness by DNA fingerprinting. *Molec. Biol. Evol.* **5**: 584–599.

Majluf, P. (1987). *Reproductive ecology of female south American fur seals at Punta San Juan, Peru*. PhD thesis: University of Cambridge.

Marlow, B. J. (1975). The comparative behaviour of the Australian sea lions *Neophoca cinerea* and *Phocartos hookeri* (Pinnipedia: Otariidae). *Mammalia* **39**: 159–230.

McCann, T. S. (1980). Territoriality and breeding behaviour of adult male Antarctic fur seal, *Arctocephalus gazella*. *J. Zool., Lond.* **192**: 295–310.

McCann, T. S. (1981). Aggression and sexual activity of male Southern elephant seals, *Mirounga leonina*. *J. Zool., Lond.* **195**: 295–310.

Merdsoy, B. R., Curtsinger, W. R. & Renouf, D. (1978). Preliminary underwater observations of the breeding behavior of the harp seal (*Pagophilus groenlandicus*). *J. Mammal.* **59**: 181–185.

Miller, E. H. (1974). Social behaviour between adult male and female New Zealand fur seals, *Arctocephalus forsteri* (Lesson) during the breeding season. *Aust. J. Zool.* **22**: 155–173.

Miller, E. H. (1975). Social and evolutionary implications of territoriality in adult male New Zealand fur seals, *Arctocephalus forsteri* (Lesson, 1828), during the breeding season. *Rapp. P.-v. Réun. Cons. perm. int. Explor. Mer* **169**: 170–187.

Morin, P. A. & Woodruff, D. S. (1992). Paternity exclusion using multiple hypervariable microsatellite loci amplified from nuclear DNA of hair cells. In *Paternity in primates: genetic tests and theories*: 63–81. (Eds Martin, R. D., Dixson, A. F. & Wickings, E. J.). Karger, Basel.

Moritz, C., Dowling, T. E. & Brown, W. M. (1987). Evolution of animal mitochondrial DNA: relevance for population biology and systematics. *A. Rev. Ecol. Syst.* **18**: 269–292.

Morton, E. S., Forman, L. & Braun, M. (1990). Extra-pair fertilizations and the evolution of colonial breeding in purple martins. *Auk* **107**: 275–283.

Odell, D. K. (1972). *Studies on the biology of the California sea lion and the northern elephant seal on San Nicolas Island, California.* PhD thesis: University of California, Los Angeles.

Oftedal, O. T., Boness, D. J. & Tedman, R. A. (1987). The behavior, physiology, and anatomy of lactation in the Pinnipedia. *Curr. Mammal.* **1**: 175–245.

Olds, J. M. (1950). Notes on the hooded seal (*Cystophora cristata*). *J. Mammal.* **31**: 450–452.

Packer, C., Gilbert, D. A., Pusey, A. E. & O'Brien, S. J. (1991). A molecular genetic analysis of kinship and cooperation in African lions. *Nature, Lond.* **351**: 562–565.

Packer, C., Lewis, S. & Pusey, A. (1992). A comparative analysis of non-offspring nursing. *Anim. Behav.* **43**: 265–282.

Pemberton, J. & Amos, B. (1990). DNA fingerprinting: a new dimension. *Trends Ecol. Evol.* **5**: 132–134.

Pemberton, J. M., Albon, S. D., Guinness, F. E., Clutton-Brock, T. H. & Dover, G. A. (1992). Behavioral estimates of male mating success tested by DNA fingerprinting in a polygynous mammal. *Behav. Ecol.* **3**: 66–75.

Perry, E. A. (1993). *Mating system of harbour seals.* PhD thesis: Memorial University.

Peterson, R. S. (1965). *Behavior of the northern fur seal.* DSc thesis: Johns Hopkins University.

Peterson, R. S. (1968). Social behavior in pinnipeds, with particular reference to the northern fur seal. In *The behavior and physiology of pinnipeds*: 3–53. (Eds Harrison, R. J., Hubbard, R. C., Peterson, R. S., Rice, C. E. & Schusterman, R. J.). Appleton-Century Crofts, New York.

Peterson, R. S. & Bartholomew, G. A. (1967). The natural history and behavior of the California sea lion. *Spec. Publs Am. Soc. Mammal.* No. **1**: 1–79.

Pierson, M. O. (1987). Breeding behavior of the Guadalupe fur seal, *Arctocephalus townsendi*. In *Status, biology, and ecology of fur seals*: 83–94. (Eds Croxall, J. P. & Gentry, R. L.). US Dept. of Commerce, Washington, D.C. (*NOAA tech. Rep. NMFS* **51**.)

Popov, L. A. (1966). On an ice floe with harp seals: ice drift of biologists in the White Sea. *Priroda* **9**: 93–101. Translation No. 814 of Fisheries Research Board of Canada (1966).

Quinn, T. W., Quinn, J. S., Cooke, F. & White, B. N. (1987). DNA marker analysis detects multiple maternity and paternity in single broods of the lesser snow goose. *Nature, Lond.* **326**: 392–394.

Rabenold, P. P., Rabenold, K. N., Piper, W. H., Haydock, J. & Zack, S. W. (1990). Shared paternity revealed by genetic analysis in cooperatively breeding tropical wrens. *Nature, Lond.* **348**: 538–540.

Rabenold, P. P., Rabenold, K. N., Piper, W. H. & Minchella, D. J. (1991). Density-dependent dispersal in social wrens: genetic analysis using novel matriline markers. *Anim. Behav.* **42**: 144–146.

Ralls, K. (1976). Mammals in which females are larger than males. *Q. Rev. Biol.* **51**: 245–276.

Rand, R. W. (1967). The Cape fur-seal (*Arctocephalus pusillus*). 3. General behaviour on land and at sea. *Investl Rep. Div. Fish. Un. S. Afr.* **60**: 1–39.

Ray, C., Watkins, W. A. & Burns, J. J. (1969). The underwater song of *Erignathus* (bearded seal). *Zoologica, N.Y.* **54**: 79–83.

Reeve, H. K., Westneat, D. F., Noon, W. A., Sherman, P. W. & Aquadro, C. F. (1990). DNA 'fingerprinting' reveals high levels of inbreeding in colonies of the eusocial naked mole-rat. *Proc. natn. Acad. Sci. USA* **87**: 2496–2500.

Ribble, D. O. (1991). The monogamous mating system of *Peromyscus californicus* as revealed by DNA fingerprinting. *Behav. Ecol. Sociobiol.* **29**: 161–166.

Riedman, M. L. (1982). The evolution of alloparental care and adoption in mammals and birds. *Q. Rev. Biol.* **57**: 405–435.

Riedman, M. L. & Le Boeuf, B. J. (1982). Mother–pup separation and adoption in northern elephant seals. *Behav. Ecol. Sociobiol.* **11**: 203–215.

Rubenstein, D. I. & Wrangham, R. W. (1986). *Ecological aspects of social evolution: birds and mammals.* Princeton University Press, Princeton.

Sandegren, F. (1970). *Breeding and maternal behavior of the Steller sea lion in Alaska.* MSc thesis: University of Alaska.

Shaughnessy, P. D. & Fletcher, I. (1987). Fur seals, *Arctocephalus* spp., at Macquarie Island. In *Status, biology, and ecology of fur seals*: 177–188. (Eds Croxall, J. P. & Gentry, R. L.). US Dept. of Commerce, Washington, D.C. (*NOAA tech. Rep. NMFS* **51**.)

Shaughnessy, P. D. & Kerry, K. R. (1989). Crabeater seals *Lobodon carcinophagus* during the breeding season: observation on five groups near Enderby Island, Antarctica. *Mar. Mamm. Sci.* **5**: 68–77.

Shutler, D. & Weatherhead, P. J. (1991). Owner and floater red-winged blackbirds: determinants of status. *Behav. Ecol. Sociobiol.* **28**: 235–241.

Siniff, D. B., Stirling, I., Bengston, J. L. & Reichle, R. A. (1979). Social and reproductive behavior of crabeater seals (*Lobodon carcinophagus*) during the austral spring. *Can. J. Zool.* **57**: 2243–2255.

Smith, H. G., Montgomerie, R., Poldmaa, T., White, B. N. & Boag, P. T. (1991). DNA fingerprinting reveals relation between tail ornaments and cuckoldry in barn swallows, *Hirundo rustica. Behav. Ecol.* **2**: 90–97.

Smith, J. N. M. & Arcese, P. (1989). How fit are floaters? Consequences of alternative territorial behaviors in a nonmigratory sparrow. *Am. Nat.* **133**: 830–845.

Smith, T. G. & Hammill, M. O. (1981). Ecology of the ringed seal, *Phoca hispida*, in its fast ice breeding habitat. *Can. J. Zool.* **59**: 966–981.

Stewart, B. S., Yochem, P. K., DeLong, R. L. & Antonelis, G. A. (1987). Interactions between Guadelupe fur seals and California sea lions at San Nicolas and San Miguel Islands, California. In *Status, biology, and ecology of fur seals*: 103–106. (Eds Croxall, J. P. & Gentry, R. L.). US Dept. of Commerce, Washington, D.C. (*NOAA tech. Rep. NMFS* **51**.)

Stirling, I. (1971). Studies on the behaviour of the South Australian fur seal, *Arctocephalus forsteri* (Lesson). I. Annual cycle, postures and calls, and adult males during the breeding season. *Aust. J. Zool.* **19**: 243–266.

Stirling, I. (1977). Adaptations of Weddell and ringed seals to exploit the polar fast-ice habitat in the absence or presence of surface predators. In *Adaptations within antarctic ecosystems*: 741–748. (Ed. Llano, G. A.). Gulf Publishing, Houston.

Stirling, I. (1983). The evolution of mating systems in pinnipeds. In *Advances in the study of mammalian behavior*: 489–527. (Eds Eisenberg, J. F. & Kleiman, D. G.). American Society of Mammalogists, Shippensburg, Pa. (*Spec. Publs Am. Soc. Mammal. No. 7.*)

Stirling, I., Calvert, W. & Spencer, C. (1987). Evidence of stereotyped underwater vocalizations of male Atlantic walruses (*Odobenus rosmarus rosmarus*). *Can. J. Zool.* **65**: 2311–2321.

Sullivan, R. M. (1981). Aquatic displays and interactions in harbor seals, *Phoca vitulina*, with comments on mating systems. *J. Mammal.* **62**: 825–831.

Thomas, J. T., Pastukhov, V. D., Elsner, R. & Petrov, E. (1982). *Phoca sibirica. Mammalian Sp.* **188**: 1–6.

Trillmich, F. (1984). The Galapagos seals. Part 2. Natural history of the Galapagos fur seal (*Arctocephalus galapagoensis*, Heller). In *Key environments: Galapagos*: 215–223. (Ed. Perry, R.). Pergamon Press, Oxford.

Trillmich, F. (1987). Galapagos fur seal, *Arctocephalus galapagoensis*. In *Status, biology, and ecology of fur seals*: 23–27. (Eds Croxall, J. P. & Gentry, R. L.). US Dept. of Commerce, Washington, D.C. (*NOAA tech. Rep. NMFS 51.*)

Trivers, R. L. (1972). Parental investment and sexual selection. In *Sexual selection and the descent of man, 1871 1971*: 136–179. (Ed. Campbell, B.). Aldine, Chicago.

Trivers, R. L. (1985). *Social evolution*. Benjamin/Cummings, Menlo Park.

van Aarde, R. J. (1980). Harem structure of the southern elephant seal *Mirounga leonina* at Kerguelen Island. *Terre Vie* **34**: 31–44.

Verrell, P. A. (1991). Male mating success in the Dusky salamander, *Desmognathus ochrophaeus*: are small, young, inexperienced males at a disadvantage? *Ethology* **88**: 277–286.

Wade, M. J. & Arnold, S. J. (1980). The intensity of sexual selection in relation to male sexual behaviour, female choice, and sperm precedence. *Anim. Behav.* **28**: 446–461.

Walker, B. G. & Bowen, W. D. (In press). Changes in body mass and feeding behaviour in male harbour seals, *Phoca vitulina*, in relation to female reproductive status. *J. Zool., Lond.*

Warneke, R. M. & Shaughnessy, P. D. (1985). *Arctocephalus pusillus*, the South African and Australian fur seal: taxonomy, evolution, biogeography and life history. In *Studies of sea mammals of south latitudes*: 53–77. (Eds Ling, J. K. & Bryden, M. M.). South Australian Museum, Adelaide.

Westneat, D. F. (1990). Genetic parentage in the indigo bunting: a study using DNA fingerprinting. *Behav. Ecol. Sociobiol.* **27**: 67–76.

Wilson, S. C. (1975). Attempted mating between a male grey seal and female harbor seals. *J. Mammal.* **56**: 531–534.

Wylie, R. H. (1973). Territoriality and nonrandom mating in sage grouse, *Centrocerus urophasianus*. *Anim. Behav. Monogr.* **6**: 87–169.

Symp. zool. Soc. Lond. (1993) No. 66: 95–114

Influence of rare ecological events on pinniped social structure and population dynamics

Fritz TRILLMICH

University of Bielefeld,
Faculty of Biology
Behavioural Ecology,
PO Box 10 01 31
D–4800 Bielefeld 1, FRG

Synopsis

The best documented 'rare' ecological events affecting pinniped populations are (1) drastic changes in physical parameters and food resources, e.g. during the 1982–83 El Niño event in the eastern Pacific, and (2) disease outbreaks such as the 1988 epizootic caused by phocine distemper virus in harbour seals. These events were particularly drastic, but smaller events seem to occur quite regularly, if unpredictably. Such events are part of the environmental variance to which most animals are exposed during their lifetime and they may have significant consequences for social and population processes or structure. El Niños can also serve as models of the potential reaction of top predators to human-induced global changes.

During both the 1982–83 El Niño and the epizootic, social patterns changed in response to reduced competition for space and food in the declining pinniped populations. Mother–young interactions and interactions among both females and males were altered, and animals were redistributed over the available habitat. Dispersal, mortality and fertility changed in the affected populations. The magnitude of the effects on sex, cohorts and (sub-) populations depended on the timing of the event in relation to the annual life cycle of the animals.

Rare events may influence differentiation among populations and reduce the long-term mean carrying capacity of the environment and/or the probability of reaching the carrying capacity. They also may create bottlenecks, thereby reducing population heterozygosity. The split of populations into more or less separated subpopulations decreases the probability of extinction. Density-dependent factors are known to influence pinniped population dynamics, but environmental stress as a density-independent, stochastic process influences the population dynamics of mammals as large as pinnipeds far more than is generally appreciated.

ZOOLOGICAL SYMPOSIUM No. 66
ISBN 0–19–854069–8

Copyright © 1993 The Zoological Society of London
All rights of reproduction in any form reserved

Introduction

... and there shall be famines, and pestilences, and earthquakes, in various places
(Matthew 24:7)

One of the major problems in ecology is the explanation of distribution,
abundance and persistence of populations. The debate about the causes of
population persistence and relative stability (DeAngelis & Waterhouse
1987) has become ever more important as conservation of species has
become a more pressing need (Simberloff 1988). The relative importance of
density-dependent regulative processes and density-independent stochastic
events for the determination of population size is still an unresolved issue.
Generally, populations of larger animals (in particular larger vertebrates)
are believed to be less influenced by stochastic events than populations of
small animals (e.g. insects).

Within the last decade, two apparently stochastic events of catastrophic
dimension befell pinniped populations: one was the Pacific-wide El Niño in
1982–83, the other the harbour seal (*Phoca vitulina*) epizootic of 1988–89
in the North Sea. Both these events had a major impact on pinniped
populations and both were documented in detail. They can serve as
examples of the influence of rare ecological events on pinniped populations
and may help to elucidate the role of stochastic processes for population
dynamics of large vertebrates. Since detailed documentation of these events
is impossible in the space available here and has been provided elsewhere
(Trillmich & Ono 1991; Dietz, Heide-Jorgensen & Härkönen 1989; Heide-
Jorgensen *et al.* 1992), this paper will concentrate on a number of more
general issues.

First, methodological problems of the definition of a 'rare' event and of
the documentation of its effects will be briefly discussed. Keeping these
problems in mind, examples will then be given of how physical or biological
stressors influence the main population parameters, i.e. mortality, fertility
and migration/dispersal, and how these effects differ between the sexes and
among age classes. Wherever possible, the more poorly documented role
that behavioural factors play in mediating the population effects will be
mentioned. Finally, the role that spatial heterogeneity and differences in
timing of major stochastic environmental disturbances play in long-term
population survival will be discussed.

What are rare ecological events?

The definition of a 'rare ecological event' depends largely on one's time
frame. A rather vague definition would be 'an event of major impact on
population structure or abundance'. 'Rare' positive events producing a

boom in population condition and subsequent reproduction have never been described for pinnipeds. However, the reduction of whale populations in the Southern Ocean may have been a rare environmental effect allowing the phenomenal rates of increase realized by the Antarctic fur seal (*Arctocephalus gazella*) population after the end of sealing (Croxall *et al.* 1988). La Niña, very cold, productive years after the warm-water catastrophes of El Niño, may provide another example of a rare positive event (see below).

Weatherhead (1986) studied the frequency of reports about unusual events. He relied on the investigators to know what was 'unusual' for their study system. Most 'unusual' events were abiotic in nature (87%) and, surprisingly, the frequency with which they were recorded declined in studies which spanned more than seven years. Apparently, what is considered rare or unusual changes somewhat with an investigator's experience in a particular study system. For Weatherhead's sample of studies on mammals of temperate and tropical ecosystems ($n = 37$), 8% reported unusual events and Weatherhead concluded from this that 'rare' ecological events are relatively frequent. Rare biotic events, in particular diseases, may be under-represented in his samples since they are harder to diagnose than abiotic changes for which weather reports often provide the evidence.

For the purpose of this paper rare ecological events are loosely defined as environmental stressors which induce major deviations of birth, death, or migration rate in one or several age/sex classes away from long-term means. The events have an influence on population structure and/or population dynamics for several subsequent years.

Such changes may be mediated primarily by abiotic factors like temperature or oceanic regime (as in the El Niño example), or by biotic factors, e.g. new parasites as in the case of the harbour seal epizootic.

Methodological problems

Census methods

Given that a rare ecological event is taking place, it is a far from trivial matter to determine population changes in the affected pinniped population(s). Most population assessments produce only relative numbers or indices of population size, whereas the real changes in numbers are needed to quantify the effects of an event. Data are also required to assess the impact of an event on different age and sex classes. Such detail is particularly important in assessing the long-term consequences of an event.

Pinnipeds are usually distributed over wide, often inaccessible areas, which makes a census of the desired accuracy problematic. Aerial censuses may often be the best tool, but they still provide only an index of

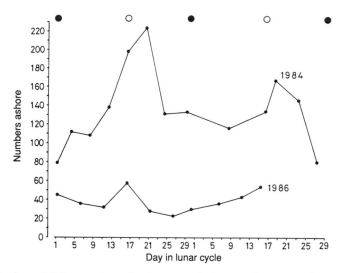

Fig. 1. Numbers of Galapagos fur seals ashore over the lunar cycle. Day 1 and 29 correspond to new moon, day 16 to full moon (see symbols above graph). The 1984 censuses were made between August 26 and October 18 during the cold season when fur seals reproduce. At this time, females spend on average about 60% of their time ashore. The 1986 censuses were made between March 10 and April 23 during the warm season, when females spend only about 20% of the time ashore.

abundance, as many seals will be at sea during the census. Furthermore, the proportions of the populations ashore vary considerably depending upon time of day, phase of the lunar cycle and season (Trillmich & Mohren 1981; Thompson & Harwood 1990; Fig. 1). To estimate population size and its changes for pinnipeds, a thorough study of behaviour over the tidal, daily, lunar, seasonal, and life cycles is needed before census numbers can be converted into population estimates that are at all reliable.

The importance of such studies for the assessment of population change was illustrated by the 1988 epizootic among harbour seals of the Wadden Sea where mortality from carcass counts was 108% of the population estimate for the Wadden Sea before the event (Dietz *et al.* 1989; Fig. 2). Consequently, numbers counted in aerial surveys prior to the epizootic did not represent 100% of the population, although additional biases may have resulted from a redistribution of seal carcasses (Heidemann & Schwarz 1990; Reijnders 1989; Reijnders, Ries & Traut 1990).

A complicating factor: habitat choice

In addition to the uncertainties surrounding most population estimates of pinnipeds, another factor complicates the correct assessment of population changes. During a population crash animals may move into the best habitat,

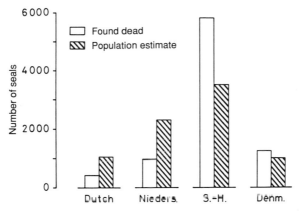

Fig. 2. Number of harbour seals found dead and previous population estimates for subsections of the Wadden Sea area (data from Dietz *et al.* 1989). Coast-section abbreviations: Dutch; Nieders. = Niedersachsen; S.-H. = Schleswig-Holstein; Denm. = Denmark.

making it appear almost as full as before the catastrophe. Such effects are expected to be widespread if individuals distribute according to an ideal free or despotic free distribution over the available habitat (Fretwell & Lucas 1970; Rosenzweig 1985; Milinski & Parker 1991). Estimates of population size in good habitat (where researchers tend to make their observations) will then underestimate the impact of an event.

Despite the reported high site fidelity of pinnipeds, such shifts among habitats of differing quality have been documented for several species. For example, in northern elephant seals (*Mirounga angustirostris*) about 32% of juvenile females gave birth away from their own birth site, and 17.6% of all females changed the location where they gave birth at least once in their lifetime (Le Boeuf & Reiter 1988). Apparently, dominant animals fill the best habitat first and less competitive (young) individuals settle elsewhere once the prime habitat is occupied. The approximation to a free distribution is normally achieved primarily by dispersal of young animals, but under catastrophic conditions all animals may take part.

The El Niño: a model event

El Niño (EN) is an unpredictably occurring but regular feature of the eastern tropical Pacific (Arntz, Pearcy & Trillmich 1991; Quinn, Neal & Antuñez de Mayolo 1987). Figure 3 shows the cumulative probability distribution of the intervals between events. Using a lognormal fit to these data Glynn (1988) determined that the mean recurrence interval for strong events was 11.95 years (± 6.6 S.D.) and for moderate events it was 6.1 (± 3.38) years. In view of the potential lifespan of an otariid pinniped (roughly

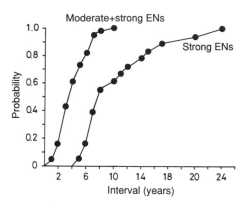

Fig. 3. Cumulative probability of a next moderate or strong, and a next strong El Niño after the end of a previous one (data from Quinn *et al.* 1987).

10–20 years), almost every animal surviving to reproductive age will experience such an event and most animals will even have to live through a strong event. In the long-term view, EN is an example of a regular, unpredictable event which may appear to be exceptional in a short study.

Biotic effects of EN

Marine ecosystems along the eastern margin of the Pacific are highly productive owing to the influence of cold currents and coastal upwellngs. El Niño-Southern Oscillation (ENSO, here called EN) refers to an oceanographic/meteorological condition causing massive advection of warm waters into the cold ocean current systems along the eastern margin of the tropical and subtropical Pacific. In very strong events, the influence of the disturbance is globally felt and influences coastal ecosystems Pacific-wide from Fireland to Alaska (Fahrbach, Trillmich & Arntz 1991, and references therein). During EN, productivity of the marine ecosystems decreases (but may increase towards subpolar latitudes), prey distributions and prey quality for pinnipeds decrease, and increased sea surface levels may cause inundation of otherwise safe haul-out and pupping sites.

These effects have been documented in detail for a number of species for the 1982–83 EN which was the strongest in the last 100 years (Trillmich & Ono 1991). Here, examples will be selected of changes caused by the disturbance. Since EN originates in the tropical Pacific all effects were generally strongest at the equator and attenuated towards higher latitudes. Later observations during other ENs demonstrated that effects observed during the strong 1982–83 EN at higher latitudes were similar to those caused by weaker ENs at low latitudes (Trillmich *et al.* 1991).

Mortality and consequent changes in age structure

By removing whole cohorts or even series of cohorts through mortality of pups and juveniles, EN 1982–83 caused a very ragged age structure in the affected populations. The most striking example of this effect is the case of the Galapagos fur seal in which all individuals of the 1980–82 cohorts died during EN (Trillmich & Limberger 1985; Trillmich & Dellinger 1991). However, loss of whole cohorts or major percentages of cohorts was also reported from the Galapagos sea lion (Trillmich & Dellinger 1991), the South American fur seal (*Arctocephalus australis*) and sea lion (*Otaria byronia*) in Peru (Majluf 1991), the California sea lion (*Zalophus c. californianus*) along the coast of California (Francis & Heath 1991a; DeLong *et al.* 1991), the northern fur seal (*Callorhinus ursinus*) on San Miguel Island (DeLong & Antonelis 1991), and the northern elephant seal (Le Boeuf & Reiter 1991; Huber, Beckham & Nisbet 1991; Stewart & Yochem 1991). Apparently there was no subsequent density-dependent compensation through increased survival of a small cohort; thus the contribution of the affected cohort to the total reproductive output of a population will be almost negligible, as is shown by data on the northern elephant seal (Fig. 4; Le Boeuf & Reiter 1991). Data on birds even show that in affected cohorts low initial survival might be coupled to survival rates which throughout life are lower than those of cohorts from more normal years (Gibbs & Grant 1987). Such effects of poor early rearing conditions on long-term 'quality' (fitness) of a cohort have not been documented for pinnipeds but could slow population recovery after an event.

Not only did fewer juveniles survive, but adults were also affected, and

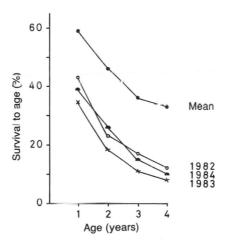

Fig. 4. Survival of elephant seal cohorts was lower for pups weaned in 1982–84: mean = mean of years 1971 to 1981 (after data in Le Boeuf & Reiter 1991).

mortality in the eastern tropical Pacific pinniped populations increased significantly (Trillmich *et al.* 1991). During the 1982–83 EN, 50–70% of adult female Galapagos fur seals died and increased adult mortality was also documented for California sea lions (DeLong *et al.* 1991) and northern fur seals on San Miguel (DeLong & Antonelis 1991). This increased adult mortality affected males and females quite differently. Many males of the Galapagos fur seal and sea lion, as well as territorial males of the South American fur seal in Peru, died during and immediately after the EN, presumably because they had been fasting during the reproductive season and returned to the sea in poor condition at a time when food availability was minimal. In Galapagos, 100% of the territorial Galapagos fur seal males died in 1982 (Trillmich & Limberger 1985). In California sea lions, females apparently suffered relatively higher mortality than males as total female numbers on the Channel Islands had not recovered by 1986 whereas male numbers remained stable throughout. Apparently males had escaped to the north from the area of major EN impact (Huber 1991). Similarly, adult female northern fur seals on San Miguel Island appeared to suffer major mortality during EN whereas males showed no indication of a decrease (DeLong & Antonelis 1991). In northern elephant seals no obvious increase in adult female mortality was noticed and there was only a slight increase in male mortality.

Considering that differences in life history and ecology are almost as great between the pinniped sexes as between pinnipeds and other species, a differential impact of a catastrophic decline in food abundance is to be expected. This is the consequence of the great sexual dimorphism in size and behaviour which characterizes the polygynously breeding species. Unfortunately, no comparative data are available for less dimorphic species like harbour seals.

Fertility

Concomitant with changes in mortality, the 1982–83 EN caused major effects on female fertility. In Galapagos fur seals, only 11% of the normal pupping rate was observed after the end of EN in 1983. Northern fur seals on San Miguel Island were similarly affected, but here it was more difficult to disentangle the effects of female mortality from reduced pupping rate (DeLong & Antonelis 1991). Among California sea lions on the Channel Islands (Francis & Heath 1991a) and the outer coast of Mexico (Aurioles & Le Boeuf 1991) fertility was reduced to 30–70% of normal.

The elephant seals on Año Nuevo and the Farallones reacted differently to EN despite the fact that elephant seal females feed in the open ocean beyond the shelf break (see Le Boeuf *et al.* this volume; R. L. DeLong, B. S. Stewart & R. Hill, pers. comm.), an area which most probably was similarly affected during EN for both populations. Nevertheless, there was a clear decrease in female fertility on the Farallones (Huber *et al.* 1991) and no noticeable change in fertility on Año Nuevo (Le Boeuf & Reiter 1991), suggesting that these animals feed in different patches of ocean.

In addition to immediate changes in fertility there was also a delayed effect on female fertility. Elephant seal cohorts that grew up under EN conditions delayed first reproduction (Le Boeuf & Reiter 1991; Huber *et al.* 1991).

The importance of movements and of spatial differences over the range of a population

Shifts in habitat may have reduced the impact of EN by improving survival chances of animals in the best available habitat. Galapagos fur seals concentrated into the best pupping habitats after EN 1982–83 whereas the coast around this colony was deserted (Trillmich & Dellinger 1991). Similarly, California sea lions shifted among the Channel Islands (DeLong *et al.* 1991).

Some South American fur seals escaped the worst impact of EN by migrating away from the highly disturbed Peruvian upwelling system. They emigrated into northern Chile during 1982, where many eventually settled permanently and bred (Guerra & Portflitt 1991).

In California, northern fur seal juveniles apparently migrated further north than in normal years, even producing the first record of an animal from San Miguel Island back on the Pribilof Islands (DeLong & Antonelis 1991). California sea lion males and juveniles migrated in greater numbers to central California in 1983 and 1984 (Huber 1991) than in other years, thus avoiding the greater EN impact in the Channel Island area of southern California. Adult females were restricted in their capacity to move out of affected areas because of the need to return to pups ashore. This may explain why their mortality increased more than that of males and their fertility was lower in the year(s) following the 1983 EN impact in California (Francis & Heath 1991a; DeLong *et al.* 1991).

Sheltered subpopulations may contribute to long-term population persistence. Interestingly, the Sea of Cortez subpopulation of the California sea lion showed no influence of EN (Aurioles & Le Boeuf 1991), whereas the population just outside the Gulf which depends on upwelling processes for its food resources was affected just as much as all other animals along the Californian coast.

On the Galapagos, the northern population of the fur seal on Genovesa came close to extinction. Before EN, in 1978, it was estimated at approximately 1000 individuals, but after EN censuses in 1988 and 1989 showed that the population had been reduced to between 10% and 20% of its former size (F. Trillmich & B. J. Le Boeuf unpubl. data). Compared to this, the population on the western Islands declined by about 50%.

Both these cases demonstrate that spatial differentiation can decrease the probability of species extinction because a subpopulation sheltered in a habitat less affected by El Niño can repopulate more severely affected areas.

Maternal behaviour

Females were clearly energy-stressed during EN. This applied to both otariids and the elephant seals. Elephant seal females stayed at sea for significantly longer periods between the end of lactation and the onset of the moult. Apparently they needed more time to recover from the stress of lactation under poor foraging conditions. While lactating on land, however, these animals only suffered from the immediate effects of storms which caused increased pup mortality (Le Boeuf & Reiter 1991; Huber *et al.* 1991; Stewart & Yochem 1991); they showed no signs that they were providing less food for the pups.

In contrast, otariid pinniped mothers, which forage during pup-rearing, needed more time and energy to replenish their reserves during foraging trips to sea. They showed more diving activity at sea to get the same food intake (Feldkamp, DeLong & Antonelis 1991; Costa, Antonelis & DeLong 1991) and spent more time away from their young (Heath *et al.* 1991). Consequently pups received less food and in many cases starved (Boness, Oftedal & Ono 1991; Iverson, Oftedal & Ono 1991). In the California sea lion, pups reduced their energy expenditure but this was not sufficient to make up for reduced milk intake when mothers were ashore (Ono & Boness 1991). Slower growth of juveniles then led to longer times to weaning, which was most pronounced in male young (Francis & Heath 1991b). Galapagos fur seal mothers proved totally unable to support the energetic demands of their pups and abandoned them.

Effects on intraspecific competition

Following the decrease in male numbers, the territorial system of Galapagos fur seals changed to a dominance system, since only very few large males were still alive after EN (Trillmich & Limberger 1985). These males claimed large areas but were unable to defend them against smaller intruding males. Reduced competition among males thus led to a major change in social pattern. The loss of three entire Galapagos fur seal cohorts and reduction of those from 1979 and 1983 may lead to similar changes about 8–10 years after the event when most males from these cohorts are expected to breed. These changes in the population's age structure will lead to reduced sexual selection through male–male competition on the cohorts which mature immediately after a major event and another set of cohorts 8–10 years later. Reduced competition within and between the sexes will reduce density-dependent effects on females. This will aid recovery of the population by decreasing age at first reproduction, increasing birth rate, and reducing mortality of pups, juveniles and perhaps adult females alike (Fowler 1990).

Weaker events and rare positive events

Weak events have much lesser effects on otariid populations in the eastern tropical Pacific (Majluf 1991). Rather than causing increased mortality, they lead to delayed weaning of young, which are very likely to be suckled into their second year of life. This lengthened lactation decreases female fertility (Trillmich 1990) and could periodically decrease population growth rate.

Another factor complicating the analysis of the effects of EN events on pinniped populations is the so-called 'La Niña'. After EN, sea surface temperatures regularly decline below the long-term mean and marine productivity increases above the mean. The effect was very strong in 1984 and 1985 in the tropical Pacific when the usual warm season on Galapagos was almost undetectable. Consequently, young Galapagos fur seals and sea lions grew extraordinarily fast and almost all young were independent of their mothers at one year of age. In addition, these cohorts have grown particularly fast and are likely to begin reproduction earlier than other cohorts. Such an upswing of recruitment rate following an EN clearly reduces the impact of the environmental disturbance.

The effects of diseases

Biotic disturbances represent the other major type of environmental stress. The recent harbour seal epizootic in the North Sea area is a good example. Diseases influence populations differently from abiotic factors mainly because transmission dynamics and development of resistance play a key role (Anderson 1991).

Frequency of diseases

Over the last 100 years epizootics have been reported for several species of seals. These have been compiled by Dietz et al. (1989), Harwood & Hall (1990) and Heide-Jorgensen et al. (1992) and are listed in Table 1. Diseases were observed on all continents, except Australia, and also appear to be recurrent events in some pinniped populations. The technical problems of investigating the occurrence and distribution of disease-related mortality in wild populations of pinnipeds are still formidable and explain why we have information only about mass mortalities (Harwood & Hall 1990). Given these limitations, the available data are a highly conservative estimate of the frequency of epidemics in pinniped populations. The mean time-period between the disease outbreaks in harbour seals of northern European waters is 34.5 years (Table 1). Given a generation time of about 5 years, a disease outbreak will influence at least part of a population every fifth to tenth generation. Diseases thus qualify as 'rare' ecological events in pinniped populations but perhaps only because of our insufficient ability to recognize such events. Alternatively, they may really be rare because the immunological memory of individuals surviving an epidemic makes another outbreak of the same disease in the same population unlikely (Heide-Jorgensen *et al.* 1992).

Background information on the harbour seal epizootic

The recent harbour seal epizootic was caused by a morbillivirus related to canine distemper virus, but identified as a separate form called phocine distemper virus (PDV; Curran *et al.* 1990). The disease began in about February to April 1988 in the Kattegat area, spread north along the Norwegian coast and south into the Wadden Sea, until it reached England and Scotland in August to September 1988. The epidemic came to an end in winter and spring 1988–89 when very little further increased mortality was noted (Dietz *et al.* 1989).

Presumably, the disease was transmitted to harbour seals by harp seals (*Phoca groenlandica*) which, in 1986–88, had invaded in great numbers along the Norwegian coast. The causes of this invasion are largely unknown, but may be related to an oceanographic perturbation in the north Atlantic. A few individuals also reached the Kattegat area. Later investigations showed that, in 1987, 40–50% of the harp seals from the West Ice and Barents Sea populations carried antibodies to PDV (Markussen & Have 1992).

Disease-related mortality

The timing of the major impact of the disease influenced its impact on different age classes in the population. The early infections in April and May of 1988 in the Skagerrak-Kattegat area led to total loss of the 1988 cohort since females either aborted or abandoned their pups. In contrast, in the United Kingdom pup mortality was minor since pups were weaned by the time the disease got to the UK. At that time of year, pups rarely come into contact with other animals because of their reduced tendency to haul out (Thompson & Harwood 1990). In contrast, moulting adults haul out gregariously during that time.

Most animals in the populations were infected even though mortality was quite different among subpopulations. Whereas the Kattegat-Skagerrak, Wadden Sea, Irish Sea and East Anglia populations of harbour seals showed a 40–60% mortality, the Scottish population suffered only a 10–20% mortality (Heide-Jorgensen *et al.* 1992; Dietz *et al.* 1989). Of the adult seals

Table 1. Mass deaths of pinnipeds most probably due to biotic factors. The two earliest records about 'harbour seals' may also refer to grey seals. In the case of the walrus the cause of mortality could not be established unequivocally. The so-called 'pox' of the Galapagos seal lion and fur seal clearly is a transmissible disease, but of unknown etiology.

Species	Year	Agent	Effects	Location	Source[a]
Phocids					
Crabeater seal	1955	Virus?	Mass mortality	Antarctica	1
Baikal seal	1987	CDV-like	Mass mortality	Lake Baikal	2, 3
Grey seal	1988/89	PDV	Fewer pups	Britain	4
Harbour seal	ca. 1780	Unknown	Mass mortality	Britain	5
	1813	Unknown	Mass mortality	Orkney	6
	1836	Unknown	Mass mortality	Orkney	6
	1869–1870	Unknown	Mass mortality	Orkney	6
	1918	Unknown	Mass mortality	Iceland	7
	1930	Unknown	Mass mortality	Shetland	8
	1975/80, 82	Avian influenza	Local mortality	New England	9
	1988/89	PDV	Mass mortality	Northern Europe	3, 7
Otariids					
California sea lion	1947	Pneumonia?	Local mortality	California	10
	1968–70	Leptospirosis SMSV?	Abortions, death of subadult males	California	10, 11
South African fur seal	1828	Red tide?	Mass mortality	South Africa	12
Galapagos sea lion	1970–71	('Pox'-virus?)	Mass mortality	Galapagos	13, 14
	1976	('Pox'-virus?)	Mass juvenile mortality	Galapagos	14
	1982–83	('Pox'-virus?)	Mass juvenile mortality	Galapagos	15
Galapagos fur seal	1970–71	('Pox'-virus?)	Small-scale mortality	Galapagos	14
	1976	('Pox'-virus)	Small-scale mortality	Galapagos	14
Odobenid					
Pacific walrus	1978	Unknown	Local mass mort., abortions	Alaska	16

[a]1, Laws & Taylor (1957); 2, Grachev et al. (1989); 3, Heide-Jørgensen et al. (1992); 4, Harwood (1990); 5, Fleming, in Harwood & Hall (1990); 6, McConnell, in Harwood & Hall (1990); 7, Dietz et al. (1989); 8, Bonner (1972); 9, Geraci et al. (1982); 10, Vedros et al. (1971); 11, Smith et al. (1973); 12, Wyatt after Morell, in Harwood & Hall (1990); 13, Rand (1975); 14, Trillmich (1979 and unpubl. obs.); 15, Robinson (1985, and pers. comm.); 16, Fay & Kelly (1980).

in Britain tested in 1989, 88% were seropositive, which indicated almost complete infection of those age-classes, whereas only 12% of tested juveniles were seropositive (Heide-Jorgensen *et al.* 1992). Since this applied also to the Scottish population it suggests that these animals were more resistant to the disease or had lower infective doses, or that the virus had mutated (Thompson *et al.* 1992).

There also appeared to be sex-differences in mortality: in the seal herds infected early 45–60% of the males died (Kattegat), whereas in the herds infected later (UK) 55–75% of the males died (Heide-Jorgensen *et al.* 1992). Perhaps this was due to the fact that males haul out to moult later than females.

The importance of population subdivision

In the Kattegat area the spread of the disease was not density-related, which suggests that social behaviour, in particular gregarious haul-out, always creates local density sufficiently high for disease transmission.

Although the disease spread fast through the harbour seal populations along the coast of the North Sea, some remote populations appeared unaffected. Within the Kattegat and Skagerrak area, the spread between colonies proved to be inversely proportional to distance between herds (Heide-Jorgensen & Härkönen in press). Obviously, migration between haul-out areas is quite frequent as otherwise the disease could not have spread so quickly around the border of the North Sea. However, populations in northern Norway as well as on Iceland and in the Baltic were never affected, which suggests relative isolation from the other populations around the North Sea (but these populations may also have been resistant to the virus).

The grey seal (*Halichoerus grypus*) population showed only limited signs of being affected by the disease. Except for a reduction of 6–24% in pup production in 1988 (Heide-Jorgensen *et al.* 1992) no major effects on grey seal populations were noted. Obviously grey seals were largely resistant to the disease, since seroconversion, i.e. the build-up of antibodies to PDV, occurred in most individuals.

Similarly, the reduced impact of PDV on the Scottish population of harbour seals might indicate that more western populations were less susceptible to the disease. Alternatively, the western stocks may have been less affected because they were hit by the disease in a different stage of their live cycle when contacts between individuals were less frequent and transmission of the disease less likely. It was postulated that seroconversion might have taken place in many cases without clinical signs, perhaps due to low-level infections (Heide-Jorgensen *et al.* 1992).

Conclusions

The effects of the documented rare, catastrophic events on pinniped populations were mainly density-independent. Moreover, the affected populations will take many years to recover from these events. Therefore, rare ecological disasters can keep even pinniped populations well below the estimated carrying capacity of the environment (Harwood & Hall 1990; Reijnders in press).

On the other hand, density-dependent changes of population parameters have been demonstrated for pinnipeds (Fowler 1990) and such effects could reduce the probability of population extinction through environmental catastrophes by increasing the population's growth rate at low densities.

For both abiotically caused reductions in food availability and disease-related mortality, the subdivision of a species into rather isolated sub-populations may be the more important parameter, permitting the species to persist in the face of catastrophic local declines or even local extinctions. Separation of subpopulations and retreat of survivors into optimal habitat reduce the probability of extinction. Sheltered populations, like the sea lions in the Gulf of Cortez during EN, may play a major role in long-term population stability because different stressors will influence each population in a unique way. This would lead to survival of local populations which could then contribute to the repopulation of harder-hit or even extinct populations elsewhere.

The impact of rare events makes migratory behaviour and breeding dispersal very important for long-term population persistence. However, knowledge about dispersal processes is very limited and more information is urgently needed to enable a better assessment of the role of dispersal in long-term population stability in pinnipeds. Distributional shifts during rare events, as exemplified by the emigration of South American fur seals from Peru into Chile, may induce permanent settlement of a new geographical range or of areas where extinction occurred previously.

Lastly, closer attention needs to be given to the investigation of potential biotic causes of population declines because many such events may have been overlooked in free-living populations in the past. To achieve such close documentation and the understanding of population processes deriving from it, long-term monitoring of populations and detailed study of their social behaviour is necessary to evaluate changes in numbers. The variable reaction of harbour seals of different subpopulations to infection with PDV raises the possibility that local populations may be sufficiently different genetically to react differently to parasite challenges. Therefore, dispersal, migration, the role social behaviour plays in the transmission of diseases, and genetic differentiation among populations as well as the role of gene flow need to be studied.

Altogether 'spreading of risk' (den Boer 1968) by the existence of many more or less isolated subpopulations may be as important for populations of large vertebrates as for insect populations. The effects of this spreading, however, will have to be measured on a different time-scale commensurate with the longer lifespan of large vertebrates. Spreading of risk throughout a large geographic range should also be a major aim of conservation strategies to avoid extinction of pinniped species, especially within the perspective of global climate changes.

Acknowledgements

I would like to thank Ian Boyd, Paul Thompson and Peter Reijnders for discussion, many useful hints and critical reading of this paper. E. Geissler kindly drew the graphs. My work in Galapagos was supported throughout by the Max-Planck Institut für Verhaltensphysiologie, Seewiesen, and I express my sincere gratitude for W. Wickler's continuous support.

References

Anderson, R. M. (1991). Populations and infectious diseases: ecology or epidemiology? *J. Anim. Ecol.* **60**: 1–50.
Arntz, W., Pearcy, W. G. & Trillmich, F. (1991). Biological consequences of the 1982–83 El Niño in the eastern Pacific. In *Pinnipeds and El Niño*: 22–42. (Eds Trillmich, F. & Ono, K. A.). Springer Verlag, Heidelberg.
Aurioles, D. & Le Boeuf, B. J. (1991). Effects of the El Niño 1982–1983 on California sea lions in Mexico. In *Pinnipeds and El Niño*: 112–118. (Eds Trillmich, F. & Ono, K. A.). Springer Verlag, Heidelberg.
Boness, D. J., Oftedal, O. T. & Ono, K. A. (1991). The effect of El Niño on pup development in the California sea lion (*Zalophus californianus*) I. In *Pinnipeds and El Niño*: 173–179. (Eds Trillmich, F. & Ono, K. A.). Springer Verlag, Heidelberg.
Bonner, W. N. (1972). The grey seal and common seal in European waters. *Oceanogr. mar. Biol.* **10**: 461–507.
Costa, D. P., Antonelis, G. A. & DeLong, R. L. (1991). Effects of El Niño on the foraging energetics of the California sea lion. In *Pinnipeds and El Niño*: 156–165. (Eds Trillmich, F. & Ono, K. A.). Springer Verlag, Heidelberg.
Croxall, J. P., McCann, T. S., Prince, P. A. & Rothery, P. (1988). Reproductive performance of seabirds and seals at South Georgia and Signy Island, South Orkney Islands, 1976–1987: implications for southern ocean monitoring studies. In *Antarctic Ocean and resources variability*: 261–285. (Ed. Sahrhage, D.). Springer Verlag, Heidelberg.
Curran, M. D., O'Loan, D., Rima, B. K. & Kennedy, S. (1990). Nucleotide sequence analysis of phocine distemper virus reveals its distinctness from canine distemper virus. *Vet. Rec.* **127**: 430–431.

DeAngelis, D. L. & Waterhouse, J. C. (1987). Equilibrium and non-equilibrium concepts in ecological models. *Ecol. Monogr.* **57**: 1–21.

DeLong, R. L. & Antonelis, G. A. (1991). Impact of the 1982–1983 El Niño on the northern fur seal population at San Miguel Island, California. In *Pinnipeds and El Niño*: 75–83. (Eds Trillmich, F. & Ono, K. A.). Springer Verlag, Heidelberg.

DeLong, R. L., Antonelis, G. A., Oliver, C. W., Stewart, B. S., Lowry, M. S. & Yochem, P. K. (1991). Effects of the 1982–1983 El Niño on several population parameters and diet of California sea lions on the California Channel Islands. In *Pinnipeds and El Niño*: 166–172. (Eds Trillmich, F. & Ono, K. A.). Springer Verlag, Heidelberg.

den Boer, P. J. (1968). Spreading the risk and stabilization of animal numbers. *Acta biotheor.* **18**: 165–194.

Dietz, R., Heide-Jorgensen, M.-P. & Härkönen, T. (1989). Mass deaths of harbor seals (*Phoca vitulina*) in Europe. *Ambio* **18**: 258–264.

Fahrbach, E., Trillmich, F. & Arntz, W. (1991). The time sequence and magnitude of physical effects of El Niño in the eastern Pacific. In *Pinnipeds and El Niño*: 8–21. (Eds Trillmich, F. & Ono, K. A.). Springer Verlag, Heidelberg.

Fay, F. H. & Kelly, B. P. (1980). Mass natural mortality of walruses (*Odobenus rosmarus*) at St. Lawrence Island, Bering sea, Autumn 1978. *Arctic* **33**: 226–245.

Feldkamp, S. D., DeLong, R. L. & Antonelis, G. A. (1991). Effects of El Niño on the foraging patterns of California sea lions (*Zalophus californianus*) near San Miguel Island, California. In *Pinnipeds and El Niño*: 146–155. (Eds Trillmich, F. & Ono, K. A.). Springer Verlag, Heidelberg.

Fowler, C. W. (1990). Density dependence in northern fur seals (*Callorhinus ursinus*). *Mar. Mamm. Sci.* **6**: 171–195.

Francis, J. M. & Heath, C. B. (1991a). Population abundance, pup mortality, and copulation frequency in the California sea lion in relation to the 1983 El Niño on San Nicholas Island. In *Pinnipeds and El Niño*: 119–128. (Eds Trillmich, F. & Ono, K. A.). Springer Verlag, Heidelberg.

Francis, J. M. & Heath, C. B. (1991b). The effects of El Niño on the frequency and sex ratio of suckling yearlings in the California sea lion. In *Pinnipeds and El Niño*: 193–204. (Eds Trillmich, F. & Ono, K. A.). Springer Verlag, Heidelberg.

Fretwell, S. D. & Lucas, H. J. Jr. (1970). On territorial behavior and other factors influencing habitat distribution in birds. 1. Theoretical development. *Acta biotheor.* **19**: 16–36.

Geraci, G. R., St. Aubin, D. J., Barker, I. K., Webster, R. G., Hinshaw, V. S., Bean, W. J., Ruhnke, H. L., Prescott, J. H., Early, G., Baker, S. S., Madoff, S. & Schooley, R. T. (1982). Mass mortality of harbor seals: pneumonia associated with influenza A virus. *Science* **215**: 1129–1131.

Gibbs, H. L. & Grant, P. R. (1987). Adult survivorship in Darwin's ground finch (*Geospiza*) populations in a variable environment. *J. Anim. Ecol.* **56**: 797–813.

Glynn, P. W. (1988). El Niño-Southern Oscillation 1982–83: nearshore population, community, and ecosystem responses. *A. Rev. Ecol. Syst.* **19**: 309–345.

Grachev, M. A., Kumarev, V. P., Mamarev, L. V., Zorin, V. L., Baranova, L. V., Denikina, N. N., Belikov, S. I., Petrov, E. A., Kolesnik, V. S., Kolesnik, R. S.,

Dorofeev, V. M., Beim, A. M., Kudelin, V. M., Nagieva, F. G. & Sidorov, V. N. (1989). Distempter virus in Baikal seals. *Nature, Lond.* **338**: 209.

Guerra, C. G. & Portflitt, G. (1991). El Niño effects on pinnipeds in northern Chile. In *Pinnipeds and El Niño*: 47–54. (Eds Trillmich, F. & Ono, K. A.). Springer Verlag, Heidelberg.

Harwood, J. (1990). The 1988 seal epizootic. *J. Zool., Lond.,* **222**: 349–351.

Harwood, J. & Hall, A. (1990). Mass mortality in marine mammals: its implications for population dynamics and genetics. *Trends Ecol. Evol.* **5**: 254–257.

Heath, C. B., Ono, K. A., Boness, D. J. & Francis, J. M. (1991). The influence of El Niño on female attendance patterns in the California sea lion. In *Pinnipeds and El Niño*: 138–145. (Eds Trillmich, F. & Ono, K. A.). Springer Verlag, Heidelberg.

Heide-Jorgensen, M. P. & Härkönen, T. (In press). Epizootiology of the seal disease in the eastern North Sea. *J. appl. Ecol.*

Heide-Jorgensen, M. P., Härkönen, T., Dietz, R. & Thompson, P. M. (1992). Retrospective of the 1988 European seal epizootic. *Dis. aquat. Organ.* **13**: 37–62.

Heidemann, G. & Schwarz, J. (1990). Das Seehundsterben im schleswig-holsteinischen Wattenmeer 1988/89. In *Warnsignale aus der Nordsee*: 325–330. (Eds Lozán, J. L. & Lenz, W.). Parey Verlag, Hamburg.

Huber, H. R. (1991). Changes in the distribution of California sea lions north of the breeding rookeries during the 1982–1983 El Niño. In *Pinnipeds and El Niño*: 129–137. (Eds Trillmich, F. & Ono, K. A.). Springer Verlag, Heidelberg.

Huber, H. R., Beckham, C. & Nisbet, J. (1991). Effects of the 1982–1983 El Niño on northern elephant seal on the South Farallon Islands, California. In *Pinnipeds and El Niño*: 219–233. (Eds Trillmich, F. & Ono, K. A.). Springer Verlag, Heidelberg.

Iverson, S. J., Oftedal, O. T. & Ono, K. A. (1991). The effect of El Niño on pup development in the California sea lion. II. In *Pinnipeds and El Niño*: 180–184. (Eds Trillmich, F. & Ono, K. A.). Springer Verlag, Heidelberg.

Laws, R. M. & Taylor, R. J. F. (1957). A mass dying of Crabeater seals, *Lobodon carcinophagus* (Gray). *Proc. zool. Soc. Lond.* **129**: 315–324.

Le Boeuf, B. J. & Reiter, J. (1988). Lifetime reproductive success in northern elephant seals. In *Reproductive success*: 344–362. (Ed. Clutton-Brock, T. H.). Chicago University Press, Chicago.

Le Boeuf, B. J. & Reiter, J. (1991). Biological effects associated with El Niño, Southern Oscillation 1982–83, on northern elephant seals breeding at Año Nuevo, California. In *Pinnipeds and El Niño*: 206–218. (Eds Trillmich, F. & Ono, K. A.). Springer Verlag, Heidelberg.

Majluf, P. (1991). El Niño effects on pinnipeds in Peru. In *Pinnipeds and El Niño*: 55–65. (Eds Trillmich, F. & Ono, K. A.). Springer Verlag, Heidelberg.

Markussen, N. H. & Have, P. (1992). Phocine distemper virus infection in Harp seals (*Phoca groenlandica*). *Mar. Mamm. Sci.* **8**: 19–26.

Milinski, M. & Parker, G. A. (1991). Competition for resources. In *Behavioural ecology*: 137–168. (Eds Krebs, J. R. & Davies, N. B.). Blackwell Sci. Publ., Oxford.

Ono, K. A. & Boness, D. J. (1991). The influence of El Niño on mother–pup behavior, pup ontogeny, and sex ratios in the California sea lion. In *Pinnipeds and El Niño*: 185–192. (Eds Trillmich, F. & Ono, K. A.). Springer Verlag, Heidelberg.

Quinn, W. H., Neal, V. T. & Antuñez de Mayolo, S. E. (1987). El Niño occurrences over the past four and a half centuries. *J. geophys. Res.* **92C**: 14449–14461.

Rand, C. S. (1975). Nodular suppurative cutaneous cellulitis in a Galapagos sea lion. *J. Wildl. Dis.* **11**: 325–329.

Reijnders, P. J. H. (1989). The recent virus outbreak amongst harbour seals in the Wadden sea: possible consequences for future population trends. *Wadden Sea Newsl.* **1989**: 10–12.

Reijnders, P. J. H. (In press). Retrospective population analysis and related future management perspectives for the harbour seal (*Phoca vitulina*) in the Wadden sea. *Neth. Inst. Sea Res. Publ. Ser.* **20**.

Reijnders, P. J. H., Ries, E. H. & Traut, I. M. (1990). Robbenbestände. In *Warnsignale aus der Nordsee*: 320–324. (Eds Lozán, J. L. & Lenz, W.). Parey Verlag, Hamburg.

Robinson, G. (1985). The influence of the 1982–83 El Niño on Galapagos marine life. In *El Niño in the Galapagos Islands: the 1982–1983 event*: 153–190. (Eds Robinson, G. & del Pino, E. M.). Fundacion Charles Darwin, Quito.

Rosenzweig, M. L. (1985). Some theoretical aspects of habitat selection. In *Habitat selection in birds*: 517–540. (Ed. Cody, M. L.). Academic Press, Orlando & London.

Simberloff, D. (1988). The contribution of population and community biology to conservation science. *A. Rev. Ecol. Syst.* **19**: 473–511.

Smith, A. W., Akers, T. G., Madin, S. H. & Vedros, N. A. (1973). San Miguel sea lion virus isolation, preliminary characterization and relationship to vesicular exanthema of swine virus. *Nature, Lond.* **244**: 108.

Stewart, B. S. & Yochem, P. K. (1991). Northern elephant seals on the southern California Channel Islands and El Niño. In *Pinnipeds and El Niño*: 234–246. (Eds Trillmich, F. & Ono, K. A.). Springer Verlag, Heidelberg.

Thompson, P. M., Cornwell, H. J. C., Ross, H. M. & Miller, D. (1992). A serological study of the prevalence of phocine distemper virus in a population of harbour seals in the Moray Firth, N. E. Scotland. *J. Wildl. Dis.* **28**: 21–27.

Thompson, P. M. & Harwood, J. (1990). Methods for estimating the population size of common seals, *Phoca vitulina*. *J. appl. Ecol.* **27**: 924–938.

Trillmich, F. (1979). Galapagos sea lions and fur seals. *Notic. Galapagos* No. 29: 8–14.

Trillmich, F. (1990). The behavioral ecology of maternal effort in fur seals and sea lions. *Behaviour* **114**: 3–20.

Trillmich, F. & Dellinger, T. (1991). The effects of El Niño on Galapagos pinnipeds. In *Pinnipeds and El Niño*: 66–74. (Eds Trillmich, F. & Ono, K. A.). Springer Verlag, Heidelberg.

Trillmich, F. & Limberger, D. (1985). Drastic effects of El Niño on Galapagos pinnipeds. *Oecologia* **67**: 19–22.

Trillmich, F. & Mohren, W. (1981). Effects of the lunar cycle on the Galapagos fur seal, *Arctocephalus galapagoensis*. *Oecologia* **48**: 85–92.

Trillmich, F. & Ono, K. A. (Eds) (1991). *Pinnipeds and El Niño. Responses to environmental stress*. Springer Verlag, Heidelberg.

Trillmich, F., Ono, K. A., Costa, D. P., DeLong, R. L., Feldkamp, S. D., Francis, J. M., Gentry, R. L., Heath, C. B., Le Boeuf, B. J., Majluf, P. & York, A. E.

(1991). The effects of El Niño on pinniped populations in the eastern Pacific. In *Pinnipeds and El Niño*: 247–270. (Eds Trillmich, F. & Ono, K. A.). Springer Verlag, Heidelberg.

Vedros, N. A., Smith, A. V., Schonewald, J., Migaki, G. & Hubbard, R. C. (1971). Leptospirosis epizootic among California sea lions. *Science* 172: 1250–1251.

Weatherhead, P. J. (1986). How unusual are unusual events? *Am. Nat.* 128: 150–154.

Symp. zool. Soc. Lond. (1993) No. 66: 115–129

Influence of maternal characteristics and environmental variation on reproduction in Antarctic fur seals

N. J. LUNN
and I. L. BOYD

*British Antarctic Survey
Natural Environment Research
Council
High Cross, Madingley Road
Cambridge CB3 0ET, UK*

Synopsis

Antarctic fur seals breed synchronously in a highly seasonal environment. There is considerable inter-annual variation in food supply, which affects breeding performance. However, food supply may also influence future performance through effects on female ovulation, implantation and/or pregnancy. We used foraging trip duration, pup growth rate and weaning mass as indicators of the food available to females during the pup-rearing period (December–April) and examined relationships between these and pup production and timing of breeding in the following year.

Productivity (pup production) was positively correlated with growth rate and weaning mass of both male and female pups in the previous season ($P < 0.05$) and negatively correlated with the variation around the mean birth mass of males ($P < 0.05$), suggesting that poor feeding conditions one season led to lower production the next. The timing of birth was positively correlated with foraging trip duration ($P < 0.05$) and negatively correlated with the birth and weaning masses of male and female pups, the duration of the perinatal period, growth of male pups, and the variation in the late-season growth (last two months) of male and female pups ($P < 0.05$ in all cases). This indicates that females give birth later in the season following a year when food resources were scarce.

These results are consistent with the hypothesis that poor feeding conditions, indicated by slow pup growth and long foraging trips, lead to lower production and delayed pupping the following year. We suggest that these effects result from females being in poorer condition in years of food shortage, so that (1) fewer of them implant (or more abort) and (2) implantation is delayed.

Introduction

Antarctic krill (*Euphausia superba*) is the main component in the diet of Antarctic fur seals (*Arctocephalus gazella*) breeding at South Georgia

ZOOLOGICAL SYMPOSIUM No. 66
ISBN 0-19-854069-8

Copyright © 1993 The Zoological Society of London
All rights of reproduction in any form reserved

(Croxall & Pilcher 1984; Doidge & Croxall 1985) and reproductive rates of both fur seals and other krill predators occurring there are sensitive to variations in krill abundance (Croxall & Prince 1979; Croxall *et al.* 1988). In particular, events which occurred in the 1977/78, 1983/84 and 1990/91 breeding seasons, when there was an apparent shortage of krill at South Georgia (Bonner, Everson & Prince 1978; Croxall & Prince 1979; Heywood, Everson & Priddle 1985; Priddle *et al.* 1988; British Antarctic Survey unpubl. data), resulted in sharp declines in the reproductive success of all krill predators (Croxall *et al.* 1988). Krill predators at South Georgia are probably highly sensitive to fluctuations in this food supply because there are few alternative exploitable prey available to these species. Reproductive success of these predators is therefore likely to be an indicator of fluctuations in the environmental carrying capacity (Croxall & Prince 1979; Croxall *et al.* 1988) and the three poor krill seasons observed to date are probably extremes in a continually fluctuating food base. For fur seals such events were easily observed in terms of low pup birth mass, long maternal foraging trips, high neonatal mortality rates and depressed pup growth during nursing (Croxall & Prince 1979; Croxall *et al.* 1988; Lunn & Boyd 1993). However, a poor krill year may have longer-term effects causing depression of breeding success in the following year. This may result from increased mortality of breeding adults, as suggested by Croxall *et al.* (1988), but it may also occur because severe food shortages can block oestrus, terminate pregnancy, prevent lactation and probably influence the timing of implantation and subsequent birth (Short 1984; Bronson 1989; Boyd 1991b).

Fur seals are likely to be particularly susceptible to fluctuations in food supply because of specific features of their reproductive cycle. While the basic reproductive cycle lasts 12 months, it takes 16 months (from conception to weaning) to rear a pup; investment in consecutive offspring thus overlaps by four months. This is because key stages of reproduction are split between two consecutive summers: ovulation, oestrus and implantation occur the first summer whereas parturition, lactation and weaning occur the second summer (Fig. 1). Events occurring during one cycle can thus readily influence the events of the next cycle.

Adult females are thought to lead a pelagic existence away from South Georgia during the austral winter (May to November) (Payne 1979a; Doidge, McCann & Croxall 1986; Duck 1990) but return to traditional breeding beaches in November or December (Doidge, McCann & Croxall 1986; Lunn & Boyd 1991) and give birth within a few days of arrival. They remain with their newborn pups for approximately seven days (the perinatal period) and fast (Costa & Trillmich 1988) which means that all nutritional requirements of mothers and pups must be met from stored maternal resources. Normally within 24 h prior to first departure, females come into

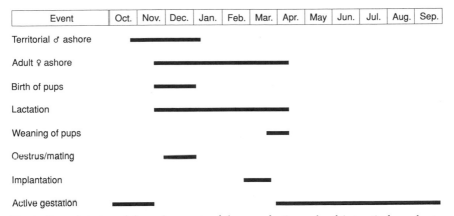

Event	Oct.	Nov.	Dec.	Jan.	Feb.	Mar.	Apr.	May	Jun.	Jul.	Aug.	Sep.

Territorial ♂ ashore

Adult ♀ ashore

Birth of pups

Lactation

Weaning of pups

Oestrus/mating

Implantation

Active gestation

Fig. 1. General timing of the major events of the reproductive cycle of Antarctic fur seals at Bird Island, South Georgia.

oestrus and are mated (McCann 1980; Doidge, McCann & Croxall 1986). The fertilized egg then enters a state of embryonic diapause and is thought to implant some time in late February or March (Boyd 1991a). Until they are weaned in late March or early April, pups are nutritionally dependent upon their mothers, which alternate 3–5-day trips to sea to feed with 1–2-day visits ashore to nurse (Bonner 1984; Doidge, McCann & Croxall 1986).

We expect that in summers when food resources are scarce, more female Antarctic fur seals will be nutritionally stressed. This should result in an increase in the number of females that fail to come into oestrus, an increase in the number of females that mate but do not implant, a delay in the timing of implantation and, as a consequence, a delay in the timing of parturition the following summer. We aim to test these hypotheses using data on annual estimates of pup production, birth date, pup birth mass, pup growth rate, and maternal behaviour patterns of Antarctic fur seals at Bird Island, South Georgia (54 °S, 38 °W) collected over nine consecutive breeding seasons.

Methods

Measurements of food abundance and nutritional condition

We use various reproductive parameters (Table 1) likely to reflect food availability and hence nutritional condition throughout lactation as indicators of maternal condition at oestrus and implantation. These are: (1) pup birth mass, which should reflect nutritional condition during the final stage of gestation (Boyd & McCann 1989; Trites 1991), (2) pup growth and (3) foraging trip duration, both of which should reflect food availability in the 3.5 to 4 months prior to implantation (Bengtson 1988;

Table 1. Reproductive parameters used as indicators of nutritional condition of female Antarctic fur seals and food availability at Bird Island

Breeding condition (year x)	Maternal condition all season (year x)	Timing of implantation (year x)
Pup birth mass (x)	Pup growth (x)	Late pup growth[a] (x)
Perinatal duration (x)	Foraging trip duration (x)	Pup weaning mass (x)
	Pup production ($x + 1$)	Birth date ($x + 1$)

[a] Late growth: pup age = 62–93 days

Croxall *et al.* 1988; Costa, Croxall & Duck 1989), (4) weaning mass, and (5) late-season growth, both of which should reflect maternal condition at implantation.

The various reproductive parameters measured only apply to females that pup and as a consequence our analyses do not take into account those females that do not pup. The latter, once mated, would be free from the constraint of having to return to Bird Island every 3–5 days and could leave an area of reduced food abundance in search of more productive areas; therefore the effects of food shortage on their future reproductive performance are likely to be less pronounced. From 1988/89 through 1991/92, the proportion of females represented by this group ranged from 0.17 to 0.35 ($\bar{x} = 0.25 \pm 0.04$).

Estimation of birth date, mass, perinatal period duration and foraging trip duration

A study beach (SB) was surveyed twice daily from mid-November through early January for the birth of pups which were assumed to have been born on the day they were first observed. Newborn pups were retrieved from the beach with the aid of a noosing pole, sexed, weighed to the nearest 100 g on a 10-kg capacity spring scale, marked with temporary serial numbers by bleaching the fur on their backs with peroxide hair dye (Clairol Born Blonde®, Bristol-Myers Company Limited, Swakeleys House, Milton Road, Ickenham, Uxbridge, UK), and returned to their mothers.

The duration of the perinatal period was defined as the interval between parturition and a mother's first departure to sea (determined by her absence ashore). For the remainder of lactation we assumed a female to have been present ashore all day if observed during a daily search of SB, adjacent beaches, and surrounding tussock (*Poa flabellata*) grassland (an area of approximately 25 ha) and this information was used to calculate foraging trip duration. Daily searches were made until it was apparent that the vast majority of pups had been weaned, as determined by their absence ashore (late March to mid-April).

Estimation of pup growth

Pup growth was estimated by weighing SB pups on day of birth and thereafter samples of pups at two and three months of age from a large adjacent beach (to avoid the disruption likely to be caused at SB). Pups were weighed (n = 100: 1983/84–1985/86, 1988/89–1990/91; n = 150: 1986/87, 1987/88) to the nearest 100 g by methods outlined by Payne (1979b) and Croxall & Prince (1979). Pups were given a temporary mark after weighing to prevent their recapture during that sampling period.

Growth rates of pups were estimated by least squares regression of weight on age.

Data analysis

A matrix of the means and standard deviations of the reproductive parameters measured over the past nine consecutive seasons was constructed (Table 2). For analysis, three parameters—date of parturition, variation around the mean date of parturition and total pup production—were considered as part of the data set for the preceding year, since changes in these parameters would be most affected by changes in the remaining parameters in the previous year. After checking the data for normality (Kolmogorov-Smirnov goodness of fit, Sokal & Rohlf 1981), Pearson's product-moment correlations were run to determine relationships between parameters measured in the current year and those measured in the previous year. Level of significance was $P = 0.05$.

Results

There appeared to be a significant overall effect of food availability in one summer, as indicated by pup growth and maternal foraging behaviour, on reproduction in the following summer, because 41.2% (14/34) of correlations with these two types of parameters were significant (Table 3). The variation around the mean pupping date was not correlated with any parameter and was therefore excluded from further analyses.

Pup production

Pup production was negatively correlated with the variation around the mean birth mass of male pups in the previous year (Table 3); the negative correlation with the variation around the mean birth mass of female pups did not quite reach significance ($r = -0.538$, $P = 0.085$). This suggests that

Table 2. Basic reproductive parameters of female Antarctic fur seals measured at Bird Island, South Georgia[a]

Parameter	Austral summer								
	1983/84	1984/85	1985/86	1986/87	1987/88	1988/89	1989/90	1990/91	1991/92
Mean birth date (Julian date)	343	347	343	343	346	339	337	345	347
S.D.[b] birth date (days)	6.1	7.0	6.7	7.2	7.1	7.9	8.0	11.6	7.5
Mean birth mass (kg) ♂	5.7	5.4	5.5	5.4	5.8	5.8	5.6	5.2	5.6
♀	5.0	4.7	4.9	4.8	5.2	5.2	4.9	4.6	5.0
S.D. birth mass (kg) ♂	0.7	0.1	0.3	0.4	0.4	0.6	0.6	0.7	0.7
♀	0.6	0.1	0.1	0.4	0.4	0.6	0.5	0.6	0.6
Duration perinatal (days)	5.6	6.6	6.2	6.6	6.8	6.7	6.3	4.8	6.4
S.D. perinatal (days)	1.5	1.2	1.4	1.1	1.4	1.9	1.2	1.3	1.3
Pup production (n)	850	664	854	818	799	757	822	545	718
Foraging trip (days)	6.8	3.1	3.7	4.8	3.3	4.1	4.4	7.3	4.1
Pup growth rate (g d⁻¹) ♂	62.8	97.7	103.7	97.3	92.6	102.7	86.5	79.9	99.3
♀	55.5	85.4	82.5	78.4	73.2	78.4	72.3	70.8	87.2
S.D. pup growth rate (g d⁻¹) ♂	1.8	2.6	1.9	1.3	1.8	1.8	1.9	2.2	2.5
♀	1.6	2.4	1.7	1.3	1.4	1.6	1.7	2.0	1.9
Late growth[c] (g d⁻¹) ♂	47.6	74.0	100.9	152.8	121.6	74.8	56.7	119.9	107.3
♀	46.8	75.8	108.3	108.5	62.2	39.8	57.1	121.8	67.2
S.D. late growth (g d⁻¹) ♂	9.7	15.6	13.2	11.4	14.8	16.9	14.3	14.0	15.9
♀	8.7	11.8	11.7	10.9	11.8	14.5	12.0	12.9	10.1
Weaning mass[d] (kg) ♂	13.2	17.1	17.8	16.9	16.8	18.1	15.9	14.6	17.3
♀	11.5	14.6	14.5	13.9	13.7	14.4	13.3	12.7	15.1

[a] Data sources: Croxall *et al.* (1988); Duck (1990); British Antarctic Survey unpubl. data
[b] S.D. = standard deviation
[c] Early growth: pup age = 0–62 days; Late growth: pup age = 62–93 days
[d] Weaning mass estimated from regression (age at weaning: ♂ = 119 days, ♀ = 116 days)

Table 3. Significant correlations[a] between parameters measured in the previous year and those measured in the current year

Parameter (previous year)	Parameter (current year)	
	Pup production	Birth date
Birth mass ♂		−0.659 *
♀		−0.737 *
S.D. birth mass ♂	−0.646 *	
Perinatal duration		−0.679 *
S.D. perinatal duration		−0.654 *
Foraging trip		0.684 *
Growth ♂	0.651 *	−0.624 *
♀	0.658 *	
S.D. late growth ♂		−0.742 *
♀		−0.621 *
Weaning mass ♂	0.629 *	−0.719 *
♀	0.670 *	

[a] Entire rows or columns with non-significant correlations have been omitted. * = $P \leqslant 0.05$

increased pup production occurred in summers following those when there was less variability in pup birth mass, at least for males. The difference between the sexes did not appear to result from greater variability in male than in female birth mass (mean C.V. males = 12.3 ± 3.28, $n = 7$; females = 11.7 ± 3.13, $n = 7$). By way of contrast, there was no relationship between pup production and the variation around the mean birth mass of male or female pups within the same year ($P > 0.3$ in both cases).

Both growth rate and weaning mass (which indicate feeding conditions through lactation and implantation) of male and female pups were positively correlated with pup production the next year (Table 3, Fig. 2) suggesting that pup production increased following a summer when local food resources appeared to be abundant. However, in the austral summer 1990/91, both growth rate and weaning mass were low (Table 2) which, we would have predicted, should have resulted in reduced pup production the next summer. However, an increase in pup production occurred. Although this increase in 1991/92 may have been related to the apparent improvement in food availability in the later stages of 1990/91 (high late-season growth of male and female pups), overall, late-season growth of male and female pups was not correlated with subsequent pup production ($P > 0.1$ in both cases).

Timing of parturition

We also predicted that, following a summer of reduced food resources, parturition would be delayed the next summer because more females would have been nutritionally stressed and hence implantation delayed. Eight parameters measured in one summer were negatively correlated with the timing of parturition the following summer (Table 3, Figs 3 and 4a),

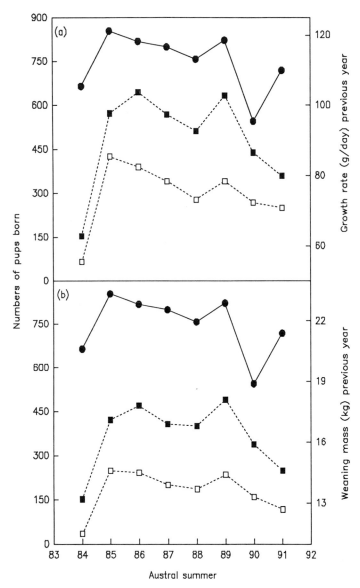

Fig. 2. Relationship between pup production (●) in current year and (a) pup growth in previous year and (b) pup weaning mass in previous year. Male pups (■), female pups (□).

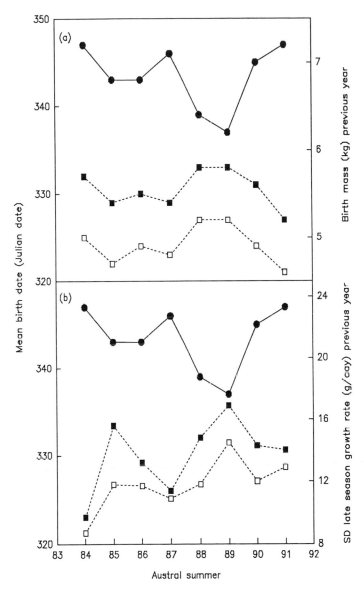

Fig. 3. Relationship between the timing of parturition (●) in current year and (a) pup birth mass in previous year and (b) variation in late-season pup growth in previous year. Male pups (■), female pups (□).

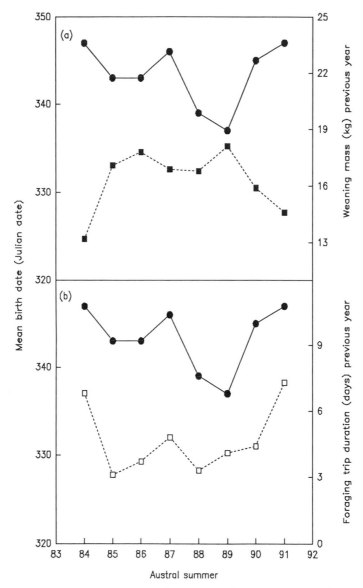

Fig. 4. Relationship between the timing of parturition (●) in current year and (a) male pup weaning mass (■) in previous year and (b) foraging trip duration (□) in previous year.

including male and female birth mass, duration of the perinatal period, variation in late-season growth of male and female pups, and male weaning mass (female weaning mass approached significance, $r = -0.613$, $P = 0.053$). In addition, foraging trip duration in the previous summer was positively correlated with the timing of parturition (Table 3, Fig. 4b). These correlations all suggest that, following summers when food resources were reduced and hence females were in poorer condition, the timing of parturition was delayed, probably owing to a delay in the timing of implantation.

Discussion

Our aim in this paper was to examine the longer-term effects of variation in food supply on reproduction of female Antarctic fur seals at Bird Island, South Georgia, by investigating the relationships between various reproductive parameters measured over nine consecutive years. We found that reduced food resources in the current year had two main effects on reproduction the following year: (1) a reduction in pup production and (2) a delay in the timing of parturition.

Pup production

Pup production depends upon the number of females that come into oestrus, mate, implant, and are successful in carrying a foetus to full term. Consequently factors that affect any of these four stages will obviously affect the numbers of pups born. As stated earlier, successful reproduction is dependent upon nutrition, females in poor condition being unlikely to reproduce (Short 1984; Lochmiller, Hellgren & Grant 1986; Bronson 1989). From this study it did not appear that the number of females entering oestrus was greatly affected by reduced food resources since, of six parameters used as indicators of maternal condition at oestrus, only one (variability in male birth mass) was significantly correlated with subsequent pup production. We had expected a greater number of females not to enter oestrus and mate when food shortages occur but it is important to note that, although years of food shortages were easily observed in terms of the parameters measured, they may not have been severe enough to prevent oestrus.

It appeared that food shortages may have had a greater impact on subsequent pup production at the time of implantation. On each foraging trip throughout lactation, females replenish reserves not only for subsequent milk production but also to meet their own requirements. It seems logical to assume that in years of high pup growth when females are producing more milk, they must have larger reserves and hence be in better condition. The positive correlation with both the growth rate and weaning mass of the current season's pup cohort (both males and females) and the numbers of pups born the following year suggests that good feeding conditions in one year promote increased pup production the next and this is most probably realized through an increase in the proportion of females that implant and carry the foetus to term.

Increased adult mortality resulting from a year of food shortage may explain decreased pup production the following year. From sightings of tagged females and taking into account that 6.4% of animals lose both tags (Lunn & Boyd 1991), the annual proportion of females which are never seen again is 0.161 ± 0.002 ($n = 6$ years). After 1983/84 and 1989/90, the proportion of females never seen again was 0.362 and 0.358 respectively, both over twice the average, which suggests that increased mortality may, in part, be responsible for decreased pup production. Using an average pup production of 800, a pupping rate for all females on SB of 0.75 (British Antarctic Survey unpubl. data), and 0.161 as the proportion of adult females never seen again, we estimated that adult mortality would have to have been 0.388 and 0.586 in 1983/84 and 1989/90 respectively in order for mortality solely to account for the observed decrease in pup production. Although increased adult mortality may therefore have been entirely responsible for decreased pup production in 1984/85, it certainly could not have been in 1990/91; another cause is indicated, which we speculate was an increase in the proportion of females failing to implant towards the end of the 1989/90 season or a decline in the proportion carrying the foetus to term.

In the austral summer of 1990 both growth rates and weaning masses were lower than in the previous year, which should have resulted in production of fewer pups in 1991 than in 1990; in fact there was an increase of 173 pups. Several mechanisms may explain this apparent discrepancy. First, there were 150 more females present on the study beach in 1991 than in 1990, so that much of the increased pup production may have been due to the presence of more females. Second, although there was no significant correlation between late-season growth in the previous year and subsequent pup production, late-season growth in 1990 was high (Table 2) and indicates that females must have been able to find high-quality krill patches. If so, more females than expected may have been able to replenish reserves and were in sufficiently good condition for implantation to occur.

In addition, it is worth keeping these types of data in perspective. Pup production in 1990/91 was the lowest recorded, foraging trips the longest, growth rates the second slowest, and weaning masses the second lowest (Table 2). Despite high late-season growth in 1990/91 and an increase in females on SB in 1991/92, pup production in 1991/92 was still the third lowest and well below the mean of 800 ± 19 pups born in average years.

Timing of parturition

Nine measures of female condition and food availability in the current season were significantly correlated with the timing of parturition the following season and suggest that food availability influenced the timing of implantation and, as a consequence, the timing of parturition the following year. We know that, even in years of 'average' food availability, younger female Antarctic fur seals tend to be in poorer condition than older females and give birth later in the season (Lunn & Boyd 1993) and that similar relationships hold in other pinnipeds (Reiter, Panken & Le Boeuf 1981; Boyd 1984). Parameters associated with male pups provided more correlations with the subsequent timing of parturition than those associated with female pups (four versus two). This difference between sexes may reflect the higher demand made on maternal resources by male compared with female pups, as males grow faster than females (Doidge, Croxall & Ricketts 1984; Table 2 this paper). As a consequence mothers of males are probably more sensitive to changes in food availability in all years than mothers of females.

Relationships involving timing of parturition with birth mass of male and female pups and with the duration and variability of the perinatal period in the previous summer probably reflect the general influence of conditions from one year to the next but they are more difficult to explain in terms of their influence on specific events in the reproductive cycle unless the timing of oestrus varies with environmental conditions. The negative relationships between the variation in late-season growth of male and female pups with the timing of parturition emphasize the importance of nutrition to females around the time of implantation. In years when food resources are scarce, we would expect less variability in individual pup growth because the foraging abilities of most mothers would be constrained and therefore more mothers would be nutritionally stressed, unable to feed their pups adequately, and likely to delay the timing of implantation as a consequence of being in poor condition. By way of contrast, when food is relatively abundant, most mothers are able to rear healthy pups but there may be more variability in individual pup growth because those mothers that are better foragers would be unconstrained by food resources.

In conclusion, the relationship between food availability and reproduction in female Antarctic fur seals is not confined to a single season but is such that a population may take several years to recover from environmental change. The long-term effect of seasons of food shortage on individuals is unknown because data at this time are insufficient. However, what is clear is that a season of reduced food resources can result in some females not raising pups for a minimum of two consecutive seasons. When one considers that three poor krill years have occurred within a period of 13 years and that peak pregnancy rates occur over a 12-year span from ages 5–16 years (Payne 1977), these lost opportunities to rear pups could represent up to half of a female's lifetime reproductive potential.

Acknowledgements

We thank all those who have worked on Bird Island and helped to collect the data on which this paper was based. J. P. Croxall and F. Trillmich provided valuable comments on earlier drafts.

References

Bengtson, J. L. (1988). Long-term trends in the foraging patterns of female Antarctic fur seals at South Georgia. In *Antarctic Ocean and resources variability*: 286–291. (Ed. Sahrhage, D.). Springer-Verlag, Berlin.

Bonner, W. N. (1984). Lactation strategies in pinnipeds: problems for a marine mammalian group. *Symp. zool. Soc. Lond.* No. 51: 253–272.

Bonner, W. N., Everson, I. & Prince, P. A. (1978). *A shortage of krill*, Euphausia superba, *around South Georgia*. Unpublished report: International Council for the Exploration of the Sea, Biological Oceanography Committee, CM 1978/L:22: 1–4.

Boyd, I. L. (1984). The relationship between body condition and the timing of implantation in pregnant Grey seals (*Halichoerus grypus*). *J. Zool., Lond.* 203: 113–123.

Boyd, I. L. (1991a). Changes in plasma progesterone and prolactin concentrations during the annual cycle and the role of prolactin in the maintenance of lactation and luteal development in the Antarctic fur seal (*Arctocephalus gazella*). *J. Reprod. Fert.* 91: 637–647.

Boyd, I. L. (1991b). Environmental and physiological factors controlling the reproductive cycles of pinnipeds. *Can. J. Zool.* 69: 1135–1148.

Boyd, I. L. & McCann, T. S. (1989). Pre-natal investment in reproduction by female Antarctic fur seals. *Behav. Ecol. Sociobiol.* 24: 377–385.

Bronson, F. H. (1989). *Mammalian reproductive biology*. University of Chicago Press, Chicago & London.

Costa, D. P., Croxall, J. P. & Duck, C. D. (1989). Foraging energetics of Antarctic fur seals in relation to changes in prey availability. *Ecology* 70: 596–606.

Costa, D. P. & Trillmich, F. (1988). Mass changes and metabolism during the perinatal fast: a comparison between Antarctic (*Arctocephalus gazella*) and Galápagos fur seals (*Arctocephalus galapagoensis*). *Physiol. Zoöl.* 61: 160–169.

Croxall, J. P. & Pilcher, M. N. (1984). Characteristics of krill *Euphausia superba* eaten by Antarctic fur seals *Arctocephalus gazella* at South Georgia. *Bull. Br. Antarct. Surv.* No. 63: 117–125.

Croxall, J. P. & Prince, P. A. (1979). Antarctic seabird and seal monitoring studies. *Polar Rec.* 19: 573–595.

Croxall, J. P., McCann, T. S., Prince, P. A. & Rothery, P. (1988). Reproductive performance of seabirds and seals at South Georgia and Signy Island, South Orkney Islands, 1976–1987: implications for Southern Ocean monitoring studies. In *Antarctic Ocean and resources variability*: 261–285. (Ed. Sahrhage, D.). Springer-Verlag, Berlin.

Doidge, D. W. & Croxall, J. P. (1985). Diet and energy budget of the Antarctic fur seal, *Arctocephalus gazella*, at South Georgia. In *Antarctic nutrient cycles and food webs*: 543–550. (Eds Siegfried, W. R., Condy, P. R. & Laws, R. M.). Springer-Verlag, Berlin.

Doidge, D. W., Croxall, J. P. & Ricketts, C. (1984). Growth rates of Antarctic fur seal *Arctocephalus gazella* pups at South Georgia. *J. Zool., Lond.* **203**: 87–93.

Doidge, D. W., McCann, T. S. & Croxall, J. P. (1986). Attendance behavior of Antarctic fur seals. In *Fur seals: maternal strategies on land and at sea*: 102–114. (Eds Gentry, R. L. & Kooyman, G. L.). Princeton University Press, Princeton.

Duck, C. D. (1990). Annual variation in the timing of reproduction in Antarctic fur seals, *Arctocephalus gazella*, at Bird Island, South Georgia. *J. Zool., Lond.* **222**: 103–116.

Heywood, R. B., Everson, I. & Priddle, J. (1985). The absence of krill from the South Georgia zone, winter 1983. *Deep-Sea Res. (A)* **32**: 369–378.

Lochmiller, R. L., Hellgren, E. C. & Grant, W. E. (1986). Reproductive responses to nutritional stress in adult female collared peccaries. *J. Wildl. Mgmt* **50**: 295–300.

Lunn, N. J. & Boyd, I. L. (1991). Pupping site fidelity of Antarctic fur seals at Bird Island, South Georgia. *J. Mammal.* **72**: 202–206.

Lunn, N. J. & Boyd, I. L. (1993). Effects of maternal age and condition on parturition and the perinatal attendance period of female Antarctic fur seals. *J. Zool., Lond.* **229**: 55–67.

McCann, T. S. (1980). Territoriality and breeding behaviour of adult male Antarctic Fur seal, *Arctocephalus gazella*. *J. Zool., Lond.* **192**: 295–310.

Payne, M. R. (1977). Growth of a fur seal population. *Phil. Trans. R. Soc. (B)* **279**: 67–79.

Payne, M. R. (1979a). Fur seals *Arctocephalus tropicalis* and *A. gazella* crossing the Antarctic Convergence at South Georgia. *Mammalia* **43**: 93–98.

Payne, M. R. (1979b). Growth in the Antarctic fur seal *Arctocephalus gazella*. *J. Zool., Lond.* **187**: 1–20.

Priddle, J., Croxall, J. P., Everson, I., Heywood, R. B., Murphy, E. J., Prince, P. A. & Sear, C. B. (1988). Large-scale fluctuations in distribution and abundance of krill—a discussion of possible causes. In *Antarctic Ocean and resources variability*: 169–182. (Ed. Sahrhage, D.). Springer-Verlag, Berlin.

Reiter J., Panken, K. J. & Le Boeuf, B. J. (1981). Female competition and reproductive success in northern elephant seals. *Anim. Behav.* **29**: 670–687.

Short, R. V. (1984). Oestrous and menstrual cycles. In *Hormonal control of reproduction*: 115–152. (Eds Austin, C. R. & Short, R. V.). Cambridge University Press, Cambridge. (*Reproduction in mammals Book 3*.)

Sokal, R. R. & Rohlf, F. J. (1981). *Biometry: the principles and practice of statistics in biolgical research*. (2nd edn). W. H. Freeman, New York.

Trites, A. W. (1991). Fetal growth of northern fur seals: life-history strategy and sources of variation. *Can. J. Zool.* **69**: 2608–2617.

Symp. zool. Soc. Lond. (1993) No. 66: 131–145

Social organization of humpback whales on a North Atlantic feeding ground

Phillip J. CLAPHAM

Department of Zoology
University of Aberdeen
Aberdeen AB9 2TN
Sootland, UK;
Cetacean Research Program
Center for Coastal Studies
Provincetown, MA 02657, USA

Synopsis

Data from a long-term study of individually identified humpback whales, *Megaptera novaeangliae*, were used to describe patterns of association and grouping of this species on one of its principal North Atlantic feeding grounds in the southern Gulf of Maine. Most groups were small and unstable, and individual whales of both sexes and all age classes associated with many conspecifics. Only six instances of stable associations were recorded. Analysis of the class composition of singles and pairs showed that: (1) among singles, juveniles of both sexes were significantly over-represented and mature females significantly under-represented; (2) male–female adult pairs were over-represented; (3) adult-juvenile pairs of any gender combination were under-represented; and (4) pairings between adult males were under-represented except during feeding. Only 12 of 2690 pairs consisted of animals that were known to be related. It is suggested that the fission–fusion sociality that characterized the study population represents a response to two ecological factors. Firstly, absence of predation nullifies the need for large groups for predator detection or communal defence. Secondly, the spatial characteristics of piscine prey favour a foraging strategy involving frequent changes in group size. In this system, kinship and dominance probably play reduced roles, while the apparent lack of territoriality is typical of taxa confronted by heterogeneously distributed and mobile resources. The apparent preference by mature males for associations with mature females may represent an attempt to establish bonds with potential future mates.

Introduction

The behaviour and social organization of mysticete cetaceans have historically proved difficult to study owing to the wide-ranging movements

ZOOLOGICAL SYMPOSIUM No. 66
ISBN 0–19–854069–8

Copyright © 1993 The Zoological Society of London
All rights of reproduction in any form reserved

of the animals and the opaque and often inhospitable nature of the environment in which they live. As a result, considerably less has been discovered of the social ecology of baleen whales than of most large mammals. Over the last two decades, the development of techniques with which to identify individual whales has led to significant advances in our knowledge of the distribution, population structure and reproductive biology of several species (recently summarized in Hammond, Mizroch & Donovan 1990), yet such investigations have generally suffered from observational sample sizes that are insufficient to permit us to place our understanding of these animals into the broader context of behavioural ecology.

Humpback whales, *Megaptera novaeangliae*, present a particularly interesting case for study. Among the most wide-ranging of all mammals (Stone, Florez-Gonzalez & Katona 1990), their annual life cycle is characterized by a distinct seasonality in both distribution and behaviour. During the months of spring, summer and autumn, humpbacks feed in productive high-latitude waters. In late autumn, they undertake long migrations to breeding and calving ranges in the tropics, where little or no feeding occurs (Baraff *et al.* 1991), and where virtually all mating activity takes place (Baker & Herman 1984a). In high latitudes, humpbacks subsist on a variety of patchily distributed and mobile prey, primarily midwater fish and euphausiid crustaceans. The occurrence of these resources is largely confined to restricted areas of relatively high biological productivity (often associated with coastal upwellings or frontal zones: Gaskin 1982), which may be separated by distances of tens or even hundreds of kilometres. This separation, together with fluctuation in local prey availability (both from day to day and year to year), frequently necessitates considerable movements by foraging whales.

Long time-series of observations have begun to provide sufficient data on the life histories of individual humpbacks for observers to investigate the social behaviour of this species in some detail (e.g. Weinrich 1991). The aim of the present study was to describe patterns of association and grouping among humpback whales on one of their principal North Atlantic feeding grounds in the Gulf of Maine, and to examine the relationship between aspects of the ecology and social organization of this species.

Methods

This study was conducted between 1979 and 1990 in the southern Gulf of Maine. The study area is described in detail in Clapham & Mayo (1987), and includes the waters of Massachusetts Bay and the Great South Channel. Humpback whales return annually to this region from the West Indies each spring, and remain resident within the Gulf of Maine and adjacent waters until late autumn (Katona & Beard 1990; Clapham, Baraff *et al.* 1993).

Since the mid 1970s, the predominant prey of humpback whales in the southern Gulf of Maine has been the American sand lance, *Ammodytes americanus* (Overholtz & Nicolas 1979; Payne, Nicolas *et al.* 1986; Payne, Wiley *et al.* 1990). During the study period, this species was identified on all but a few of the many occasions on which the prey upon which humpbacks were feeding could be observed (either at the surface or in the whale's open mouth).

The majority of observations were made from 30 m commercial whalewatching vessels in Massachusetts Bay. Additional directed cruises to this area, and to the Great South Channel, were made on a 14 m auxiliary ketch. A total of 9532 cruises were made during the study period; 97% of these were 4 h trips aboard whalewatching vessels operating daily from Provincetown, Massachusetts, between mid-April and the end of October each year. Because of the commercial nature of these cruises, behavioural observations of groups of whales varied in duration from a few minutes to more than an hour, and systematic sampling was impossible. Consequently, with the exception of surface feeding (defined below), behaviour is not evaluated here as a potential correlate of group size and composition.

Although information on group size frequencies was available from 1979 to 1990, data on identified individuals were available only from 1979 to 1988. Individual humpback whales were recognized from variations in the ventral fluke pattern (Katona, Kraus *et al.* 1980), the shape and size of the dorsal fin, and prominent scars.

Where possible, gender was determined from photographs of the genital area, exposed above the surface during certain behaviours. As reported by Glockner (1983), a hemispherical lobe is present in the genital region of females, but not in males. A whale was also considered female if it was in repeated close association with a first-year calf. The sex of 278 individuals (116 males, 162 females) was determined during this study.

Definitions
Association and group
While two or more humpback whales separated by considerable distances may be in acoustic contact and therefore associated, such affiliations (if they exist) are impossible to determine in the field. Consequently, I have limited my definition of a group to obvious affiliations in which two or more animals were swimming side by side during the observation period and were generally co-ordinating at least their surfacing and diving activities and their speed and direction of movement. An association is an affiliation between two individuals; consequently, in a group of three whales, each animal is considered to have two associations. Lone animals are referred to here as

singles. A stable association was defined as one in which two whales were together, continuously or repeatedly (more than five times), for a minimum period of two weeks.

Age class
Individual whales were classified into one of two age classes, juvenile and adult. Since most animals were observed in several years, separate classifications were made for each individual for each year of the study. A juvenile whale was one which had been observed first as a calf, which was no longer with its mother, and which was known to be less than five years old at the time of observation (separation between mother and calf occurs in most cases after the calf's natal year: Clapham & Mayo 1987). At five years a whale was considered to be an adult, on the basis of the average age at attainment of sexual maturity in this population (Clapham 1992). If a whale had not been observed as a calf, but had been recorded in the population for at least four years at the time of observation, or was known to have had a calf, it was considered an adult. Whales not observed as calves and seen for less than four years were not assigned to an age class, irrespective of how small or large they appeared to be.

Surface feeding
In order to examine whether the composition of groups was influenced by feeding, analyses of group composition were stratified into surface-feeding and 'other' behaviours. Surface feeding is defined as a behaviour where the mouth was agape at the surface with the ventral pleats extended. Occasionally, prey could be seen in the open mouth. Surface feeding was accompanied in most cases by the production of bubble clouds or bubble nets; these facilitate the capture of schooling fish or euphausiids (Hain *et al.* 1982), and are never observed with non-feeding behaviours. Surface feeding represented the only occasion when observers could positively state that whales were feeding; many other instances of feeding take place below the surface, but its occurrence is generally impossible to verify.

Calculation of expected values for group composition analysis
In analyses of group composition by sex and age class, expected values for each category were calculated separately for each year of the study and based upon the number of individuals in each class observed in that year; expected and observed values were then summed for all years to give the results represented in Tables 2 and 3, and the differences tested with chi-square. For example, in 1985, of 99 observed individuals of known sex and age class, 49 were adult females. Thus, adult females represented 0.495 $(1/(99/49))$ of the sample of animals of known sex and age class observed that year. As a result, the expected value for their representation in the 85

singles of known sex and age class observed during 1985 was 42.1 (= 85 ×
0.495).

Calves were not included in any assessments of group composition on the
assumption that calves do not exercise independent choice in associations
but are instead present in the group solely because of maternal bonds.

Rate of change of associates

Because of the non-systematic nature of most observations, it was not
possible to measure directly the time that groups of whales remained
together. However, as a measure of the degree to which an animal's
associates changed, the following ratio (R) was calculated for each individual:

$$R = T_a/T_i$$

where T_a = total number of associations; T_i = total number of individual
whales associated with.

For example, for an individual humpback observed to have 20 associations
(T_a) with 10 individuals (T_i), $R = 2.0$ (20/10). A whale with a low rate of
change of associates might be observed in 20 associations with two
individuals, giving $R = 10.0$ (20/2).

Mean ratios were calculated for all known adult females, adult males,
juvenile females and juvenile males.

Results

Group size

In total, 27 252 groups observed during the study period were used in
analysis of group size frequency. While many other groups were recorded in
the field, I analysed only those for which the group size was in no doubt (in

Table 1. Observed size frequencies for calf and non-calf groups, 1979–1990

	Group size						
	1	2	3	4	5	6+	Total
Calf absent							
n	13244	6829	1820	489	146	144	22672
%	58.4	30.1	8.0	2.2	0.7	0.6	
Calf present							
n	—	3491	737	240	68	44	4580
%	—	76.2	16.1	5.2	1.5	1.0	
Total							
n	13244	10320	2557	729	214	188	27252
%	48.6	37.8	9.4	2.7	0.8	0.7	

practice, this was when the group was observed at close enough range for individual identifications to be made). Table 1 summarizes observed group-size frequencies for both calf and non-calf groups. The great majority of groups in both categories were small; among groups that did not contain a calf, 88.5% were either singles or pairs. Large groups (more than five whales) were observed infrequently (0.6% of non-calf groups, 1.0% of calf groups); the majority (81%) were associated with surface-feeding behaviour.

Rate of change of associates

The overall mean ratio R for all individuals ($n = 91$) in the four sex and age classes was 1.897 (range = 1.0 to 7.5, S.D. = 0.902). There was no significant difference between classes (1-way ANOVA: $f = 2.108$, $P = 0.104$, $d.f. = 3,86$). The consistently low ratios for all classes reflect the high turnover of associates observed.

A randomly chosen example of the constant flux that characterized the associations of individuals is a mature female (known to us as 'Janus') who was observed on 19 separate occasions over a 14 h period in a single day. 'Janus' was alone in only two sightings; in the remaining 17 groups (11 of them involving surface feeding), a total of 29 associations with 10 individuals was recorded. Associated group size ranged from one to five animals, and changed frequently. 'Janus's' associations during this period are broadly representative of those of most whales.

Stable associations

Stable associations were rare: only six were observed during the study. These involved four pairs, and are described below:

Pair 1: 'Nurse' (adult female) and 'Silver' (adult female). Stable associations in two separate years, with both animals pregnant in both cases. In the first, in 1982, the two animals were observed together on nine of 10 days over a 33-day period. In the second, in 1984, they were seen together on 13 of 18 days over a 42-day period.

Pair 2: 'Binoc' (adult female) and 'Stub' (adult male). Stable associations in two separate years. In the first, in 1983 (with 'Binoc' pregnant), they were observed together on 18 of 19 days over a 24-day period. In the second, in 1985 (with 'Binoc' neither pregnant nor lactating), they were seen together on nine out of 11 days over a 36-day period.

Pair 3: 'Crown' and 'Appaloosa' (both females; age class unknown, but small size strongly suggested that both were juveniles). Observed together in 1983 on 42 of 68 days over a 172-day period.

Pair 4: 'Cats Paw' (adult female, neither pregnant nor lactating) and 'Cluster' (male; age class unknown, but adult the following year). Observed together in 1985 on 14 of 38 days over a 59-day period.

Table 2. Observed and expected values for the composition of singles by animals of known age and gender. Results for each class are divided into surface feeding and all other behaviours. All χ^2 values are based on one degree of freedom, alpha = 0.05

Class	Behaviour	Number of singles		χ^2	$P <$	Representation
		Observed	Expected			
Adult females	Feeding	508	646	46.23	0.001	Under-represented
	Other	1037	1648	226.35	0.001	Under-represented
Adult males	Feeding	554	364	99.84	0.001	Over-represented
	Other	784	793	0.10	n.s.	
Juvenile females	Feeding	90	86	0.21	n.s.	
	Other	541	250	339.09	0.001	Over-represented
Juvenile males	Feeding	121	137	1.96	n.s.	
	Other	690	362	297.72	0.001	Over-represented

n.s., not significant

Table 3. Observed and expected values for the composition of pairs by animals of known age and gender. Results for each class are divided into surface-feeding and other behaviours. All χ^2 values are based upon one degree of freedom, alpha = 0.05

Class[a]	Behaviour	Number of pairs		χ^2	$P <$	Representation
		Observed	Expected			
AF-AF	Feeding	41	65	8.80	0.005	Under-represented
	Other	263	279	0.91	n.s.	
AF-AM	Feeding	90	68	7.42	0.01	Over-represented
	Other	389	265	58.02	0.001	Over-represented
AF-JF	Feeding	9	17	3.62	n.s.	
	Other	48	88	18.25	0.001	Under-represented
AF-JM	Feeding	25	26	0.06	n.s.	
	Other	102	125	4.23	0.05	Under-represented
AM-AM	Feeding	29	18	6.83	0.01	Over-represented
	Other	21	65	29.61	0.001	Under-represented
AM-JF	Feeding	7	8	0.18	n.s.	
	Other	24	40	6.53	0.025	Under-represented
AM-JM	Feeding	18	13	1.75	n.s.	
	Other	41	58	5.08	0.025	Under-represented
JF-JF	Feeding	0	1	1.21	n.s.	
	Other	11	8	1.63	n.s.	
JF-JM	Feeding	3	4	0.17	n.s.	
	Other	38	21	13.54	0.001	Over-represented
JM-JM	Feeding	1	3	1.16	n.s.	
	Other	27	15	9.60	0.005	Over-represented

[a] AF, adult female; AM, adult male; JF, juvenile female; JM, juvenile male
n.s., not significant

Group composition

The composition of singles and pairs (the most common group sizes) by sex and age class is summarized in Tables 2 and 3. Results for each class are separated into two categories: surface feeding, and all other behaviours. Significant differences between observed and expected values were found for many classes, and show that: (1) among singles, juveniles of both sexes were over-represented, and adult females under-represented; (2) male–female adult pairs were over-represented; (3) adult–juvenile pairs of any gender combination were under-represented; and (4) pairings between adult males were under-represented except during feeding.

Associations between relatives

In total, 210 individuals were known to have one or more relatives (grandmother, mother, calf, grandcalf or sibling) in the population, yet of the 2690 pairs (excluding mother–calf pairs) which included individuals from this subset, only 12 consisted of animals that were known to be related. The maximum number of associations observed between any pair of known relatives was four.

Discussion

Ecological determinants of group size

The social organization of humpback whales during this study was characterized by small, unstable groups. This confirms results reported for this species from the same area (Weinrich & Kuhlberg 1991), and from other high-latitude feeding grounds, including Newfoundland (Whitehead 1983), south-east Alaska (Baker & Herman 1984a; Perry, Baker & Herman 1990), and west Greenland (F. Larsen unpubl. data). Two ecological variables, absence of predation and spatial dispersal of prey, probably represent the major determinants of these group characteristics.

Humpback whales do not appear to live under threat of predation whilst in the southern Gulf of Maine (Katona, Beard *et al.* 1988), a contention which is well supported by the virtual absence of killer whales during the period of this study (five sightings and no attacks on mysticetes were recorded in 12 years). Since defence against predators is a major effector of group size in a wide variety of taxa (Hamilton 1971; Bertram 1978; Pulliam & Caraco 1984), the absence of predation on humpbacks in the study area presumably nullifies the need for large groups for communal defence or predator detection.

It is likely that spatial characteristics of prey, particularly school size, represent the other major determinant of group size and instability. Using data from pulse depth recorders (PDR), Whitehead (1983) suggested that humpback whale group size was positively correlated with the horizontal size of the school of fish being exploited. Estimates made during this study of the approximate horizontal extent of visible prey schools, and occasional PDR traces taken in the vicinity of feeding whales, provided support for Whitehead's (1983) conclusion. Large co-ordinated groups of whales (the largest recorded was an unstable aggregation of up to 18 animals) were always observed surface-feeding, and were often associated with extensive patches of surface prey, which suggests that the upper limit of group size is determined by the quantity of prey immediately available. This is in agreement with Gosling & Petrie (1981), who reviewed the economics of social organization and predicted that, if group size is limited by the costs of competition between individuals for food, large groups should form if the feeding constraint is removed.

Schools of *Ammodytes* vary considerably in horizontal extent. Meyer, Cooper & Langton (1979), working in the area where most of these observations of humpbacks were made, quantified the size of offshore sand lance schools and reported that they varied in size from about 500 to tens of thousands of fish, a finding that is similar to the variation reported by Kühlmann & Karst (1967) for another species of *Ammodytes* in the North Sea. The constant change in the size of the fish schools encountered by feeding humpbacks, and the consequent need for a foraging strategy involving patch-by-patch decisions on the energetic cost or benefit of co-operative feeding, explains the fluctuating group sizes that are often characteristic of humpback feeding aggregations. This phenomenon of so-called fission-fusion sociality (Kummer 1971) has been observed for other animals confronted with patchily distributed resources of varying quantity; examples include spider monkeys (Symington 1988; Chapman 1990), deer (Schaller 1967) and hyaenas (Kruuk 1972). The lack of stable groups in this study contrasts with observations from south-east Alaska (Perry *et al.* 1990) of large stable groups of humpbacks, apparently associated with consistently large schools of herring (*Clupea harengus*).

Territoriality and dominance

Another consequence of the distribution of prey is the lack of territoriality observed in humpback whales. Defence of a specific area becomes pointless if the resources within that area are fluctuating and unpredictable, and sometimes completely absent; as noted by Pulliam & Caraco (1984) and others, a high degree of heterogeneity in resource distribution renders local concentrations economically indefensible. In this, humpbacks are typical of other animals that subsist on mobile resources with similar characteristics (Davies & Houston 1984).

Furthermore, within this framework of shifting associations and small unstable groups, dominance plays no obvious role. Neither aggressive interactions nor contests of any kind were observed during this study, nor have they been reported from other feeding-ground investigations of this species. In many cases, co-operation rather than competition appeared to characterize intraspecific interactions, an observation which agrees with results reported by Baker & Herman (1984a) and Perry *et al.* (1990). Given the low rates of association between particular individuals, humpbacks have fewer opportunities, and less need, to establish a dominance system than animals whose ecology requires or allows them to coexist for extended periods in stable groups (e.g. elephants: Moss & Poole 1983; baboons: Hamilton & Bulger 1990; reindeer: Hirotani 1990). This situation is further reinforced by the absence of breeding behaviour, and consequently of direct competition for mates, during the summer.

However, in a species whose resources are neither infinite nor homo-geneously distributed, it is unrealistic to presume that dominance and competition play no role simply because they are not obviously manifest. As noted by Weinrich *et al.* (1985), apparently prime feeding areas (those hosting large aggregations of whales and prolonged periods of surface feeding) appear to be colonized primarily by adult animals; juveniles, by contrast, are often found in peripheral areas where resource abundance appears to be lower. It is unclear whether this situation represents active exclusion of juveniles by adults (and therefore reflects the existence of some form of age-related dominance), or differing energetic requirements of the two age-classes.

The lack of a dominance hierarchy among individuals on a feeding ground may also be an important consequence of factors operating several months later during the winter breeding season. Mature male humpbacks fight for access to mature females (Tyack & Whitehead 1982; Baker & Herman 1984b; Clapham, Palsbøll *et al.* 1992). Given that animals which are sympatric during the summer (e.g. Gulf of Maine whales) will be spatially mixing in the West Indies mainly with unfamiliar conspecifics from other feeding grounds in the western North Atlantic in the winter (Mattila *et al.* 1989; Mattila & Clapham 1989), most contests between mature males for females will not be between animals from the same feeding area. In light of this, it will not be worthwhile for males to invest time and energy into establishing what Barnard & Burk (1979) have termed 'assessment hierarchies' during the summer, since these would be of little value in the assessment of largely unfamiliar competitors on the breeding grounds.

Group composition by sex and age class

It is difficult to interpret differences in the composition of groups between surface feeding and other behaviours. The data suggest that associations are often more random during feeding than at other times, although there were apparent exceptions to this. Furthermore, the problem is complicated by the fact that the 'other behaviours' category undoubtedly included many events where subsurface feeding occurred but could not be observed. It is possible that, as in many other taxa, significant social interactions among humpback whales occur principally during time not allotted to feeding, but data are currently insufficient to address this question.

Some trends in group composition are rather more clear. Juveniles of both sexes are alone a great deal, a phenomenon which other data link to a process of social development (P. J. Clapham unpubl. data). A related finding from the group composition analysis is that animals tend to associate most with conspecifics of their own age class. The tendency for mother–calf pairs to be alone for much of the time has been noted previously (Clapham & Mayo 1987) and may simply reflect the relative infrequency with which humpbacks form groups larger than two, for reasons discussed above.

Weinrich & Kuhlberg (1991) have suggested that male humpbacks on a feeding ground tend to avoid each other once they reach sexual maturity because by associating (particularly when feeding) they would be assisting potential reproductive competitors. The results of this study suggest otherwise. While male–male pairs are indeed under-represented, they are not uncommon, and are actually marginally over-represented in surface-feeding bouts. While the role of kinship in humpback social organization has yet to be elucidated, the idea that all but a few of a male's male associates are close relatives (and therefore that co-operation might somehow result in enhancement of inclusive fitness) is untenable in view of the large number of individuals with which males associate.

I suggest that the low incidence of reproductive competition between Gulf of Maine males on the breeding grounds and the virtual absence of any breeding behaviour during the summer account for the apparently tolerant and sometimes co-operative coexistence that characterizes male–male relationships in high latitudes. Males incur low costs by associating with other males, and do not avoid them; rather, the disproportionately low frequency of (apparently non-feeding) male–male pairs may simply reflect the fact that males prefer to associate with females. This is suggested by the pair composition analysis presented here; furthermore, an analysis of the association patterns of mature males from this population showed that 75.9% of their associations were with adult females (P. J. Clapham, unpubl. data). This may represent a reproductive strategy in which males establish bonds with many females in the summer with a possible payoff on the breeding grounds during the winter. This is conceivable, since a male meeting and courting a female from the Gulf of Maine in the West Indies will generally not be in competition for her with other males from the same feeding ground, but existing data are insufficient to test this explanation of the high frequency of associations between adult males and females.

Role of kinship

Weinrich (1991) has hypothesized that at least some long-term associations are based upon kinship, although data on the degree of relatedness between individuals in such groups are currently lacking. While future molecular studies may elucidate the composition of these groups, the paucity of associations between relatives during this study suggests that kinship plays little or no role in the social organization of this species, at least in the southern Gulf of Maine. Given the unstable group structure resulting from the ecological factors discussed above, it is difficult to see the value of stable associations with relatives. If the upper limit of group size is indeed determined by the size of a prey school, stable pairs (whether composed of relatives or not) will be at a disadvantage if most of the schools that they encounter are too small for more than one whale to exploit efficiently. It is possible that pairs could outcompete other whales for the most profitable schools, but this idea cannot be tested with existing data.

Conclusions

The observed patterns of social organization of humpback whales in the southern Gulf of Maine conform well to ecological predictions derived from knowledge of their life cycle as well as associated resource and predation characteristics. However, a more comprehensive understanding of the ecological determinants of social behaviour requires accurate quantitative data on prey abundance in relation to group sizes of foraging whales, information about the degree of relatedness between associated individuals, and investigation of whether humpbacks show any higher order of community structure beyond the obvious affiliations described here.

Acknowledgements

This study would not have been possible without the efforts of the many observers from the Center for Coastal Studies who have recorded data on humpback whales since 1979. Particular thanks are due to Lisa Baraff, Carole Carlson and Peggy Christian for the photographic analysis which underpinned much of this work, and to Margaret Murphy and Sharon Pittman for careful computer management of the daunting volume of data involved. I am grateful also to the captains and crew of the Dolphin Fleet for their assistance in the field, to the Cetacean Research Unit for supplying additional data on the sex of individuals, and to Phil Hammond (Sea Mammal Research Unit, Cambridge), Paul Racey and Roger Thorpe (University of Aberdeen), and Lisa Baraff, Laurie Goldman and Karen Steuer (Center for Coastal Studies) for their helpful comments on the manuscript. This study was funded in part by the U.S. National Marine Fisheries Service (contract number 50–EANF–8–00054), by the Marine and Estuarine Management Division (U.S. Department of Commerce), and by grants from the Seth Sprague Educational and Charitable Foundation.

References

Baker, C. S. & Herman, L. M. (1984a). Seasonal contrasts in the social behaviour of the humpback whale. *Cetus* 5: 14–16.

Baker, C. S. & Herman, L. M. (1984b). Aggressive behavior between humpback whales (*Megaptera novaeangliae*) wintering in Hawaiian waters. *Can. J. Zool.* 62: 1922–1937.

Baraff, L. S., Clapham, P. J., Mattila, D. K. & Bowman, R. (1991). Feeding behaviour of humpback whales in low-latitude waters. *Mar. Mamm. Sci.* 7: 49–54.

Barnard, C. J. & Burk, T. (1979). Dominance hierarchies and the evolution of 'individual recognition'. *J. theor. Biol.* 81: 65–73.

Bertram, B. C. R. (1978). Living in groups: predators and prey. In *Behavioural ecology: an evolutionary approach*: 64–96. (Eds Krebs, J. R. & Davies, N. B.). (1st edn). Blackwell Scientific Publications, Oxford.

Chapman, C. A. (1990). Association patterns of spider monkeys: the influence of ecology and sex on social organization. *Behav. Ecol. Sociobiol.* 26: 409–414.

Clapham, P. J. (1992). Age at attainment of sexual maturity in humpback whales, *Megaptera novaeangliae*. *Can. J. Zool.* 70: 1470–1472.

Clapham, P. J., Baraff, L. S., Carlson, C. A., Christian, M. A., Mattila, D. K., Mayo, C. A., Murphy, M. A. & Pittman, S. (1993). Seasonal occurrence and annual return of humpback whales in the southern Gulf of Maine. *Can. J. Zool.* 71.

Clapham, P. J. & Mayo, C. A. (1987). Reproduction and recruitment of individually identified humpback whales, *Megaptera novaeangliae*, observed in Massachusetts Bay, 1979–1985. *Can. J. Zool.* 65: 2853–2863.

Clapham, P. J., Palsbøll, P. J., Mattila, D. K. & Vasquez, O. (1992). Composition and dynamics of humpback whale competitive groups in the West Indies. *Behaviour* 122: 182–194.

Davies, N. B. & Houston, A. I. (1984). Territory economics. In *Behavioural ecology: an evolutionary approach*: 148–169. (Eds Krebs, J. R. & Davies, N. B.). (2nd edn). Blackwell Scientific Publications, Oxford.

Gaskin, D. E. (1982). *The ecology of whales and dolphins*. Heinemann, London.

Glockner, D. A. (1983). Determining the sex of humpback whales (*Megaptera novaeangliae*) in their natural environment. In *Behaviour and communication of whales*: 447–464. (Ed. Payne, R. S.). Westview Press, Boulder. (*AAAS sel. Symp.* No. 76.)

Gosling, L. M. & Petrie, M. (1981). The economics of social organization. In *Physiological ecology: an evolutionary approach to resource use*: 315–345. (Eds Townsend, C. R. & Calow, P.). Blackwell Scientific Publications, Oxford.

Hain, J. H. W., Carter, G. R., Kraus, S. D., Mayo, C. A. & Winn, H. E. (1982).

Feeding behavior of the humpback whale, *Megaptera novaeangliae*, in the western North Atlantic. *Fish. Bull. U.S. Fish Wildl. Serv.* 80: 259–268.

Hamilton, W. D. (1971). Geometry for the selfish herd. *J. theor. Biol.* 31: 295–311.

Hamilton, W. J. & Bulger, J. B. (1990). Natal male baboon rank rises and successful challenges to resident alpha males. *Behav. Ecol. Sociobiol.* 26: 357–362.

Hammond, P. S., Mizroch, S. A. & Donovan, G. P. (Eds) (1990). Individual recognition of cetaceans: use of photo-identification and other techniques to estimate population parameters. *Rep. int. Whal. Commn spec. Issue* No. 12: 1–440.

Hirotani, A. (1990). Social organization of reindeer (*Rangifer tarandus*), with special reference to relationships among females. *Can. J. Zool.* 68: 743–749.

Katona, S. K. & Beard, J. A. (1990). Population size, migrations and feeding aggregations of the humpback whale (*Megaptera novaeangliae*) in the western North Atlantic Ocean. *Rep. int. Whal. Commn spec. Issue* No. 12: 295–305.

Katona, S. K., Beard, J. A., Girton, P. E. & Wenzel, F. (1988). Killer whales (*Orcinus orca*) from the Bay of Fundy to the Equator, including the Gulf of Mexico. *Rit Fiskideildar* 11: 205–224.

Katona, S. K., Kraus, S. D., Perkins, J. & Harcourt, P. (1980). *Humpback whales: a catalogue of individuals from the western North Atlantic identified by fluke photographs.* College of the Atlantic, Bar Harbor, Maine.

Kruuk, H. (1972). *The spotted hyena: a study of predation and social behaviour.* University of Chicago Press, Chicago & London.

Kühlmann, D. H. H. & Karst, H. (1967). Freiwasserbeobachtungen zum Verhalten von Tobiasfischschwärmen (Ammodytidae) in der westlichen Ostsee. *Z. Tierpsychol.* 24: 282–297.

Kummer, H. (1971). *Primate societies: group techniques of ecological adaptation.* Aldine-Atherton, Chicago.

Mattila, D. K. & Clapham, P. J. (1989). Humpback whales, *Megaptera novaeangliae*, and other cetaceans on Virgin Bank and in the northern Leeward Islands, 1985 and 1986. *Can. J. Zool.* 67: 2201–2211.

Mattila, D. K., Clapham, P. J., Katona, S. K. & Strone, G. S. (1989). Population composition of humpback whales, *Megaptera novaeangliae*, on Silver Bank, 1984. *Can. J. Zool.* 67: 281–285.

Meyer, T. L., Cooper, R. A. & Langton, R. W. (1979). Relative abundance, behavior, and food habits of the American sand lance, *Ammodytes americanus*, from the Gulf of Maine. *Fish. Bull. U.S. Fish Wildl. Serv.* 77: 243–253.

Moss, C. J. & Poole, J. H. (1983). Relationships and social structure in African elephants. In *Primate social relationships: an integrated approach*: 315–325. (Ed. Hinde, R. A.). Blackwell Scientific Publications, Oxford.

Overholtz, W. J. & Nicolas, J. R. (1979). Apparent feeding by the fin whale, *Balaenoptera physalus*, and humpback whale, *Megaptera novaeangliae*, on the American sand lance, *Ammodytes americanus*, in the northwest Atlantic. *Fish. Bull. U.S. Fish Wildl. Serv.* 77: 285–287.

Payne, P. M., Nicolas, J. R., O'Brien, L. & Powers, K. D. (1986). The distribution of the humpback whale, *Megaptera novaeangliae*, on Georges Bank and in the Gulf of Maine in relation to densities of the sand eel, *Ammodytes americanus*. *Fish. Bull. U.S. Fish Wildl. Serv.* 84: 271–277.

Payne, P. M., Wiley, D., Young, S., Pittman, S., Clapham, P. J. & Jossi, J. W. (1990). Recent fluctuations in the abundance of baleen whales in the southern Gulf of Maine in relation to changes in selected prey. *Fish. Bull. U.S. Fish Wildl. Serv.* **88**: 687–696.

Perry, A., Baker, C. S. & Herman, L. M. (1990). Population characteristics of individually identified humpback whales in the central and eastern North Pacific: a summary and critique. *Rep. int. Whal. Commn spec. Issue* No. 12: 307–317.

Pulliam, H. R. & Caraco, T. (1984). Living in groups: is there an optimal group size? In *Behavioural ecology: an evolutionary approach*: 122–147. (Eds Krebs, J. R. & Davies, N. B.) (2nd edn). Blackwell Scientific Publications, Oxford.

Schaller, G. B. (1967). *The deer and the tiger: a study of wildlife in India.* University of Chicago Press, Chicago & London.

Stone, G. S., Florez-Gonzalez, L. & Katona, S. (1990). Whale migration record. *Nature, Lond.* **346**: 705.

Symington, M. M. (1988). Food competition and foraging party size in the black spider monkey (*Ateles paniscus* Chamek). *Behaviour* **105**: 117–134.

Tyack, P. & Whitehead, H. (1982). Male competition in large groups of wintering humpback whales. *Behaviour* **83**: 132–154.

Weinrich, M. T. (1991). Stable associations among humpback whales (*Megaptera novaeangliae*) in the southern Gulf of Maine. *Can. J. Zool.* **69**: 3012–3019.

Weinrich, M. T., Belt, C. R., Schilling, M. R. & Cappellino, M. E. (1985). Habitat use patterns as a function of age and reproductive status in humpback whales. *Abstr. bienn. Conf. Biol. mar. Mammals* 6. Society for Marine Mammalogy. Unpublished.

Weinrich, M. T. & Kuhlberg, A. E. (1991). Short-term association patterns of humpback whales (*Megaptera novaeangliae*) on their feeding grounds in the southern Gulf of Maine. *Can. J. Zool.* **69**: 3005–3011.

Whitehead, H. (1983). Structure and stability of humpback whale groups off Newfoundland. *Can. J. Zool.* **61**: 1391–1397.

Foraging ecology of marine mammals

Symp. zool. Soc. Lond. (1993) No. 66: 149–178

Sex differences in diving and foraging behaviour of northern elephant seals

B. J. LE BOEUF,
D. E. CROCKER,
S. B. BLACKWELL,
P. A. MORRIS and
P. H. THORSON

Department of Biology
University of California
Santa Cruz
California 95064, USA

Synopsis

Sex differences in the foraging behaviour of adult northern elephant seals, *Mirounga angustirostris*, are predicted from the great disparity in size between the sexes, males being 1.5–10 times larger than females. Males must consume approximately three times more prey per day than females. By examining the diving behaviour, during which all foraging occurs, our aim was to elucidate how males do this and when the strategy develops. Dive data were collected by microcomputer time-depth recorders attached to the backs of free-ranging seals (nine adult males, 10 adult females, seven juvenile males and six juvenile females) during periods at sea ranging from one to three months.

The sexes foraged in different locations and exhibited differences in foraging-type dives, suggesting different foraging strategies and, possibly, different prey. Females moved steadily across the north-eastern Pacific from the coast to as far as 150 °W, in the range 44–52 °N, foraging daily *en route*. Males migrated to areas along continental margins off the state of Washington, to as far as the northern Gulf of Alaska and the eastern Aleutian Islands, where they exhibited concentrated foraging.

Female foraging was exclusively pelagic, with dive depth varying with diel vertical movements of prey, such as squid, in the deep scattering layer. Males exhibited two types of foraging dives, neither of which followed a diel pattern: pelagic dives like those of females, observed mainly during transit, and flat-bottomed dives (which accounted for over 40% of their dives), which occurred near continental margins. The characteristics of these dives, such as occurring in a long series to a uniform depth, and their location, suggest the pursuit of benthic prey such as skates, rays or small sharks on seamounts, guyots or the edge of the continental shelf or, alternatively, the pursuit of prey in the water column by means of a sit-and-wait strategy. The sex differences observed in adults were evident in juveniles less than 18 months old, suggesting that a different foraging strategy of males is important for attaining as well as maintaining large size.

ZOOLOGICAL SYMPOSIUM No. 66
ISBN 0–19–854069–8

Copyright © 1993 The Zoological Society of London
All rights of reproduction in any form reserved

Introduction

When males fight over large groups of females (Darwin 1871; Trivers 1985), as in some pinnipeds, ungulates and primates, large male size is associated with disproportionate gains in reproductive success and sexual dimorphism is most prominent (Alexander *et al.* 1979). Selection for great size in the sexual arena has important implications for performance in other contexts such as foraging. The male form, being modified for another purpose, may not be as optimal as the female form for foraging, at least in the manner in which females do it (e.g. African lions: Schaller 1972). Males must eat more than females to attain and sustain their greater size. The higher energy requirements of males may make them more vulnerable to starvation when food is scarce (Wegge 1980). This logic predicts that where sexual size dimorphism is extreme, there will be sex differences in foraging.

Sex differences in foraging are most common in those pinnipeds, primates and birds that are sexually dimorphic with males being larger than females. It is the male that takes supplementary warm-blooded prey such as birds or other seals in Steller sea lions, *Eumetopias jubatus* (Gentry & Johnson 1981), southern sea lions, *Otaria byronia* (S. Sommerhays quoted in Gentry & Johnson 1981), Antarctic fur seals, *Arctocephalus gazella* (Bonner & Hunter 1982), sea otters, *Enhydra lutris* (Riedman & Estes 1988) and walruses, *Odobenus rosmarus* (Chapskii 1936; Mansfield 1958; Fay 1960; Lowry & Fay 1984; Fay 1990). Male predation of this kind is often specific to certain areas and certain individuals (Riedman 1990). Females rarely supplement their diet in this way. This male bias is observed in other mammals such as olive baboons, *Papio anubis* (Harding 1973), chimpanzees (Teleki 1973; Goodall 1986), which capture and eat birds and other mammals such as red colobus monkeys, bush pigs and baboons. In most animals, sex differences in diet or foraging tactics can be attributed either to the different requirements or to the different capabilities of the two sexes (Baker 1978).

Male northern elephant seals, *Mirounga angustirostris*, are 1.5 to 10 times larger than females; both sexes lose 36% of their mass over the course of the breeding season, with a three- to sixfold dimorphism in mass being most common when they return to sea to forage (Deutsch, Haley & Le Boeuf 1990; Deutsch, Crocker *et al.* in press). As mass increases, absolute energy requirements increase, so we expect males to consume more than females. How much more? During a 70-day foraging trip following the breeding season, females increase their mass at departure by 25%, or at the rate of 1 kg/day (Le Boeuf, Costa, Huntley & Feldkamp 1988). From water influx data (Costa 1991; D. Costa & B. Le Boeuf unpubl. data), it is estimated that females consume 6.2% of their mass daily. Therefore, they would consume prey such as squid at the rate of about 20 kg/day at sea.

Assuming that males also increase their mass at departure by 25%, and given that metabolic rate is proportional to body mass raised to the 0.75 power (Kleiber 1932; Brody, Procter & Ashworth 1934; Benedict 1938; Kleiber 1961), we calculate that males increase their mass by 2.92 kg/day and ingest about 63.9 kg/day of prey, or about three times as much as females, if they are preying on the same animals. Males might also consume more energy by being more efficient in capturing prey than females or by consuming prey with a higher energy density. In any case, we expect the male and female patterns of diving, during which all foraging occurs, to reflect these allometric relationships.

By examining the free-ranging diving pattern, we aimed to shed light on how males accomplished this task. We ruled out the possibility that males simply spend longer periods at sea, since males, in fact, spend less time at sea during the year (two trips totalling eight months) than females (two trips totalling 10 months) and therefore have less time for foraging than females (Le Boeuf in press). Our analysis is indirect because diving records do not tell us when a diving seal catches prey, the species caught, or the amount consumed. We infer foraging from mass gain over the period at sea (Le Boeuf, Costa, Huntley, Kooyman et al. 1986; Le Boeuf, Costa, Huntley & Feldkamp 1988; Le Boeuf, Naito, Huntley & Asaga 1989).

We know, from independent observations and analysis of stomach contents of living and dead animals, that adult northern elephant seals eat as many as 28 species of squid and octopus, cartilaginous fishes, cyclostomes, a few teleosts such as Pacific hake, *Merluccius productus*, and some rockfish, *Sebastes*, and occasional crustaceans and tunicates (Condit & Le Boeuf 1984; Antonelis, Lowry, DeMaster et al. 1987; Stewart & DeLong 1991; Antonelis, Lowry, Fiscus et al. in press). Prey remains in the lavaged stomachs of seals shortly after returning to land revealed no sex differences in diet (Stewart & DeLong 1991; Antonelis, Lowry, Fiscus et al. in press) but these data are biased to prey consumed in the last few days at sea (Harvey, Antonelis & Casson 1989).

Our approach is based on comparison of the sexes with respect to (1) migratory path, speed and distance of travel, (2) indices of diving performance such as mean dive depth and mean dive duration and (3) analysis of the types of dives exhibited. For the latter, we rely heavily on arguments developed in a previous study that measured swim speed as a function of six dive types distinguishable in dive records (Le Boeuf, Naito, Asaga et al. 1992; see also Asaga et al. in press). This procedure yielded ascent and descent angles and distance travelled for each dive type, which, in combination with other attributes such as depth and activity at the dive bottom, provided information for a functional classification of dive types. In the present study, we assumed that one of the dive types observed in elephant seal dive records (A dives) represents transit and two of them (D and E dives) represent foraging. Le Boeuf, Naito, Asaga *et al.* (1992) and Hindell (1990) argued further that the two foraging dives represent pelagic and benthic foraging; the present study presents data bearing on this distinction.

We tested several hypotheses regarding sex differences. The general hypothesis is that the dive pattern of the sexes differs, reflecting the need for males to acquire more resources or to acquire them more efficiently than females. More specific hypotheses tested were: (1) males forage in different geographical locations than females; (2) males spend more time per day foraging than females; (3) males spend part of the time feeding on different prey than females (as reflected by the depth and type of dives and their peak occurrence during the day); and (4) sex differences in foraging begin early in life, being evident in the diving pattern of juveniles.

Methods

This analysis is based on 32 diving records obtained from elephant seals at Año Nuevo, California, during the years 1989–91. Instruments (Wildlife Computers, Woodinville, Washington) were attached to a radio transmitter (Advanced Telemetry Systems, Minnesota) and the package was glued to the pelage on the dorsal midline of the seal behind the shoulders with marine epoxy during immobilization with drugs (see Le Boeuf, Costa, Huntley & Feldkamp 1988; Le Boeuf, Naito, Huntley & Asaga 1989).

Instruments began collecting dive data as soon as the seals entered the water, recording dive depth every 30 s until the memory capacities of the computers were full. Mean record duration was 59.8 ± 25.1 days (range = 8–94 days).

Subjects and summary dive data

The sample included nine males recorded during the period at sea following the breeding season: six adult males, eight or more years old, and three subadult males, six to seven years old; 10 females recorded during the period at sea following breeding (post-breeding females); and seven juvenile males and six juvenile females, 1.4–1.8 years old, recorded during their third and fourth trips to sea. Summary dive statistics—dive rate, percentage of time submerged, mean depth $\pm$ one standard deviation, maximum depth, mean dive duration $\pm$ one standard deviation, maximum dive duration, and mean surface interval (excluding surface intervals exceeding 10 min) $\pm$ one standard deviation—were calculated for all animals.

Dive classification

The dives of all adult males and females, six juvenile males and six juvenile females (144 431 dives) were classified individually.

Six distinguishable dive types have been described in northern elephant seal diving records (Le Boeuf, Costa, Huntley & Feldkamp 1988; Asaga *et al.* in press) as well as in those of southern elephant seals, *M. leonina* (Hindell 1990; Hindell, Slip & Burton 1991). From strip-chart representations of the time-depth profile, two of us used the basic method of Le Boeuf, Naito, Asaga *et al.* (1992) to classify dives into four major dive types and variations on these types, and two minor categories. The major dive types were: A dives—direct descent to a sharp or rounded inflection point followed by direct ascent to the surface, C dives—direct descent to a depth at which point the descent rate decreased then continued at a slower rate to the bottom of the dive, followed by direct ascent to the surface, D dives—direct descent to a depth at which point there occurred two to 12 vertical excursions or 'wiggles', followed by nearly vertical ascent to the surface, and E dives—direct descent to the bottom of the dive, which was flat, ending in direct, nearly vertical ascent to the surface.

Dives which exhibited slightly more variation in shape, but which retained the basic characteristics of the main dive types—lack of bottom time for A dives, distinct 'wiggles' for D dives and a flat bottom for E dives—were classified as variants of types A, D, and E, i.e., A_v, D_v and E_v (Fig. 1). Unusually shallow (less than 100 m) dives of short duration, usually occurring during departure from the rookery or return to it, were classified as I dives. Dives which did not fit into the aforementioned categories were rare and classified as X dives. C dives and their variants are not discussed further in this paper but will be treated in a later paper; there were no sex differences in the frequency of C dives in adults or in juveniles.

To assess the reliability of dive-type classification, four complete dive records, which included 10 483 dives, were classified independently by two of us. This yielded a $91.3 \pm 1.3\%$ agreement on a dive-by-dive comparison. The majority of discrepancies involved variants of the major dive types. Disagreement about one of the major four dive types was rare.

Migratory path and travel speed

Additional channels on the diving instruments recorded ambient water temperature at 10-min intervals and light levels at the surface. The light-level data provided estimates of the time of dawn and dusk from which an algorithm calculated position to within ± 1 degree in latitude and longitude (Delong, Stewart & Hill 1992). Ambiguities in latitude, which are especially large near equinoxes, were resolved by matching sea-surface temperature recorded by the diving instrument to mean sea-surface temperature locations compiled semi-monthly from satellites by the National Meteorological Center of the National Weather Service (Ashville, North Carolina).

Position, which we plotted at two-day intervals, yielded partial migratory paths for six males, nine adult females and five juveniles. Minimum travel

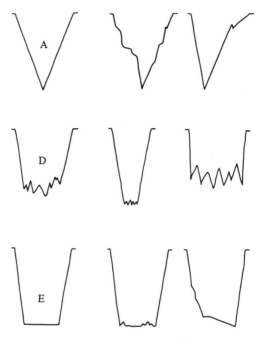

Fig. 1. Three principal dive types observed in northern elephant seal diving records and two examples of variations on each type. The top row shows an A dive, or transit dive, on the left followed by two common variations on this type. The second row shows a D dive, or pelagic foraging dive, followed by two variations. The third row shows an E dive, or flat-bottomed dive, followed by two variations on this type.

speed was calculated as the distance in kilometres between two points, one at the beginning and one at the end of a segment of the track where the seal was in transit along a great circle, divided by the number of days separating the two points. 'In transit' was defined as moving in a more or less straight line at a steady rate without lingering in an area. The longest migration distance for a seal was calculated as the distance in kilometres along a great circle between Año Nuevo (37.1 °N, 122.2 °W) and the point furthest away on the animal's track. 'Foraging area' was defined as a geographical location where the seal reduced its horizontal travel speed substantially and was relatively stationary in its migration for about eight or more days.

Time of the year

We recorded the period at sea between the breeding season and the annual moult. This is roughly the period from March to July for males of breeding age and March to mid May for adult females. Thus, the only time that males were at sea and females were not was during June and July.

The diving behaviour of five male and three female juveniles, 1.4 years old, was recorded during the months of May, June and July, and that of two male and three female juveniles, 1.8 years old, was recorded during the months of November, December and January.

Results

Migratory path, foraging areas and speed of travel

Adult males

The three adult males migrated northward along fairly direct routes (Fig. 2). One male, 'Tyke', reached the eastern Aleutians in 34 days, another, 'Zilla', was near Kodiak Island in the northern Gulf of Alaska in 28 days, and the third, 'Quake', reached the Queen Charlotte Islands off British Columbia in 18 days. All three were still in transit when the geolocation record ended. The subadult male tracks were more revealing because the records lasted 93–95 days (Fig. 2). One male, 'Joe', travelled directly to an area in the northern Gulf of Alaska near Kodiak Island in 32 days where he apparently foraged for 48 days before beginning his return along the same route. The two other males, 'Pico' and 'C508', spent most of their time foraging off the coasts of Oregon, Washington and Vancouver Island. Males, more so than females (see below), exhibited repeated diving in rather narrowly confined areas lasting from several days to two months.

Adult females

Females ranged as far as 56 °N and 150 °W (Fig. 3). No female ranged as far north or as far west as some males. The furthest locations reached by most females were in the range 44–52 °N. Four females reached the end of their migration and were returning to the rookery ('Quebec', 'Lilly', 'Renee' and 'Sydney'), three females reached a point where, given the time at sea, they were about to turn back ('Glori', 'C115' and 'D318'), and two females were still on the outward leg of their journey when the record ended ('D197' and 'YB143'). One female, 'D318', spent most of her time in the same location as two subadult males. The tracks show that most animals took a similar route on the outward leg of their migration from Año Nuevo.

Juveniles

Juveniles migrated northward along the same general route to distances comparable to those of adult females; most of them were still in transit when the records ended (Fig. 4). The longest migration was by the male 'Opus', whose route to an area south-east of Kodiak Island was almost identical to that of the subadult male 'Joe' (Fig. 2). The male 'Magus' and the females 'Gaia' and 'Arwen' lingered over certain areas, which suggests that these were foraging areas.

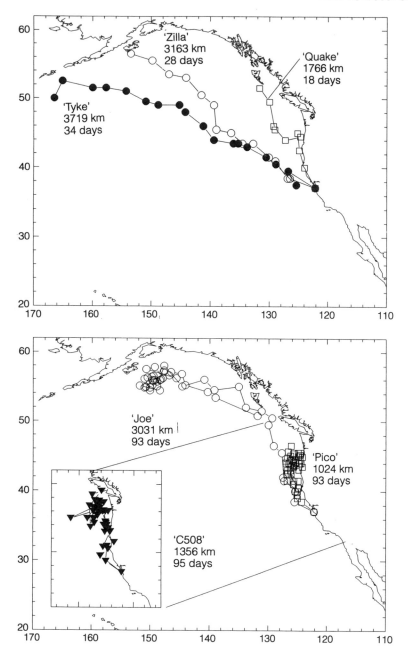

Fig. 2. Migratory paths of three adult males, eight years old or older (top), and three subadult males, six to seven years old (bottom) from Año Nuevo, California, following the breeding season. The positions were plotted every two days. Longitude and latitude are represented on the *x* and *y* axes, respectively.

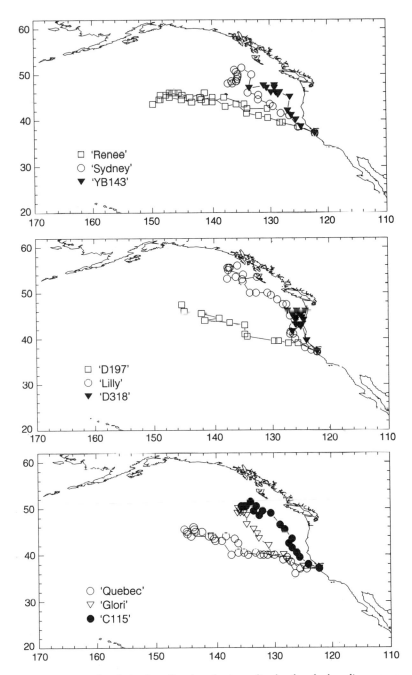

Fig. 3. Migratory paths of nine breeding females immediately after the breeding season.

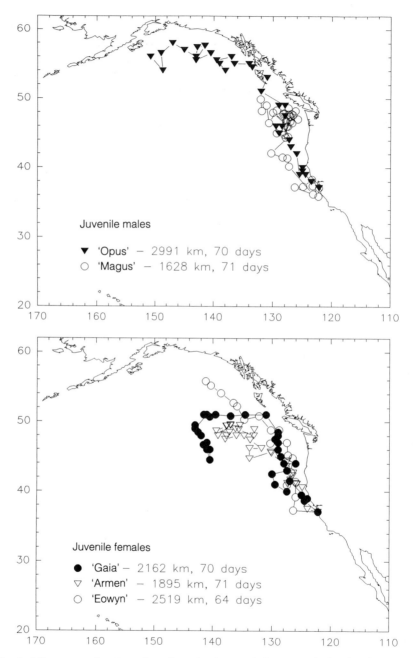

Fig. 4. Migratory paths of two juvenile males (top) and three juvenile females (bottom) on their third trip to sea at 1.4 years of age.

Distance and speed of travel

Although the longest migrations among adults and juveniles were by males, the mean distance travelled by each sex within each group was not significantly different. Moreover, the mean distance travelled by juveniles of both sexes (2239 ± 534 km) was not significantly different from that of adults of either sex: the mean for breeding males was 2353 ± 1118 km and for adult females 1910 ± 436 km.

Similarly, among both adults and juveniles the highest speed of travel while in transit was achieved by males, but the mean speed of travel of each sex within each group was not significantly different. The mean speed of in-transit travel by the group of juveniles (58.3 ± 16.6 km/day) was significantly slower than that of breeding age males (89.4 ± 25.2 km/day) ($t = 2.45$, d.f. $= 9$, $P < 0.05$) and adult females (77.2 ± 11.0 km/day) ($t = 2.28$, d.f. $= 12$, $P < 0.05$).

Diving pattern

Diving indices

Summary statistics reveal similarities in key elements of the diving pattern of both sexes (Table 1). For the data presented in Table 1, no statistically significant sex differences are evident in the diving performance of adults or juveniles.

Males did not appear to spend more time per day foraging than females, i.e., their dive rate, time at sea submerged and dive durations were not significantly different from the values for females. There was a strong trend for males to dive less deep than females, which is in part indicative of different foraging strategies (see below).

Dive type frequency

Foraging-type dives (D + E_v combined) accounted for the majority of dives in both sexes (Table 2). Males did not exhibit a higher percentage of foraging-type dives than females and there were no sex differences in the percentage of transit dives displayed. However, Table 2 reveals sex differences in the percentage of the two types of foraging dives displayed. Breeding-age males exhibited significantly fewer D-type dives than females ($t = 8.32$, d.f. $= 17$, $P < 0.05$)—D was the most common type of dive in female records—and significantly more E-type dives than females ($t = 5.03$ and 5.10 for E and E_v, respectively, d.f. $= 17$, $P < 0.05$).

Male E dives had a mean depth of 331 ± 243 m and a mean duration of 24.2 ± 4.11 min. They occurred in series with a mean length of 6.41 ± 9.01 dives ($n = 2170$ series); some series were as long as 60 dives and lasted over 24 h. Within individual male records, E dives predominated in foraging areas. This was most clear for the male 'Joe'. In the foraging area, 'Joe' exhibited 57.8% E dives, 13.5% D dives and 16.5% A dives in comparison

Table 1. Diving performance of breeding-age males and females, and juveniles of both sexes, 1.4 to 1.8 years of age. The column mean surface interval excludes rare extended surface intervals that exceed 10 min.

Sex and age	Total dives	Mean dives/h	Mean % time on surface	Mean duration (min)	Maximum duration (min)	Mean depth (min)	Maximum depth (m)	Mean surface interval (min)
Adults								
Breeding age males $n = 9$	38 043	2.45 ± 0.24	13.50 ± 5.80	21.25 ± 4.63	90.0	330 ± 222	1503	2.73 ± 0.88
Adult females $n = 10$	28 483	2.57 ± 0.22	9.70 ± 0.80	20.82 ± 4.14	55.0	509 ± 147	1273	2.08 ± 0.47
Juveniles								
Males $n = 7$	45 398	3.66 ± 0.36	10.27 ± 0.91	14.67 ± 3.74	47.0	345 ± 136	1011	1.43 ± 0.59
Females $n = 6$	32 507	3.53 ± 0.68	11.95 ± 4.70	15.54 ± 3.88	53.5	390 ± 140	870	1.70 ± 0.55

Table 2. Mean percentage of free-ranging dives of breeding-age males, post-breeding females and juveniles of both sexes according to dive type. The subscript 'v' refers to variations on the dive type, e.g. A_v refers to A dives and variations of A dives

Sex and age	Dive types					
	Transit			Forage		
	A	A_v	D	E	E_v	$D + E_v$
Adults						
Breeding-age males $n = 9$	25.7 ± 14.3	32.3 ± 18.9	17.9 ± 13.2	39.3 ± 21.1	41.4 ± 21.6	59.3 ± 18.5
Post-breeding females $n = 10$	13.2 ± 6.6	24.0 ± 8.2	60.4 ± 8.2	3.7 ± 2.3	4.5 ± 2.4	64.8 ± 8.2
Juveniles						
Males $n = 6$	21.7 ± 14.9	30.8 ± 17.4	25.7 ± 13.5	24.9 ± 17.2	28.1 ± 18.5	53.6 ± 19.2
Females $n = 6$	36.4 ± 15.6	52.9 ± 17.0	29.6 ± 15.6	3.1 ± 6.1	4.6 ± 9.4	34.2 ± 15.0

with 21.3%, 20.6% and 53.5% of these dives, respectively, while in transit. The mean ambient temperature at the bottom of E dives by 'Joe' in the foraging area was not significantly different from the temperatures at the surface (4.7 ± 1.1 vs. 4.8 ± 0.7 °C, $n = 1043$).

When the relative percentage of the different dive types was analysed on a daily basis, a strong sex difference in apparent foraging behaviour emerged. Foraging in males varied from days with 100% foraging and few or no transit dives to days with no foraging dives, dominated by transit dives (Fig. 5a, c). The occurrence of male transit dives dropped dramatically after Julian day 110 (April 20), by which time males had reached their foraging areas (Fig. 5c). In contrast, females foraged consistently within a more narrow range every day, devoting approximately 35–90% of their dives to foraging (Fig. 5b). They showed a lower daily percentage of transit dives than males, especially during the first 30 days at sea (Fig. 5d). Females exhibited far fewer days of few or no transit dives than males did (Fig. 5c, d). The daily percentage of D dives differed greatly between the sexes (Fig. 5a, b).

To assess foraging variability between the sexes, the mean of the mean depths of foraging dives was calculated for both sexes. There was little interfemale variation in mean depth of type D foraging dives (mean = 545 ± 33 m). In contrast, there was wide variation among males in mean depth for both type D (mean = 688 ± 207 m) and type E dives (331 ± 243 m).

Similar sex differences were evident among juveniles (Table 2). Males displayed significantly more E dives and E_v dives than females ($t = 2.93$ and 2.77, respectively, *d.f.* = 10, $P < 0.05$). Five of the six females had less than 1% E-type dives in their records but one juvenile female, 'Eowyn', had 23.8%, all of which occurred at the end of her record when she was high up in the Gulf of Alaska north of 50 °N. Unlike adults, both sexes displayed a similar percentage of D dives. The two types of foraging dives combined accounted for more than half of the dive types in male juvenile records but only about a third of the dive types for female juveniles; this difference was not statistically significant. Sex differences in the percentage of transit dives displayed were not statistically significant.

The differences in foraging strategy revealed by a daily dive frequency analysis in adults were also evident in juveniles. Juvenile males exhibited distinct transit phases (maximum frequency = 83.4 ± 3.3% dives per day). Juvenile males also exhibited distinct D-type foraging (maximum frequency = 68.8 ± 16.2% dives per day) *en route* to days of predominantly E-type foraging (maximum frequency = 63.5 ± 14.2% dives per day). Juvenile females were more variable than juvenile males in their behaviour. Two of the yearling females, 'Arwen' and 'Gaia', exhibited periods that were predominantly transit which are reflected in the reduced foraging percentages of females in Table 2.

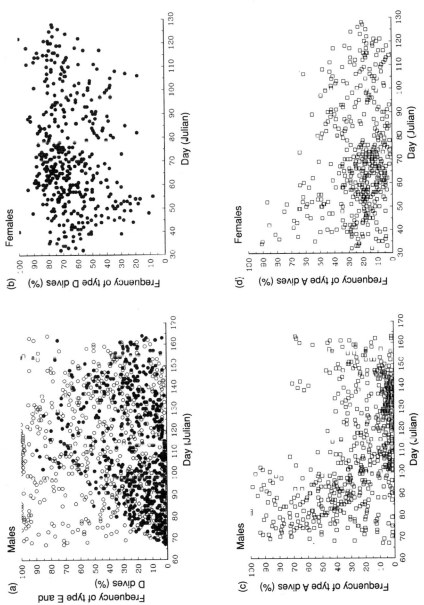

Fig. 5. Percentage of daily foraging dives of males (type E = open circles and type D = closed circles) are shown in (a) for comparison with foraging dives of females in (b). The percentage of daily transit dives (type A = open squares) of males and females are shown in (c) and (d).

Predictions and data bearing on the putative hunting strategy reflected by E dives

Sex differences in the relative percentage of foraging dives displayed provide indirect support for the third hypothesis, that males spend part of the time feeding on different prey than females. This is largely a matter of interpreting the function of male E dives. We consider evidence linking flat-bottomed E dives to benthic foraging on the ocean bottom and an alternative hypothesis that these dives reflect a sit-and-wait hunting strategy in the water column. We tested three predictions derived from reasoning that flat-bottomed E dives reflect foraging on benthic animals:

Prediction 1: E dives occur over the continental shelf, seamounts or guyots. The shape of E dives, the stability of maximum dive depth in a series, and the shape of the dives that precede and follow a series of E dives, suggest movement along the bottom of the ocean floor (Fig. 6). If this is the case, the appearance of E dives should be limited to areas with a depth of approximately 1000 m or less.

Days in which 10 or more type E dives occurred are shown in relation to seamounts, guyots and the 1000 m isobath in Fig. 7. Type E dives occurred near coastal margins; they did not occur across a wide range of the Pacific along migratory paths of males or females. For example, the male 'Tyke'

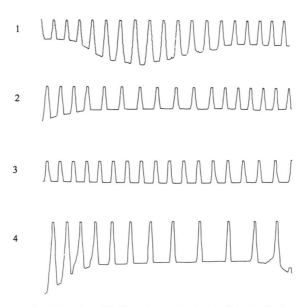

Fig. 6. Four examples of a series of E dives. Examples 1, 2 and 4 give the impression that the animal is moving along a sloping seafloor.

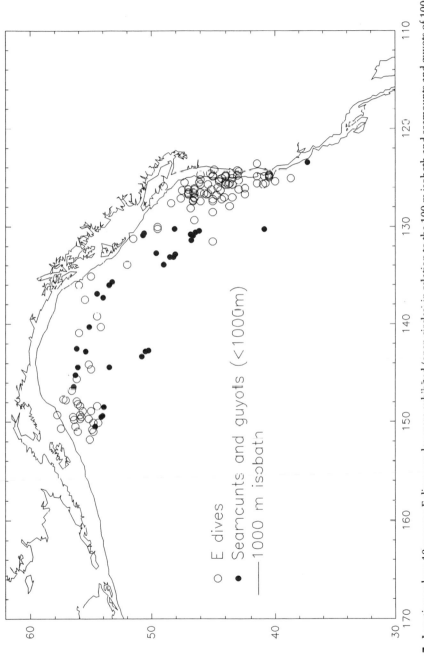

○ E dives
● Seamounts and guyots (<1000m)
——— 1000 m isobath

Fig. 7. Locations where 10 or more E dives per day were exhibited (open circles) in relation to the 100 m isobath and seamounts and guyots of 1000 m or less (closed circles). The 1000 m isobath is from General Bathymetric Chart of the Oceans (GEBCO) and seamounts and guyots are from T. E. Chase, *Bathymetry of the North Pacific*, Scripps Institution of Oceanography, 1970. All but three E-dive locations are for males; only one female, 'Lilly', met the criterion and for only three days.

showed no E dives along his entire migratory path until he was near the eastern Aleutians whereupon he exhibited them daily for five days until the diving record ended. E dives represented 70.5% of 'Quake's' dives while he was near the coast (see Fig. 2); however, these dives disappeared abruptly when he headed west into deeper waters. E dive distribution was proximal to areas of 1000 m or less and sometimes coincident with these areas but the majority of E dives occurred in adjacent deeper-water zones. These data do not provide unequivocal support for the prediction but there may be mitigating circumstances which we discuss later.

Prediction 2: The occurrence of E dives does not follow a diel pattern. Benthic animals do not migrate vertically, as do animals in the deep scattering layer. Some of these animals are most active at dawn and dusk (Woodhead 1966). If seals are feeding on benthic animals during E dives, these dives should be constant throughout the day and night or they should be most prevalent at dawn and dusk. They should not show peak frequencies near midnight and be least frequent near midday as one would predict if the males were pursuing vertically migrating prey.

This prediction is borne out by the data in Fig. 8 showing that E dive frequency did not follow a diel pattern, with fewer dives during the day (because they are deeper and hence longer) than at night, but rather, followed a crepuscular pattern to varying degrees in eight of the nine males. E dives of the exceptional male, 'C508', did not vary systematically with time of day. In addition, there was no diel pattern in frequency of E dives in juvenile males. There were too few E dives by females to analyse.

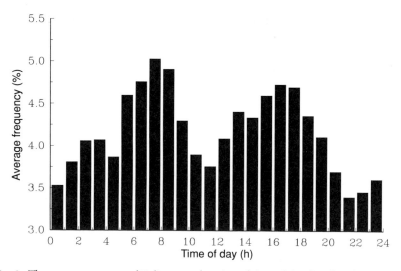

Fig. 8. The mean percentage of E dives as a function of time of day for all males.

Prediction 3: Variation in depth of E dives does not follow a diel pattern as does variation in depth of D dives. The reasoning here is similar to the previous prediction. Benthic prey are not moving down with increasing light or up with decreasing light so the dive depth of benthic foragers should not vary with peaks and troughs in light levels. In contrast, if D dives are pelagic and reflect foraging in the deep scattering layer, their depths should be deeper during midday than around midnight.

As predicted, E dive depth of males did not follow a diel pattern (Fig. 9). D dive depth of females exhibited the expected diel pattern (Fig. 10) but the D dive depths of males did not (Fig. 11). This unexpected difference was also evident in juveniles (Fig. 12). This is additional information suggestive of males and females exploiting different prey.

An alternative hypothesis to flat-bottomed E dives serving benthic foraging is that they reflect a sit-and-wait strategy in the water column (see Thompson & Fedak, this volume, for an extended discussion). Males may target rich areas where the encounter rate with prey is improved by a sit-and-wait strategy which might depend on the prey moving to them. It is unlikely that males do this for vertically moving prey since our data show that the depths of neither D dives nor E dives of males track the deep scattering layer. If males adopt a sit-and-wait strategy, it is more likely that it is to intercept horizontally moving prey. We tested one prediction from this reasoning.

Prediction 4: E dives should be of longer duration than D dives if the two dive types are matched for depth. The logic is simply that the seal sitting and waiting at depth uses less energy than the seal that is making multiple vertical excursions in the water column. If E dives serve foraging, as their high frequency suggests, we would expect the seals to maximize time at the bottom of the dive pursuing prey (Kooyman *et al.* 1980).

The mean durations of 24 312 E and D dives of six males, matched for depth at increments of 50 m, yielded 47 paired comparisons. Contrary to the prediction, E dives were of significantly shorter duration (mean = 24.20 ± 4.11 min) than D dives (mean = 26.48 ± 4.74 min) ($t = 4.3$, $d.f. = 45$, $P < 0.05$).

Discussion

Our analysis provides support for the general hypothesis that there are sex differences in the diving and foraging pattern of elephant seals that reflect the greater energy requirements of males. Some males travelled further north than females and all males migrated to areas near continental margins where concentrated foraging occurred. In contrast, females ventured across a broad expanse of the north-eastern Pacific. These data, bolstered by similar data on the movements and foraging locations of elephant seals from

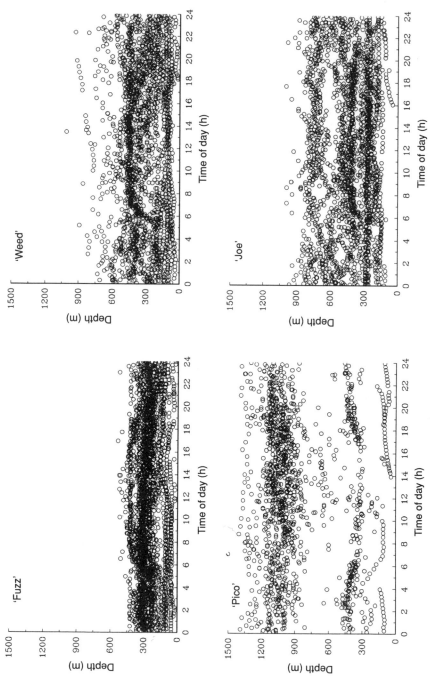

Fig. 9. Depth of E dives did not follow a diel pattern, as is shown for four males.

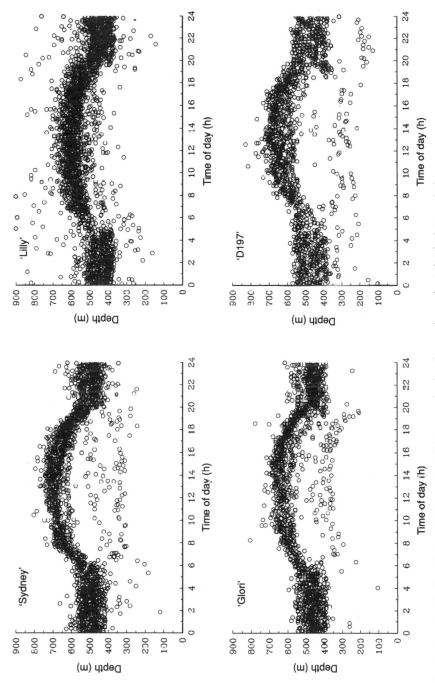

Fig. 10. Depth of D dives followed the expected diel pattern for females, as shown for individuals in each diagram.

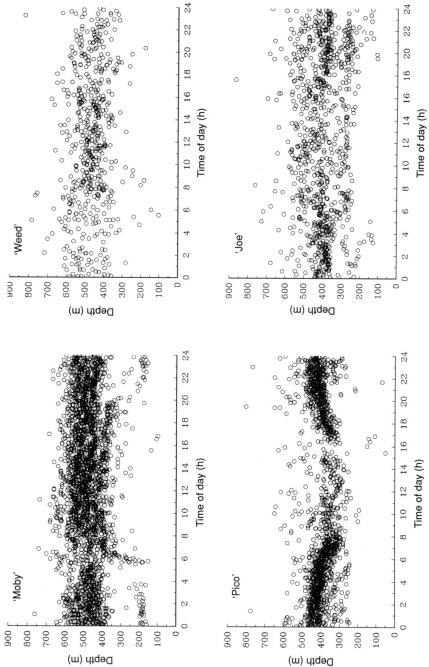

Fig. 11. Depth of D dives for males did not exhibit a diel pattern, as is evident from the dive depths of four males illustrated.

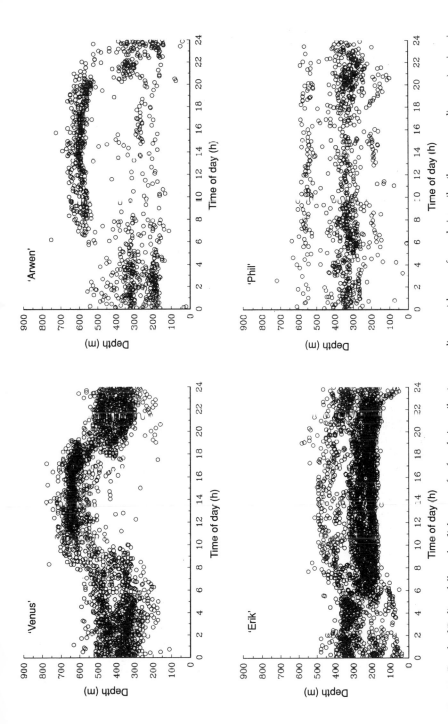

Fig. 12. Depth of D dives followed a diel pattern for female juveniles (top two diagrams) but not for male juveniles (bottom two diagrams), mirroring the sex difference seen in adults.

San Miguel Island in southern California (DeLong *et al.* 1992; Stewart & DeLong in press), support the hypothesis that males forage in different locations than females.

Sex differences in the types of dives displayed and the frequency of their occurrence suggest that males adopt a different foraging strategy than females which may be associated with capturing different prey. Over 40% of male dives were flat-bottomed E dives whereas less than 5% of female dives were of this type. The frequency and depth of male E dives did not follow a diel pattern, which suggested that males were not pursuing prey in the deep scattering layer. This, plus the presence of a diel pattern in the depth of D dives by females and the absence of one in males further suggests sexual segregation in prey pursued. Stewart & DeLong (1991) reported a diel pattern in the depth of male dives during the northward migration from San Miguel Island in early spring and the return migration in early summer, and no diel pattern in foraging areas; however, they lumped all male dives together, unlike our analysis by dive types.

Males put a premium on concentrated transit until reaching a foraging area where they remained in a relatively fixed area for 1.5 to 2 months (see also DeLong *et al.* 1992; Stewart & DeLong in press). During transit, pelagic foraging dives were interspersed with transit dives but once the foraging area was reached, repetitive, uniform, flat-bottomed E dives predominated. Thus, males appear to pursue a strategy characterized by increased relative investment in transit to a preferential foraging area, with a relatively high rate of energy acquisition in that area. Males seem to minimize the energetic cost and reduced foraging time due to transit by opportunistically foraging on pelagic prey.

Females, on the other hand, pursue a strategy of constant moderate energetic investment in transit to prey, and foraging to achieve energy acquisition that is constant and more moderate than that of males. They foraged consistently, alternating foraging with transit, and rarely stayed long in the same area. Despite dispersing over a broad geographic area, all females foraged at approximately the same mean depth and showed strong diel variation in dive depth. This suggests that females are utilizing a food source that is not determined so much by geographical boundaries, as is the case with males, as by a fixed cyclical pattern of vertical prey movement in the pelagic and mesopelagic environment.

We conclude that E dives serve foraging rather than transit or rest because these dives predominated in the dive records, were most frequent when the animal was on station foraging, were less frequent or absent in transit, and were often displayed in long series lasting over 24 h.

The foraging tactic and type of prey consumed during E dives is not clear from our analysis. Arguments can be made, and some data brought to bear, on two hypotheses: that E dives serve active benthic foraging (Hindell 1990;

Le Boeuf, Naito, Asaga *et al.* 1992) and that E dives reflect a sit-and-wait foraging strategy. We review the merits of each hypothesis briefly.

Benthic foraging

The case for E dives serving benthic foraging begins with a consideration of their flat-bottomed shape and the uniform depth of multiple dives in a series. These characteristics suggest that the animal is moving over a relatively flat surface. In some records, transit or pelagic foraging dives grade into increasingly shallower E dives as if the seal approached and moved over a seamount or was foraging on and off the continental slope. Swim speed as measured in a post-breeding adult female (Le Boeuf, Naito, Asaga *et al.* 1992) is consistent with the idea of movement at the bottom of E dives. During the bottom segment of 23 E dives, this female was always moving (mean rate = 0.91 ± 0.48 m/s).

An advantage of benthic foraging is that males could trap prey against the ocean bottom, possibly enhancing prey capture rate and reducing pursuit effort. Compared with squid found in the water column, benthic-dwelling animals such as small sharks, skates, rays, ratfish and hagfish are large and energy-dense. There is evidence suggesting that males prey on these animals and females only rarely do. Antonelis, Lowry, Fiscus *et al.* (in press) report that over 20% of the stomachs of subadult males contained the remains of cyclostomes. In contrast, cyclostomes were absent in female stomachs in four of five years and rare (less than 5% of stomachs) in the exceptional year. Subadult males also consumed more elasmobranchs in two of the three years reported. These observations are consistent with reports of males capturing sharks and rays (Condit & Le Boeuf 1984). Most of these animals are relatively slow-moving and abundant in eastern Pacific waters from Baja California to Alaska. The frequency and depth of E dives as a function of time of day is consistent with males feeding on benthic animals rather than preying on vertically moving prey in the deep scattering layer, as females appear to do. Lastly, E dives occur near benthic areas within diving distance of males; most of them do not occur in the unreachable depths of the open ocean which they cross in transit.

An argument against the benthic foraging hypothesis is that E dives are not distributed precisely in areas where the dives could reach bottom (Fig. 7), e.g. the highest concentrations of E dives are approximately two degrees distant from the 1000 m isobath. This argument depends on precision of bathymetric measurement and animal location. The likelihood of errors of this magnitude in charting depths, even in the more general maps, is unlikely. On the other hand, the error in geolocating by light levels can be large, especially near the equinoxes and at lower latitudes (Hill in press). It is standard practice to use surface temperature from other sources as an aid in locating the animal, a procedure which can be subjective. An attempt to validate location provided by a geographic location time depth recorder (of the type used in this study) attached to a ship's mast with the ship's LORAN yielded less than 50% agreement within a 110 × 110 km grid (DeLong *et al.* 1992). Clearly, geolocation derived in this way may provide a general estimate of location but it is not accurate enough to pinpoint bathymetric features.

A final point against the benthic foraging hypothesis has to do with E dive shape and uniformity. Cursory inspection suggests movement along the ocean bottom, as argued above, but the dive bottoms are so uniform in depth and over time, in many cases, that they would have to occur over an area resembling the abyssal plain. That is, they appear too flat to resemble the surface of seamounts, guyots and the continental shelf in the north-eastern Pacific Ocean.

Sit-and-wait foraging

The location of E dives along the continental margin, an area characterized by strong currents, upwelling and high biomass, is consistent with a sit-and-wait strategy. Either benthic or pelagic prey may be pursued, as this hypothesis makes no distinction in this regard. The stereotyped characteristics of E dives, especially with respect to depth, are consistent with the idea of males diving repetitively to a precise depth to capture prey that congregate at boundary layers or some other physical feature of the water column. Prey might accumulate at the discontinuities between water masses or the seals might use changes in water temperature as a cue to finding prey as Boyd & Arnbom (1991) have suggested for southern elephant seals, *M. leonina.*

Temperature change, however, may not be an optimal signal for northern elephant seals, for as they move north in their migrations, the thermocline, which is pronounced at 100–150 m in central California waters, fades quickly to a point where temperature decreases uniformly with depth (H. Hakoyama, Y. Naito & B. J. Le Boeuf unpubl. data). Moreover, the difference between surface temperature and temperature at depth attenuates as the seal moves north, offering fewer cues for identifying temperature layers at depth. For example, the mean temperature difference between the surface and bottom of E dives along the state of Washington was 3.3 °C and only 0.08 °C in the upper Gulf of Alaska.

Two additional points create difficulties for the sit-and-wait strategy. Swim speed measurement indicates movement at the bottom of E dives (Le Boeuf, Naito, Asaga *et al.* 1992); this is not, however, particularly damaging to the hypothesis, for two reasons. The measurement was made on a female and E dives in a female may have little relationship to those in males. Secondly, a slow move-and-pounce strategy would not differ much from a sit-and-wait strategy. Perhaps more damaging is the finding that E dives are significantly shorter in duration than D dives. This implies that the seal is not maximizing its time underwater during E dives and suggests that it may be using more energy than during a D dive which appears to involve active searching or pursuit of prey at the dive bottom. That is, why are E dives shorter in duration if the seal is simply sitting and waiting? It does not seem to be because the seal captures prey and moves to the surface to eat it, because the dives are of uniform duration.

General conclusions and implications

Our analysis is not conclusive with respect to the role of E dives in foraging. Further studies will be required to test the hypotheses we have considered. What is important is that males spend a significant part of their time at sea exhibiting dives rarely displayed by females. These dives have a specific pattern and are specific to place. This sex difference implies that males acquire different prey which may be larger, have higher energy value, are easier to catch or are more abundant, or can be caught by males but not by females. The data on elephant seal diet in the literature do not shed light on this prediction regarding sex differences in foraging. Prey remains identified in dead animals (Condit & Le Boeuf 1984) or lavaged stomachs of animals returning to land (Stewart & DeLong 1991; Antonelis, Lowry, Fiscus et al. in press) bias the results to pelagic foraging during the last few days in transit (Helm 1984; Harvey et al. 1989).

The principal sex differences we have described in northern elephant seals are also evident in southern elephant seals. Hindell (1990) reports that after the breeding season, adult males from Macquarie Island migrated to foraging areas within the 1000 m depth contour off the Antarctic coast while adult females foraged offshore in deep oceanic waters. Flat-bottomed dives, corresponding to our E dives, were almost exclusively a characteristic of males; they were the most common dive type in the foraging area, alternating with pelagic dives, corresponding to our D dives, and they exhibited no diel pattern in depth. Pelagic foraging dives predominated in the records of females (and were observed in the records of males in transit) and they showed marked diel fluctuation in dive depth. Hindell concludes that southern elephant seal females are exclusively pelagic foragers in the water column while males employ both a pelagic and benthic foraging strategy.

Finally, juveniles less than two years old going to sea for only the third and fourth time revealed sex differences in the frequency and depth of E dives as a function of time of day, as well as in the variability of foraging dive types exhibited per day, similar to the pattern observed in adults. Thus, the sex differences in diving behaviour and foraging strategies we have described appear early in life, as predicted, and may be essential to the growth spurt seen in males (Laws 1956) and to the attainment and maintenance of a size allowing potential reproduction for males.

Acknowledgements

We thank Guy Oliver and John Sanders and numerous others for field assistance, Clairol Inc. for marking solutions, and Ian Boyd for a thorough critical review of the manuscript. This research was funded in part by grants from the National Science Foundation, Minerals Management Service, Myers Foundation, and gifts from George Malloch and the G. MacGowan estate.

References

Alexander, R. D., Hoogland, J. L., Howard, R. D., Noonan, K. M. & Sherman, P. W. (1979). Sexual dimorphisms and breeding systems in pinnipeds, ungulates, primates and humans. In *Evolutionary biology and human social behavior: an anthropological perspective*: 402–603. (Eds Chagnon, N. A. & Irons, W.). Wadsworth, Belmont, California.

Antonelis, G. A. Jr., Lowry, M. S., DeMaster, D. P. & Fiscus, C. H. (1987). Assessing northern elephant seal feeding habits by stomach lavage. *Mar. Mamm. Sci.* 3: 308–322.

Antonelis, G. A., Lowry, M. S., Fiscus, C. H., Stewart, B. S. & DeLong, R. L. (In press). Diet of the northern elephant seal. In *Elephant seals*. (Eds Le Boeuf, B. J. & Laws, R. M.). University of California Press, Los Angeles.

Asaga, T., Naito, Y., Le Boeuf, B. J. & Sakurai, H. (In press). Functional analysis of dive types of female northern elephant seals. In *Elephant seals*. (Eds Le Boeuf, B. J. & Laws, R. M.). University of California Press, Los Angeles.

Baker, R. R. (1978). *The evolutionary ecology of animal migration*. Hodder and Stoughton, London.

Benedict, F. G. (1938). *Vital energetics: a study in comparative basal metabolism*. Washington, D.C.: Carnegie Institute of Washington.

Bonner, W. N. & Hunter, S. (1982). Predatory interactions between Antarctic fur seals, Macaroni penguins and giant petrels. *Bull. Br. Antarct. Surv.* No. 56: 75–79.

Boyd, I. L. & Arnbom, T. (1991). Diving behaviour in relation to water temperature in the southern elephant seal: foraging implications. *Polar Biol.* 11: 259–266.

Brody, S., Procter, R. C. & Ashworth, U. S. (1934). Basal metabolism, endogenous nitrogen, creatinine and neutral sulphur excretions as functions of body weight. *Res. Bull. Mo. agric. Exp. Stn* 220: 1–40.

Chapskii, K. K. (1936). The walrus of the Kara Sea. *Trudÿ vses. arkt. Inst.* 67: 1–124.

Condit, R. & Le Boeuf, B. J. (1984). Feeding habits and feeding grounds of the northern elephant seal. *J. Mammal.* 65: 281–290.

Costa, D. P. (1991). Reproductive and foraging energetics of pinnipeds; implications for life history patterns. In *The behaviour of pinnipeds*: 300–344. (Ed. Renouf, D.). Chapman and Hall, New York.

Darwin, C. (1871). *The descent of Man, and selection in relation to sex*. New York: D. Appleton and Company.

DeLong, R. L., Stewart, B. S. & Hill, R. D. (1992). Documenting migrations of northern elephant seals using day length. *Mar. Mamm. Sci.* **8**: 155–159.

Deutsch, C. J., Haley, M. P. & Le Boeuf, B. J. (1990). Reproductive effort of male northern elephant seals: estimates from mass loss. *Can. J. Zool.* **68**: 2580–2593.

Deutsch, C. J., Crocker, D. E., Costa, D. P. & Le Boeuf, B. J. (In press). Sex and age related variation in reproductive effort of the northern elephant seal. In *Elephant seals*. (Eds Le Boeuf, B. J. & Laws, R. M.). University of California Press, Los Angeles.

Fay, F. H. (1960). Carnivorous walrus and some Arctic zoonoses. *Arctic* **13**: 111–122.

Fay, F. H. (1990). Predation on a ringed seal, *Phoca hispida*, and a black guillemot, *Cepphus grylle*, by a Pacific walrus, *Odobenus rosmarus divergens. Mar. Mamm. Sci.* **6**: 348–350.

Gentry, R. L. & Johnson, J. H. (1981). Predation by sea lions on northern fur seal neonates. *Mammalia* **45**: 423–430.

Goodall, J. (1986). *The chimpanzees of Gombe: patterns of behavior*. Harvard Univ. Press, Belknap Press, Cambridge, Mass.

Harding, R. S. O. (1973). Predation by a troop of olive baboons (*Papio anubis*). *Am. J. phys. Anthrop.* **38**: 587–591.

Harvey, J. T., Antonelis, G. A. & Casson, C. J. (1989). Quantifying errors in pinniped food habit studies. *Abstr. bienn. Conf. Biol. mar. Mammals* **8**: 26. Society for Marine Mammalogy. Unpublished.

Helm, R. C. (1984). Rate of digestion in three species of pinnipeds. *Can. J. Zool.* **62**: 1751–1756.

Hill, R. (In press). The theory of geolocation. In *Elephant seals*. (Eds Le Boeuf, B. J. & Laws, R. M.). University of California Press, Los Angeles.

Hindell, M. A. (1990). *Population dynamics and diving behavior of southern elephant seals*. PhD thesis. University of Queensland.

Hindell, M. A., Slip, D. J. & Burton, H. R. (1991). The diving behaviour of adult male and female southern elephant seals, *Mirounga leonina* (Pinnipedia: Phocidae). *Aust. J. Zool.* **39**: 595–619.

Kleiber, M. (1932). Body size and metabolism. *Hilgardia* **6**: 315–353.

Kleiber, M. (1961). *The fire of life: an introduction to animal energetics*. Wiley, New York & London.

Kooyman, G. L., Wahrenbrock, E. A., Castellini, M. A., Davis, R. W. & Sinnett, E. E. (1980). Aerobic and anaerobic metabolism during voluntary diving in Weddell seals: evidence of preferred pathways from blood chemistry and behavior. *J. comp. Physiol.* **138**: 335–346.

Laws, R. M. (1956). Growth and sexual maturity in aquatic mammals. *Nature, London.* **178**: 193–194.

Le Boeuf, B. J. (In press). Variation in the diving pattern of northern elephant seals with age, mass, sex and reproductive condition. In *Elephant seals*. (Eds Le Boeuf, B. J. & Laws, R. M.). University of California Press, Los Angeles.

Le Boeuf, B. J., Costa, D. P., Huntley, A. C. & Feldkamp, S. D. (1988). Continous, deep diving in female northern elephant seals, *Mirounga angustirostris. Can. J. Zool.* **66**: 446–458.

Le Boeuf, B. J., Costa, D. P., Huntley, A. C., Kooyman, G. L. & Davis, R. W.

(1986). Pattern and depth of dives in Northern elephant seals, *Mirounga angustirostris*. *J. Zool., Lond.* **208**: 1–7.

Le Boeuf, B. J., Naito, Y., Asaga, T., Crocker, D. E. & Costa, D. P. (1992). Swim speed in a female northern elephant seal: metabolic and foraging implications. *Can. J. Zool.* **70**: 786–795.

Le Boeuf, B. J., Naito, Y., Huntley, A. C. & Asaga, T. (1989). Prolonged, continuous, deep diving by northern elephant seals. *Can. J. Zool.* **67**: 2514–2519.

Lowry, L. F. & Fay, F. H. (1984). Seal eating by walruses in the Bering and Chukchi seas. *Polar Biol.* **3**: 11–18.

Mansfield, A. W. (1958). The biology of the Atlantic walrus, *Odobenus rosmarus rosmarus* (Linnaeus) in the eastern Canadian Arctic. *Fish. Res. Bd Can. MS Rep. (Biol.)* No. 653.

Riedman, M. L. (1990). *The pinnipeds: seals, sea lions and walruses.* University of California Press, Los Angeles.

Riedman, M. L. & Estes, J. A. (1988). Predation on seabirds by sea otters. *Can. J. Zool.* **66**: 1396–1402.

Schaller, G. B. (1972). *The Serengeti lion: a study of predator–prey relations.* The University of Chicago Press, Chicago & London.

Stewart, B. S. & DeLong, R. L. (1991). Diving patterns of northern elephant seal bulls. *Mar. Mamm. Sci.* **7**: 369–384.

Stewart, B. S. & DeLong, R. L. (In press). Post-breeding foraging migrations of northern elephant seals. In *Elephant seals.* (Eds Le Boeuf, B. J. & Laws, R. M.). University of California Press, Los Angeles.

Teleki, G. (1973). *The predatory behavior of wild chimpanzees.* Bucknell University Press, Lewisburg, Pennsylvania.

Trivers, R. L. (1985). *Social evolution.* The Benjamin/Cummings Publishing Company, Menlo Park, California.

Wegge, P. (1980). Distorted sex ratio among small broods in a declining capercaillie population. *Ornis scand.* **11**: 106–109.

Woodhead, P. M. J. (1966). The behaviour of fish in relation to light in the sea. *Oceanogr. mar. Biol.* **4**: 337–403.

Symp. zool. Soc. Lond. (1993) No. 66: 179–194

Seasonal dispersion and habitat use of foraging northern elephant seals

Brent S. STEWART

Hubbs-Sea World Research Institute
1700 South Shores Road
San Diego, California 92109, USA

and Robert L. DELONG

National Marine Mammal Laboratory
NMFS, NOAA
7600 Sand Point Way
Seattle, Washington 92115, USA

Synopsis

Elephant seals are at sea and dive continously for 8–10 months each year. They range widely in the North Pacific; their dispersion and migratory patterns correspond with three major North Pacific oceanographic features where enhanced biological productivity supports deep-water assemblages of elephant seals' primary cephalopod prey. Throughout the year adult males are dispersed farther north than females. Both sexes forage primarily on squid that reduce their specific gravity to maintain neutral buoyancy either by storing fat in their bodies or by actively replacing heavy metal ions in their body fluids with lighter ammonium ions. Squid with these adaptations are slower swimmers, and therefore probably more susceptible to predation by elephant seals, but are of poorer nutritional quality than are negatively buoyant squid species. Adult elephant seal males may gain access to higher-quality prey than females by travelling farther north to capitalize on larger fat-rich gonatid squid, the dominant cephalopod fauna in the Subarctic water mass. It is not clear why females do not forage in those northern areas. But reduced intraspecific competition for food resources because of such sexual segregation of foraging adults has important implications for demographic studies of the rapidly growing northern elephant seal population.

Introduction

The diving patterns of northern elephant seals (*Mirounga angustirostris*) have been documented in striking detail in recent years (e.g., Le Boeuf, Naito *et al.* 1989; DeLong & Stewart 1991). Integration of information on elephant seal diet (Antonelis, Lowry, DeMaster & Fiscus 1987; Antonelis, Lowry, Fiscus *et al.* in press), seasonal movements (DeLong, Stewart & Hill 1992; Stewart & DeLong in press) and diving behaviour has greatly

ZOOLOGICAL SYMPOSIUM No. 66
ISBN 0–19–854069–8
Copyright © 1993 The Zoological Society of London
All rights of reproduction in any form reserved

improved basic knowledge of the foraging ecology of northern elephant seals.

From summaries of diving and location records we describe here the year-round geographic and vertical dispersion of northern elephant seals that breed and moult on the Southern California Channel Islands. We interpret these distribution patterns in relation to general oceanographic character-istics in the North Pacific and to what is known of the spatial distribution and ecology of elephant seals' primary cephalopod prey.

Materials and methods

We glued microprocessor-based dive and location recorders (= GLTDRs; DeLong *et al.* 1992) to the backs of adult males and females at the ends of their breeding (February) and moult (females: May; males: August) seasons and recovered the instruments three to nine months later when the seals returned to land. All deployments were at San Miguel Island (34°02'N, 120°23'W), California, but some instruments were recovered at San Nicolas Island (33°15'N, 119°30'W). We programmed the GLTDRs to store, in electronic memory (256 kbytes available), measurements of depth at 30-, 60- or 90-s intervals continually and sea-surface temperature (SST) and ambient light (SSL) whenever seals were within 30 m of the sea-surface. Some instruments that were used for deployments lasting more than four months were programmed to alternate between measurement periods of 7–10 days and shorter intervals (3–5 days) when no measurements were made, so that seals' entire periods at sea could be sampled.

We summarized depth data to describe general features of each dive (e.g. dive duration, maximum dive depth etc.; see DeLong & Stewart 1991; Bengston & Stewart 1992, for dive variable definitions). We determined seals' daily locations by using daylength, local apparent noon time, and SST as described by DeLong *et al.* (1992). We confirmed or corrected all daylength-based estimates of latitude by comparing GLTDR-stored measure-ments of sea-surface temperature with sea-surface temperature data that had been collected from ship- and satellite based sensors and summarized by the National Oceanic and Atmospheric Administration (e.g. NOAA 1990, 1991).

Statistical analyses of dive parameters were made using SYSTAT (Wilkinson 1988). We report sample statistics (i.e. $\bar{x}$ = sample mean, S.D. = standard deviation of the sample mean) as summary statistics for all dives of each seal and population statistics (i.e. μ = population mean, S.E.M. = standard error of the population mean) as summary statistics for all dives of all seals. We used a Kolmogorov-Smirnov two-sample test (Sokal & Rohlf 1981) to compare the dive depth distributions of males and females and a one-way analysis of variance (Zar 1974) to determine the influence of time of day on maximum dive depth; the latter analyses were performed on random samples of 20% of each seal's dive records.

Results

We recovered instruments that contained complete dive and location records from 21 females (post-breeding = 13; post-moult = 8) and 15 males (post-breeding = 9; post-moult = 6). Collectively 116 555 dives were recorded for females (post-breeding = 54 889; post-moult = 61 662) and 100 662 for males (post-breeding = 61 658; post-moult = 39 504). Females were at sea for an average of 72 days (S.D. = 7 days) in spring (= post-breeding migration) and 224 days (S.D. = 30 days) in summer and autumn (post-moult migration). Males were at sea for 119 days (S.D. = 9 days) between breeding and moult and 122 days (S.D. = 5 days) between moult and breeding (Fig. 1).

Adult males and females were ashore simultaneously only briefly during the year, when breeding; then, a typical female was ashore for about 34 days continuously whereas an adult male may have been ashore for 30 to 90 days (Stewart 1989; Fig. 1). Seals began travelling north immediately upon entering the water after breeding and moult seasons and were dispersed widely in the eastern North Pacific during the next several months or more (Figs 2–5). Females remained south of 50° N latitude during post-moult and post-breeding seasons but ranged farther west during the post-moult season (Figs 2, 3). Post-breeding females were dispersed farther north in 1991 than in 1990 (Fig. 2). Males migrated farther north than females throughout the year and most travelled to the northern Gulf of Alaska and the eastern Aleutian Islands (Figs 4, 5).

Females were submerged approximately 91% (S.D. = 1%) and males 88% (S.D. = 2%) of the time that they were at sea. While at sea seals dived continually with rare interruption. On average, dives lasted 23.6 min (S.E.M. = 0.4 min) for females and 23.1 min (S.E.M. = 0.7 min) for males. Interdive surface intervals were brief, approximately 2 min (S.E.M. = 0.1 min) for females and 3 min (S.E.M. = 0.1 min) for males. During virtually all dives seals descended rapidly to a specific depth, remained at or within 30 m of that depth for several minutes or more and then ascended rapidly to the surface. We use the maximum depth of each dive as an index of predominant vertical dispersion of seals because approximately 35% (S.E.M. = 9.6%) of the duration of each dive was spent at or near maximum depth. Few dives were shallower than 150 m (Figs 1, 6) and few exceeded 800 m (females: 0.16% > 800 m; males 0.18% > 800 m). The average depth of females' dives was 517.5 m (S.E.M. = 11.05) during the post-breeding season and 478.8 m (S.E.M. = 7.8 m) during the post-moult season (Fig. 6). The average depth of males' dives was 364.0 m during the

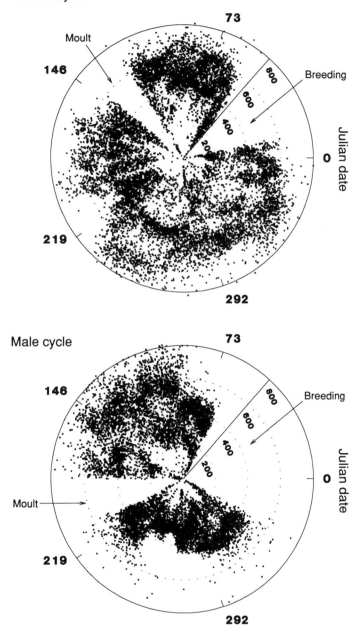

Fig. 1. Typical annual cycle of northern elephant seal adults showing periods on land (breeding and moult), periods at sea (scatterplots), and vertical dispersion (dive depths). Radius is maximum depth of each dive from 0 m at the centre to 800 m at the perimeter. Julian date from 0 (1 January) through 365 (31 December) is the perimeter axis.

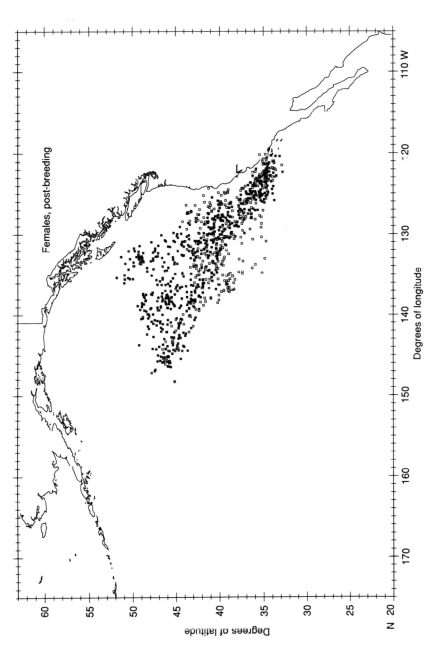

Fig. 2. Daily locations of 13 adult female northern elephant seals during the post-breeding season foraging period (February–May). Open squares indicate female dispersion in 1990 and filled squares female dispersion in 1991.

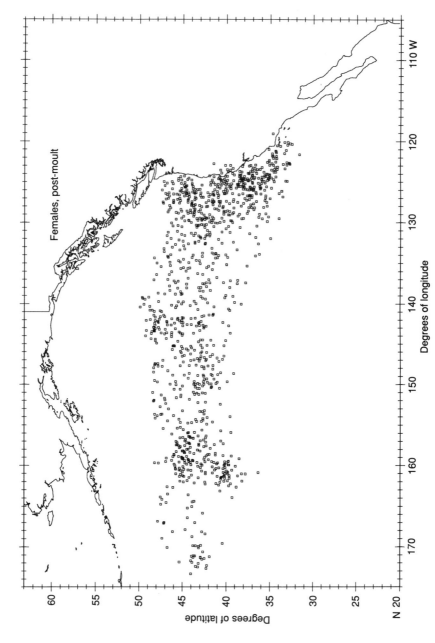

Fig. 3. Daily locations of eight adult female northern elephant seals during the post-moult season foraging period (May–January).

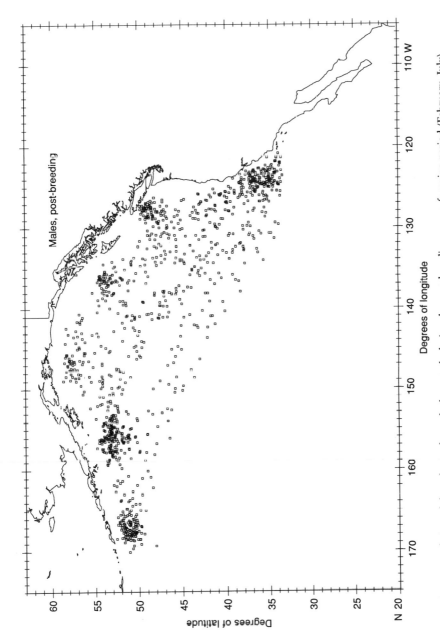

Fig. 4. Daily locations of nine adult male northern elephant seals during the post-breeding season foraging period (February–July).

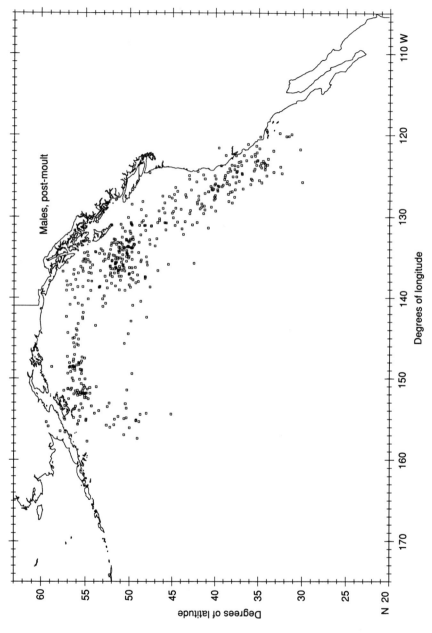

Fig. 5. Daily locations of six adult male northern elephant seals during the post-moult season foraging period (August–December).

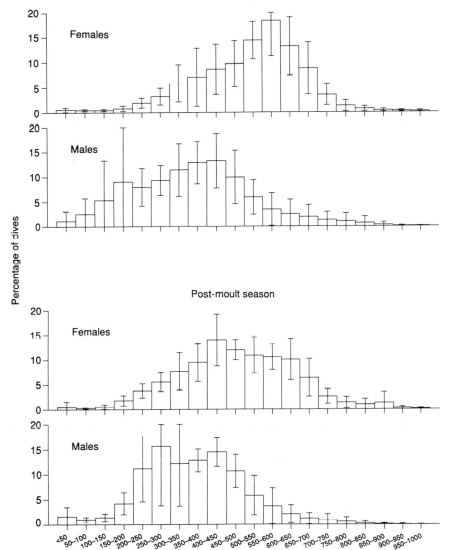

Fig. 6. Frequency distribution of maximum dive depths of adult male and female northern elephant seals during post-breeding and post-moult seasons. Histogram values are population averages (μ) of all seals and error bars are standard errors (S.E.M.) of the population averages.

post-breeding season and 366.2 m (S.E.M. = 15.3) during the post-moult season (Fig. 6). The differences in dive-depth distributions between males and females were significant for each season ($P < 0.05$). With rare exceptions, there was significant ($P < 0.01$ for each seal) diel variation in dive depth throughout the year for each seal; nighttime dives were 100–200 m shallower than daytime dives.

Discussion

Northern elephant seals are at sea for 8–10 months each year (Fig. 1). During that time they make two migrations between breeding and moulting sites on the Southern California Channel Islands and pelagic foraging grounds (Stewart & DeLong 1990; DeLong et al. 1992; Stewart & DeLong in press). During those migrations individuals are widely dispersed in the eastern North Pacific Ocean (Figs 2–5). Our records indicate that geographic dispersion of adult males differs from that of females for much of the year. Females remain south of 50° N latitude whereas males are dispersed several hundred or more kilometres farther north along the Eastern Aleutian Islands and the Gulf of Alaska. Geographic overlap of males and females occurs primarily in the California Current offshore northern and central California, Oregon and Washington; these areas are migratory corridors for both sexes and migratory destinations of females.

The patterns of seasonal geographic dispersion of foraging northern elephant seals that we describe here correspond with three North Pacific water masses and with the distribution of cephalopods, which appear to be elephant seals' predominant prey (cf. Antonelis, Lowry, DeMaster et al. 1987; DeLong & Stewart 1991; Antonelis, Lowry, Fiscus et al. in press). Those water masses are (1) the Subarctic Current (= West Wind Drift), an eastward-flowing current between 40° and 50° N latitude that separates the cooler Subarctic Water Mass to the north from the warmer Subtropical Transition Zone and the Equatorial Water Mass to the south, (2) the California Current which splits off from the Subarctic Current as it approaches the North American coast and flows south along the California coast and (3) the Alaska Stream, which also splits off from the western reach of the Subarctic Current but flows northward through the Gulf of Alaska and then continues west along the eastern Aleutian Islands (Favorite, Diomead & Nasu 1976; Reed & Schumacher 1985).

Female elephant seals tended to aggregate in the eastern North Pacific Transitional Domain between the Subarctic Domain and the North Pacific Transition Zone and especially in the Subarctic Frontal Zone (for a comprehensive description of these areas see Favorite et al. 1976; Pearcy 1991; Roden 1991). We are unable to explain the differences in post-breeding dispersion of elephant seal females between 1990 and 1991.

Although sea-surface temperatures (SSTs) were above normal during that season in both years, SSTs in 1990 did not differ significantly from SSTs in 1991 during that season. Males aggregated along the offshore boundary of the Alaska Stream, particularly along the eastern Aleutian Islands (Figs 2–5) during both post-breeding and post-moult seasons. Open-ocean frontal regions and water mass boundaries, like the Subarctic Frontal Zone and bordering areas, are known to be unique areas of biological enhancement and have been postulated to play important roles in localized biological productivity (e.g. Lutjeharms, Walters & Allanson 1985). Highly productive, near-surface fish and squid communities have been identified in or near some of these areas, particularly in the North Pacific Transitional Region (e.g. Gould & Platt in press; Pearcy 1991). However, there are evidently rich biological communities at depth in those areas, too, as suggested by the presence of northern elephant seals and other deep diving predators (e.g., sperm whales); little is known about the composition or trophic interactions of these communities. Northern elephant seals apparently spend little or no time pursuing prey in the euphotic zone (< 200 m depth). The vast bulk of their prey are probably mesopelagic squid (Antonelis, Lowry, DeMaster et al. 1987; Antonelis, Lowry, Fiscus et al. in press) that live at depths of 450 m or greater during the day and ascend to feed nearer the surface at night (Roper & Young 1975; Summers 1983). The diel dive patterns of northern elephant seals (see also Le Boeuf, Costa et al. 1988; Le Boeuf, Naito et al. 1989; DeLong & Stewart 1991; Stewart & DeLong in press) correspond strikingly with what is known about the diel migratory behaviour of these squid and of the deep scattering layer biological communities in general (e.g. Davidson & Pearcy 1972; Roper & Young 1975; Forward 1988; Brooks & Saenger 1991).

Thus the vertical dispersion of northern elephant seals appears to be intimately related to diel variations in the vertical concentration of squid and mesopelagic fish which, in turn, are responding to diel variations in the vertical dispersion of other zooplankton prey. It may also be strongly influenced by variations in squid behaviour. Common features of most of the squid that elephant seals eat when in the California Current (Antonelis, Lowry, DeMaster et al. 1987; DeLong & Stewart 1991; Antonelis, Lowry, Fiscus et al. in press) are (1) the presence of ventral photophores that are used by the squid as camouflage (countershading) or for attracting prey or mates in the dysphotic zone (e.g. Young 1977) and (2) the replacement of heavy metal ions in body fluids of the squid with lighter fat (i.e. oily squid, especially gonatid squid) or ammonium ions (ammoniacal squid) to achieve neutral buoyancy (e.g. Denton 1974; Clarke, Denton & Gilpin-Brown 1979; Clarke 1986).

The squids' bioluminescence makes them susceptible to larger predators like elephant seals, sperm whales and beaked whales that are capable of deep, long-duration diving. These neutrally buoyant squid are probably easier to catch, although they may be of poorer nutritional quality, than fast-swimming negatively buoyant squid (cf. Clarke 1986). While in the California Current elephant seals prey predominantly on ammoniacal species of squid and, to some extent, oily squid (*Gonatopsis borealis*; Antonelis, Lowry, Fiscus *et al.* in press), even though more muscular and nutritious negatively buoyant species are common in that area (Young 1972; Roper & Young 1975; Jefferts 1983, 1988a, b). They may therefore be selecting prey on the basis of energetic tradeoffs between pursuit and capture time and prey energy content and nutrition.

It is likely that elephant seals are preying mostly on squid when they are dispersed away from the California Current as do many North Pacific cetaceans that have similar functional adaptations in dentition (Gaskin 1982). Dietary studies of other deep-diving marine mammals (e.g. by Okutani & Nemoto 1964; Kajimura, Fiscus & Stroud 1980; Clarke 1986; Fiscus, Rice & Wolman 1989) indicate that the squid preferred by northern elephant seals when they are in southern California waters (see, e.g., Antonelis, Lowry, DeMaster *et al.* 1987; Antonelis, Lowry, Fiscus *et al.* in press) are also common where elephant seals concentrate their foraging effort off the coasts of central and northern California, Oregon, Washington and British Columbia, in the Gulf of Alaska and along the eastern Aleutian Islands. Although methodological constraints have limited thorough study of deep-water squid communities in the Pacific, some general information has been summarized by a few authors. Neutrally buoyant squid are common in the three water mass zones in which elephant seals are dispersed. Oily squid (primarily gonatids) dominate the northern Subarctic Water Mass (Kubodera & Jefferts 1984; Jefferts 1988a, b), where adult male northern elephant seals forage, but they are less common and smaller further south where female elephant seals forage (Roper, Sweeney & Nauen 1984; Jefferts 1988a, b). Smaller, ammoniacal enoploteuthid squid are more dominant in the latter area. The eastern reach of the North Pacific Transition Domain supports a mixture of these two squid groups, although oily squid are less common there than they are further north and the gonatid squid that occur in the Transition Domain are generally smaller than those of the Subarctic Water Mass (Jefferts 1988a, b). We think that the seasonal migrations and the geographic and vertical dispersion of foraging northern elephant seals are primarily linked to the geographic and vertical dispersion of squid that have the characteristics identified above.

The diversity of fish detected in elephant seals' diet from lavage studies is substantially less than that of cephalopods (cf. Antonelis, Lowry, Demaster *et al.* 1987; Antonelis, Lowry, Fiscus *et al.* in press). We suspect, however, that mesopelagic fish may be important prey for elephant seals in some areas of the North Pacific where mesopelagic fish and squid distributions are strongly correlated (e.g. Jefferts 1988b). Our understanding of the foraging ecology of northern elephant seals is limited by a lack of information on their diet during most of the year when seals are foraging far away from the California Islands rookeries and haulouts to which previous dietary studies have been confined. Additional studies of diet are needed in those distant foraging areas to clarify the relationships between elephant seal diving patterns and foraging ecology.

The differences in male and female elephant seal geographic dispersion may be linked to different energetic requirements of these sexually size-dimorphic seals or to differences in their abilities to find and capture prey of various sizes and nutritional content. Seals would certainly seem to benefit by travelling to the Subarctic Water Mass to capitalize on large, oily gonatid squid and this may explain the attraction of the area to adult male elephant seals. We cannot explain why females apparently do not forage in that area. Perhaps more puzzling is that males simply transit through the Subarctic Frontal Zone and the eastern North Pacific Transition Zone and Domain where females concentrate their foraging effort, despite the apparent abundance of prey there. Size, mobility, and nutritional and energetic content of prey may be important considerations that constrain elephant seal males to abandon those areas in favour of more distant foraging grounds. Regardless of the reasons for the temporal and spatial segregation of foraging elephant seal males and females, the consequent reduction in intraspecific competition for food resources has important implications for demographic analysis of the rapidly growing northern elephant seal population (cf. Stewart 1992; Stewart, Le Boeuf et al. in press).

Sexual segregation of foraging animals has been reported for other marine mammals (e.g. by Baker 1978; Gaskin 1982). The most cogent comparison for northern elephant seals is with sperm whales (*Physeter macrocephalus*) in the north Pacific Ocean. Seasonal differences in latitudinal distributions of males and females of that sexually dimorphic species are remarkably similar to those of northern elephant seal males and females (cf. Rice 1989); both species prey mostly on mesopelagic squid and data indicate great similarity in dietary composition in the California Current at least (cf. Fiscus et al. 1989). Comparative studies of patterns of geographic and vertical dispersion of these and other deep-diving, teuthophagous marine mammals could provide important insights into the factors that favoured the evolution of sexual dimorphism and geographic segregation in this predator guild.

References

Antonelis, G. A., Jr., Lowry, M. S., DeMaster, D. P. & Fiscus, C. H. (1987). Assessing northern elephant seal feeding habits by stomach lavage. *Mar. Mamm. Sci.* 3: 308–322.

Antonelis, G. A., Lowry, M. S., Fiscus, C. H., Stewart, B. S. & DeLong, R. L. (In press). Diet of the northern elephant seal. In: *Elephant seals*. (Eds Le Boeuf, B. J. & Laws, R. M.). University of California Press, Berkeley.

Baker, R. R. (1978). *The evolutionary ecology of animal migration.* Holmes & Meier Publ., N.Y.

Bengtson, J. L. & Stewart, B. S. (1992). Diving and haulout behavior of crabeater seals in the Weddell Sea, Antarctica, during March 1986. *Polar Biol.* **12**: 635–644.

Brooks, A. L. & Saenger, R. A. (1991). Vertical size-depth distribution properties of mid-water fish off Bermuda, with comparative reviews for other open ocean areas. *Can. J. Fish. aquat. Sci.* **48**: 694–721.

Clarke, M. R. (1986). Cephalopods in the diet of odontocetes. In *Research on dolphins*: 280–321. (Eds Bryden, M. M. & Harrison, R.). Clarendon Press, Oxford.

Clarke, M. R., Denton, E. J. & Gilpin-Brown, J. B. (1979). On the use of ammonium for buoyancy in squids. *J. mar. biol. Ass. U.K.* **59**: 259–276.

Davidson, H. A. & Pearcy, W. G. (1972). Sound-scattering layers in the northeastern Pacific. *J. Fish. Res. Bd Can.* **29**: 1419–1423.

DeLong, R. L. & Stewart, B. S. (1991). Diving patterns of northern elephant seal bulls. *Mar. Mamm. Sci.* **7**: 369–384.

DeLong, R. L., Stewart, B. S. & Hill, R. D. (1992). Documenting migrations of northern elephant seals using day length. *Mar. Mamm. Sci.* **8**: 155–159.

Denton, E. J. (1974). On buoyancy and the lives of modern and fossil cephalopods. *Proc. R. Soc. (B)* **185**: 273–299.

Favorite, F., Diomead, A. J. & Nasu, K. (1976). Oceanography of the Subarctic Pacific Region, 1960–1971. *Int. north Pacif. Fish. Commn Bull.* **33**: 1–187.

Fiscus, C. H., Rice, D. W. & Wolman, A. A. (1989). Cephalopods from the stomachs of sperm whales taken off California. *NOAA tech. rep. NMFS* No. 83: 1–12.

Forward, R. B. (1988). Diel vertical migration: zooplankton photobiology and behaviour. *Oceanogr. mar. Biol.* **26**: 361–393.

Gaskin, D. (1982). *The ecology of whales and dolphins.* Heinemann, London.

Gould, P. J. & Platt, J. F. (In press). Marine birds in the Central North Pacific. In *Biology, distribution and stock assessment of species caught in the high seas driftnet fisheries in the North Pacific Ocean. (Int. north Pacif. Fish. Commn Bull.)*

Jefferts, K. (1983). *Zoogeography and systematics of cephalopods of the northeastern Pacific Ocean.* PhD diss.: Oregon State University, Corvallis, OR.

Jefferts, K. (1988a). Zoogeography of cephalopods from the northeastern Pacific Ocean. *Bull. Ocean Res. Inst. Univ. Tokyo* **26**: 123–157.

Jefferts, K. (1988b). Zoogeography of northeastern Pacific cephalopods. In *Cephalopods—present and past*: 317–339. (Eds Wiedmann, J. & Kullmann, J.). Schweizerbart'sche Verlagsbuchhandlung, Stuttgart.

Kajimura, H., Fiscus, C. H. & Stroud, R. K. (1980). Food of the Pacific white-sided dolphin, *Lagenorhynchus obliquidens*, Dall's porpoise, *Phocoenoides dalli*, and northern fur seal, *Callorhinus ursinus*, off California and Washington with appendices on size and food of Dall's porpoise from Alaskan waters. *NOAA tech. Mem. NMFS F/NWC-2*: 1–30.

Kubodera, T. & Jefferts, K. (1984). Distribution and abundance of the early life stages of squid, primarily Gonatidae (Cephalopoda, Oegopsida), in the northern North Pacific. *Bull. natn. Sci. Mus., Tokyo Ser. A (Zool.)* 10: 165–193.

Le Boeuf, B. J., Costa, D. P., Huntley, A. C. & Feldkamp, S. D. (1988). Continuous, deep diving in female northern elephant seals, *Mirounga angustirostris. Can. J. Zool.* 66: 446–458.

Le Boeuf, B. J., Naito, Y., Huntley, A. C., & Asaga, T. (1989). Prolonged, continuous, deep diving by northern elephant seals. *Can. J. Zool.* 67: 2514–2519.

Lutjeharms, J. R. E., Walters, N. M. & Allanson, B. R. (1985). Ocean frontal systems and biological enhancement. In *Antarctic nutrient cycles and food webs:* 11–21. (Eds Siegfied, W. R., Condy, P. R. & Laws, R. M.). Springer-Verlag, Berlin etc.

NOAA (1990). *Oceanographic monthly summary* 10. National Ocean Service, National Weather Service, Washington, D.C.

NOAA (1991). *Oceanographic monthly summary* 11. National Ocean Service, National Weather Service, Washington, D.C.

Okutani, T. & Nemoto, T. (1964). Squids as the food of sperm whales in the Bering Sea and Gulf of Alaska. *Scient. Rep. Whales Res. Inst., Tokyo* 18: 111–121.

Pearcy, W. G. (1991). Biology of the transition zone. *NOAA tech. Rep. NMFS* 105: 39–55.

Reed, D. K. & Schumacher, J. D. (1985). On the general circulation in the subarctic Pacific. In *Proceedings of the workshop on the fate and impact of marine debris*: 483–496. (Eds Shomura, R. S. & Yoshida, H. O.). *NOAA tech. Mem. NMFS* SWFC-54.

Rice, D. W. (1989). Sperm whale, *Physeter macrocephalus* Linnaeus, 1758. In *Handbook of marine mammals* 4. *River dolphins and the larger toothed whales*: 177–233. (Eds Ridgway, S. H. & Harrison, R.). Academic Press, London etc.

Roden, G. I. (1991). Subarctic-subtropical transition zone of the North Pacific: large-scale aspects and mesoscale structure. *NOAA tech. Rep. NMFS* 105: 1–38.

Roper, C. F. E., Sweeney, M. J. & Nauen, C. E. (1984). FAO species catalogue 3. Cephalopods of the world. An annotated and illustrated catalogue of species of interest to fisheries. *FAO Fish. Synop.* No. 125 (3): 3–277.

Roper, C. F. E. & Young, R. E. (1975). Vertical distribution of pelagic cephalopods. *Smithson. Contr. Zool.* No. 209: 1–51.

Sokal, R. R. & Rohlf, F. J. (1981). *Biometry: the principles and practice of statistics in biological research.* (2nd edn). W. H. Freeman and Company, New York.

Stewart, B. S. (1989) *The ecology and population biology of the northern elephant seal,* Mirounga angustirostris *Gill 1866, on the Southern California Channel Islands.* PhD diss.: University of California, Los Angeles, CA.

Stewart, B. S. (1992). Population recovery of northern elephant seals on the Southern California Channel Islands. In *Wildlife 2001: populations*: 1075–1086. (Eds McCullough, D. R. & Barrett, R. H.). Elsevier Applied Science Press, New York.

Stewart, B. S. & DeLong, R. L. (1990). Sexual differences in migrations and foraging behavior of northern elephant seals. *Am. Zool.* 30: 44A.

Stewart, B. S. & DeLong, R. L. (In press). Post-breeding foraging migrations of northern elephant seals. In *Elephant seals*. (Eds Le Boeuf, B. J. & Laws, R. M.). University of California Press, Berkeley.

Stewart, B. S., Le Boeuf, B. J., Yochem, P. K., Huber, H. R., DeLong, R. L., Jameson, R., Sydeman, W. & Allen, S. G. (In press). History and present status of the northern elephant seal population. In *Elephant seals*. (Eds Le Boeuf, B. J. & Laws, R. M.). University of California Press, Berkeley.

Summers, W. C. (1983). Physiological and trophic ecology of cephalopods. In *The Mollusca* 6: *Ecology*: 261–279. (Ed. Russell-Hunter, W. D.). Academic Press, London.

Wilkinson, L. (1988). *SYSTAT: the system for statistics*. SYSTAT, Inc., Evanston, IL.

Young, R. E. (1972). The systematics and areal distribution of pelagic cephalopods from the seas off southern California. *Smithson. Contr. Zool.* No. 97: 1–159.

Young, R. E. (1977). Ventral bioluminescent countershading in midwater cephalopods. *Symp. zool. Soc. Lond.* No. 38: 161–190.

Zar, J. (1974). *Biostatistical analysis*. Prentice Hall, Inc., Englewood Cliffs, N.J.

Symp. zool. Soc. Lond. (1993) No. 66: 195–210

Studying the behaviour and movements of high Arctic belugas with satellite telemetry

A. R. MARTIN[1], T. G. SMITH[2] and O. P. COX[1]

[1]Sea Mammal Research Unit
Natural Environment Research Council
c/o British Antarctic Survey
High Cross, Madingley Road
Cambridge CB3 0ET, UK

[2]Department of Fisheries and Oceans
Pacific Biological Station
3190 Hammond Bay Road
Nanaimo, B.C. V9R 5K6, Canada

Synopsis

Eighteen belugas (*Delphinapterus leucas*) of the Canadian High Arctic/Baffin Bay stock were captured around Somerset Island, N.W.T., Canada, in July of the years 1988–92 and released with satellite-monitored radio transmitters. Signals from the radio packs permitted calculation of the animals' location up to 26 times per day and also carried detailed information on diving and surfacing behaviour. Whales were followed for up to 75 days and for distances of up to 3250 km, throughout the time spent in the Arctic archipelago and the subsequent eastward migration at the end of the summer.

The general pattern of behaviour of this stock is a westwards movement from Lancaster Sound and Prince Regent Inlet through Barrow Strait in late July, occupation of Peel Sound and Franklin Strait in August, followed by a usually rapid transit to Baffin Bay via southern Devon Island in early/mid September. Although much time is spent in coastal shallows, especially freshwater outlets, all the belugas followed in this study spent considerable time offshore, and often in the deepest water available. Here, they carried out prolonged bouts of intensive diving, invariably to the seabed. Dives of up to 18.3 min (average 12.9 min) and to depths of 440 m were recorded. During this activity, typically 73.5% of the time is spent in diving and 26.5% at or near the surface in ventilation sequences. Some 33–44% of the duration of the diving sequence is actually spent on the sea bed. Deep dives are usually of a square profile, with a constant rate of descent to the sea bed followed by several minutes at constant depth. Rates of ascent sometimes diminished as the surface was approached. On migration, sustained swim speeds of up to 5.8 km/h

ZOOLOGICAL SYMPOSIUM No. 66
ISBN 0–19–854069–8

Copyright © 1993 The Zoological Society of London
All rights of reproduction in any form reserved

were recorded. Difficulties encountered with application of satellite-linked radio telemetry to the study of cetaceans, and their possible solutions, are discussed.

Introduction

Progress in the study of cetaceans has always been hampered by the sheer inaccessibility of these animals to terrestrially-based researchers. The majority of whales, dolphins and porpoises spend their lives far out at sea and most of their time below the water surface, both characteristics serving to limit, and bias, our knowledge of this important component of the marine fauna.

In the 1970s cetologists began to use miniature VHF radio transmitters, attached to their study animals, to help follow them from land, ships or aircraft when they were out of visual range (e.g. Mate & Harvey 1984; Würsig & Bastida 1986). However, despite advances in technology, the success of this method has been limited by the requirement to maintain at least one, and preferably many, receivers within range of the study animal. Normally only one animal can be followed at a time, and costs become prohibitive before a meaningful sample of individual animals can be studied.

In the early 1980s a new type of radio-tracking technique became available to wildlife biologists. Instead of being picked up by earth-based receivers, radio signals were detected by polar-orbit communications satellites and subsequently relayed to the scientist via a processing centre. The advantages of this technique, including a global positioning ability and a facility to send sensor data, prompted widespread usage, especially on terrestrial mammals (Fancy et al. 1988) and more recently on seals (e.g. McConnell et al. 1992; Stewart et al. 1989). Despite early technical difficulties with this research tool, which has been termed 'satellite telemetry', by the end of the decade it had been used with varying degrees of success on at least five cetacean species varying in size from the bottlenose dolphin Tursiops truncatus (Tanaka 1987), to the northern right whale Eubalaena glacialis (Mate et al. 1991).

This paper presents results from the first five years of a study of white whales or belugas Delphinapterus leucas using satellite telemetry. It was carried out on animals of the Canadian high Arctic or Baffin Bay stock, which has been estimated to comprise between 6300 and 18 600 belugas (Smith, Hammill et al. 1985), and is subject to an average annual harvest of about 40–60 animals by Canadian Inuit settlements (D.F.O. 1992). The stock occupies waters around Somerset Island in the Canadian Arctic archipelago in summer, principally Peel Sound, Barrow Strait and Prince Regent Inlet, and was assumed to move eastward through Lancaster Sound into Baffin Bay as new ice re-covers the area in early autumn (Figs 1 and 2).

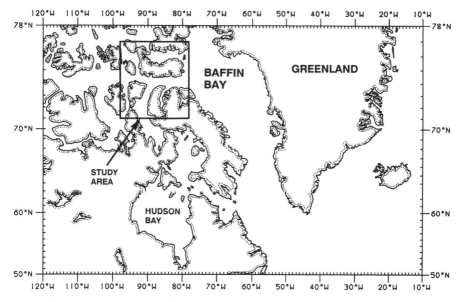

Fig. 1. Map showing location of the study area.

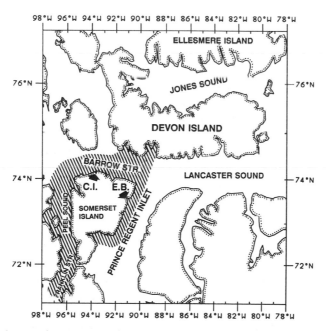

Fig. 2. Map showing the principal summering grounds of the Baffin Bay beluga stock (hatched area) as identified by the tracks of animals fitted with transmitter packs in the years 1988–1992 and concurrent aerial reconnaissance.

Across Baffin Bay, along the coast of north-west Greenland (Fig. 1), belugas occur between late September and May, and have been subject to high catches by local Inuit hunters for many years. The relatively frequent occurrence of ice entrapments (Savssats) around Disko Island often results in annual catches of over 1000 animals (Kapel 1977). If these belugas are of the same stock as those occurring in the Canadian high Arctic in summer, as seems likely, this level of hunting is above even the most optimistic yield sustainable by a population of this estimated size.

The aims of this study were twofold. Firstly, to provide information that would help lead to sensible management of the stock, not least by facilitating a more accurate assessment of stock size by aerial survey. Such surveys could be more effectively designed and their results more correctly interpreted with data on the animals' short-term movements, migrations, habitat usage and geographical distribution. Secondly, to increase our understanding of the physiology and natural history of the species by examining in detail its diving, swimming and foraging behaviour.

Methods

Eighteen sub-adult or adult belugas were captured with an inflatable boat and hoop nets, and then restrained in shallow water, in the manner described by Martin & Smith (1992), either in Cunningham Inlet or Elwin Bay, Somerset Island (Table 1). Each was fitted with a transmitter pack held in place with a saddle of 3mm-thick pliable PVC-impregnated webbing. This was laid snugly over the whale's dorsal ridge and secured with 2–4 flexible pins of nylon, passing through the fibrous ridge tissue. Plastic implants react less with beluga tissue than metallic ones (Geraci & Smith 1990) and their flexibility is probably important in allowing the dorsal tissue to flex without hindrance. A long-acting antibiotic (Amoxycillin traded as Betamox L.A. Norbrook Laboratories, London, U.K.) was administered intramuscularly to reduce risks of infection. The entire package eventually falls off the animal when the pins migrate through the ridge tissue.

A slightly different and smaller transmitter pack was used in each successive year of the study, but all essentially comprised a transmission unit (Toyocom Ltd or Mariner Radar Ltd), a programmable controller/ datalogger and a power supply, housed in one or two tubes of anodized aluminium, with an external antenna and one or more externally-mounted sensors. Power was supplied by two or three lithium thionyl chloride 'D' cells. The packs, designed and built at the Sea Mammal Research Unit, were pressure-proofed to a depth in excess of 1000 m in seawater. All were fitted with optical emergence sensors to detect surfacing and most had a pressure sensor to determine depth (see Table 1). Tests of pressure sensors in a

Table 1. Details of transmitter pack deployment on belugas at two sites on Somerset Island, N.W.T., Canada. Whales were captured and released at the freshwater outlets of Cunningham Inlet (C.I.) (74°05′N,93°49′W) and Elwin Bay (E.B.) (73°33′N,90°57′W) between 1988 and 1992.

Release					Details of animal			Tag longevity (days)
Date	Local time	Place	Tag no.	Transmitter type	Length (m)	Sex	Age	
22 July 1988	1358	C.I.	8753	Location	3.35	F	Adult	10.6
27 July 1989	1733	C.I.	5802	Location	3.20	F	Adult	7.0
30 July	1837	C.I.	5800	Location	2.50	F	Subadult	0.01
17 July 1990	1207	C.I.	5801	Location and depth	2.90	F	Subadult	9.0
17 July	1722	C.I.	8751	Location	3.05	F	Adult	37.6
17 July	1722	C.I.	8750	Location	3.24	F	Subadult	29.6
18 July	1536	C.I.	5803	Location and depth	4.04	F	Adult	20.5
23 July	1300	C.I.	5806	Location and depth	4.16	F	Adult	44.9
21 July 1991	1224	C.I.	5801	Location and depth	3.81	F	Adult	27.6
27 July	2145	C.I.	5804	Location and depth	2.64	F	Adult	0.05
28 July	2125	C.I.	5805	Location and depth	3.40	F	Subadult	24.6
31 July	1600	C.I.	8750	Location and depth	3.88	F	Adult	47.2
1 August	1250	C.I.	8757	Location and depth	3.71	F	Adult	2.5
26 July 1992	1106	E.B.	8752	Location and depth	3.88	F	Adult	75.0
23 July	1500	E.B.	8753	Location and depth	4.64	M	Adult	72.7
23 July	1410	E.B.	8754	Location and depth	4.18	F	Adult	69.8
25 July	2040	E.B.	8755	Location and depth	4.31	F	Adult	64.7
21 July	1330	E.B.	8756	Location and depth	4.50	M	Adult	0.1

calibration tank indicated that depth could be determined to 500 m with an accuracy of 2.5 m. The typical weight of a pack in seawater was 438 g (1620 g in air).

Full details of the specification and usage of Service ARGOS, the satellite-based data-reception system used in this study, are provided elsewhere (CLS/Service ARGOS 1989; Fancy et al. 1988), so only a brief summary will be given here. The two polar-orbiting satellites used by the ARGOS system provided up to 28 short (< 15 min) data-reception 'passes' in each 24-h period at the latitudes of this study. If the number and quality of signals received in a pass satisfied certain criteria, ARGOS calculated the geographical location of the animal concerned. The error of such calculations, i.e. the distance between an animal's real and estimated location, was variable. Service ARGOS assigns a 'location quality' index to each, on a scale of 0–3 where 3 is the best. Users of the system for animal-tracking purposes rarely achieve the accuracy 'guaranteed' by Service ARGOS. Keating, Brewster & Key (1991) calculated 68 percentile errors of 1188 m, 903 m and 361 m for qualities 1, 2 and 3 respectively in a study of terrestrial mammals, and these figures are likely to be more appropriate for the current study. We 'filtered' all ARGOS-derived locations by removing those that would have involved impossibly fast swimming speeds from reliably known positions.

Signals of 401.65 MHz were transmitted immediately an animal broke the surface, and only when it was at the surface, subject to an ARGOS-imposed minimum interval of 40 s between signals. Each transmission carried an identifying number unique to the individual pack and up to 256 data bits derived from the onboard sensors. Because signals were only receivable for a small proportion of the time, all sensor information was deliberately transmitted many times in the hope that it would eventually be received successfully during a satellite pass. The transmitter controllers were programmed to take data from the various sensors at frequent intervals throughout the day, condense and then store them ready for the next transmission opportunity.

True water depths were determined by reference to the most recent charts of the Canadian Hydrographic Service.

Results

Effects of transmitter attachment on animal behaviour

Three of the study animals remained near to the site of capture and were observed closely for at least 24 h after initial attachment of their transmitters. A further five were re-sighted between five days and 21 days after attachment. All of these whales appeared to be behaving normally and we could not detect any behaviour indicating irritation with the device, even in the case of an adult female (No. 8751 in 1990) where the package was beginning to work loose. Social bonding seemed to be unimpaired; two animals (one adult, one sub-adult) which were captured together and equipped with transmitters in 1990 remained together for the lifetime of the units (30 days), during which time they travelled over 1000 km.

Further confidence that the devices do not affect behaviour has been drawn from the evidence of deep diving (see below) which demonstrates routinely long and deep submergences by animals with transmitter packages. We conclude that capture, and attachment and subsequent carrying of these devices, does not materially influence behaviour in these whales.

Movements and distribution

The paths of the 15 belugas tracked successfully and the concurrent observations made each year from the air demonstrated a general fairly slow movement from Prince Regent Inlet westwards through Barrow Strait and as far south into Peel Sound as open water permitted (Fig. 2). In 1991, when Peel Sound became almost ice-free, all the tagged animals and (as confirmed visually from the air) probably the bulk of the entire stock of whales penetrated Franklin Strait and remained within a relatively small area throughout August (Fig. 2). During this period waters off the northern coast of Somerset Island, including Cunningham Inlet, Barrow Strait and northern Peel Sound, were almost devoid of whales. The preferred focus of activity thus seemed to be Peel Sound and, if it could be reached, Franklin Strait.

Five transmitters functioned long enough to demonstrate the end-of-season migration in advance of re-forming sea ice. This proved to be a very rapid movement in late August/early September northward out of Peel Sound and then eastward through Barrow Strait and Lancaster Sound, most often along the southern shoreline of Devon Island (Fig. 3). The journey was usually broken by stops of 0.5–10 days at inlets along the Devon Island coast, where high-quality locations (class 2 or 3) were obtained. After reaching Baffin Bay, the animals moved north to Jones Sound before heading eastward towards Greenlandic waters in late September (Fig. 3).

An average of 9.3 and a maximum of 26 locations per whale were obtained each day. In most cases these provided a reasonably detailed description of the movements of each whale. Belugas spent on average 68% of their time ($n = 3914$ h of track from nine animals) within 2 km of the coast. Variation in habitat preference was marked, with individual whales spending between 18.6% and 85.7% of their time near the coast ($n = 9$), and much of this in one or more freshwater outlets such as bays or estuaries. Whales sometimes paused at one of these features for a few hours or could remain there for up to five days. Many whales spent time at several of the

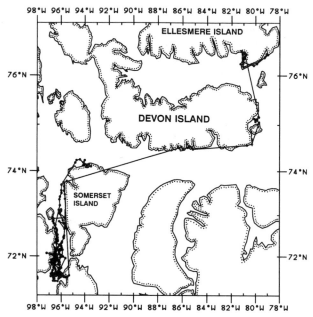

Fig. 3. The track of adult female No. 8750 between 31 July and 16 September 1991 showing occupation of Franklin Strait throughout much of August and rapid transit to Ellesmere Island in September. The transmitter pack exhausted its power supply after 47 days of operation as the animal was heading towards Greenland across Baffin Bay.

approximately ten such sites which were available at Somerset Island and south-east Prince of Wales Island. Directed travel between freshwater outlets, and more generally, was usually within a few kilometres and often within 200 m of the coast. However, contrary to expectations based on anecdotal evidence and earlier aerial surveys (Smith, Hammill *et al.* 1985), these results demonstrate that most animals spent considerable time in deep water. The reason for the earlier observational bias is clear from the dive records (see below); in deeper water a greater proportion of time is spent out of visual range below the surface.

Examination of dive records for animals Nos. 5801, 5805 and 8750 in 1991, for which we know the maximum depth achieved in consecutive 22-h periods, showed that dives to 100 m or more were made in 78 out of 85 such periods (91.8%). Dives to 300 m or more were made in 71 periods (83.5%) and no more than two consecutive periods were spent in shallow water (< 50 m) by any of these three whales. For most of this time, the belugas were in Franklin Strait where water depth increases rapidly away from land, and deep dives can be made within a few kilometres of the Prince of Wales Island coast.

Although locations of animals derived from satellite passes did not reveal every detail of an animal's movements, conditions sometimes permitted an accurate determination of a beluga's track. Since the time of each location was known, velocities could be accurately calculated. For this analysis, potential errors due to inaccurate reporting of the whale's position were reduced by choosing only locations of qualities 2 or 3 and intervals of at least 5 h between such locations. Over these periods, straight-line speeds of 2–4 km/h were commonly recorded. The fastest sustained swimming speed was accomplished on 7 and 8 September 1991 by adult female No. 8750, which covered 145.5 km in 25.3 h (an average speed of 5.8 km/h), during the easterly migration through Lancaster Sound.

Diving behaviour

We examined depth-with-time data from 10 individual belugas in one of four seasons. A typical record for a 24-h period is shown in Fig. 4. Sample sizes varied between the different analyses because of the varying levels of recording detail obtained in different years (a single transmitter could be programmed to provide high-resolution data for a short period of time or a continuous picture in less detail).

The diving behaviour of belugas monitored in this study could be divided into 'shallow water' or 'deep water' categories. The 'shallow' category includes all the time spent in coastal habitats, where submergence times were usually short. Horizontal speeds were either extremely low, when the whales were in bays or estuaries, or quite rapid as whales moved between such sites.

The 'deep' category covered the time that the belugas were outside coastal waters, usually for periods of more than a day, and is characterized by

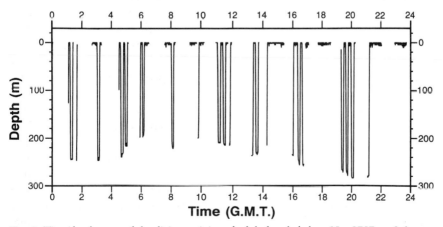

Fig. 4. Time/depth trace of the diving activity of adult female beluga No. 8757 on 3 August 1991. Time shown is G.M.T., 5 h ahead of local time. Gaps in the trace are when data were not received because satellites were not within range.

sequences of prolonged dives which were usually to the seabed. Often these dives were carried out in the deepest water available, and had a very characteristic profile.

Submergences in shallow water were invariably of short duration (< 120 s). Concurrent visual observation of whales which stayed in Cunningham Inlet for many hours or days after release with a transmitter pack showed they had a behavioural cycle linked to the tide. At high water, the whales swam closer to the freshwater outlet and even penetrated up into the river itself, often remaining in water of < 2 m depth for many hours. As the tide receded, the belugas retreated back into the Inlet proper and milled around in small social groups in water depths of 5–25 m. Although the whales mostly remained near the surface during these periods, dives were made to the bottom on occasion. This behaviour was also seen at Elwin Bay in 1992 and probably occurs at all the freshwater-outlet aggregation sites.

Subsurface activity in deep water could be divided into three types:

1. 'V' shaped or spike dives, where the animal swam down at a fairly constant rate to a certain depth (usually in mid-water) and immediately returned to the surface.

2. Prolonged shallow dives, in which the whale remained submerged for at least five minutes at relatively shallow depths.

3. Prolonged and deep square profile dives, where the animal dived at a uniform rate to the seabed and remained at an almost constant depth for many minutes before returning to the surface (Fig. 5).

Spike dives comprised fewer than 5% of dives to 50 m or more. By their nature they are of fairly short duration, and none appeared to be made to depths of > 150 m.

Prolonged shallow dives, defined as those of $\geq$ 300 s duration and to a depth of $\leq$ 50 m, comprised about 9% of our sample of submergences to $\geq$ 15 m ($n = 629$). Most were made in shallow waters, but some occurred offshore in water of $\geq$ 100 m depth. About 28% of all prolonged dives were of this type.

In a sample of 629 submergences to $\geq$ 15 m by three animals in 1991, 141 (22.4%) were to depths of at least 200 m and all of these were of a characteristic 'square' profile, involving time spent on the seabed. Whales normally surfaced near the point of submergence after this type of dive, so little horizontal movement was involved. Many consecutive dives to a similar depth ($\pm$ 5 m) were usually recorded when whales were diving in this manner. Dive duration was in the range 522–1098 s (8.7–18.3 min) and averaged 773 s (12.9 min). The regression of maximum depth of dive on dive duration was significant and positive ($r^2 = 8.9\%$, $P < 0.001$). There was no significant relationship between dive duration and the duration of the preceding surface interval (P.S.I.) ($P > 0.4$). Information on the length of the succeeding surface interval was not available. The median P.S.I. was 278 s (4.6 min) and the mean was 392 s (6.5 min). Thus, in a typical bout of deep diving, about 73.5% of the time was spent below the surface.

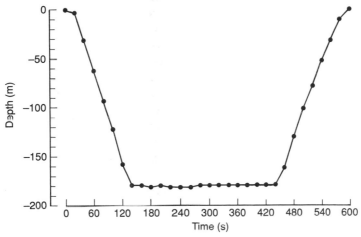

Fig. 5. Time/depth trace of a prolonged 'square' profile dive to the seabed. This dive had an estimated bottom time of 5.3 min, a mean descent rate of 1.55 m/s and an ascent rate decreasing from 1.55 to 1.15 m/s. It was made by a 2.9 m sub-adult female in 180 m of water at position 74°33'N, 92°42'W.

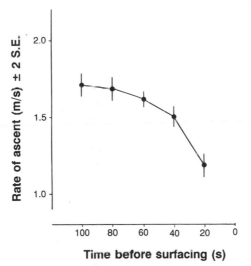

Fig. 6. Average rate of ascent from deep dives as the surface is approached. Values indicate the rate over consecutive periods of 20 s. The right-most point is that of the last 20 s interval completed before the whale surfaced; this ended randomly between 0.1 and 19.9 s before surfacing. The value on the x-axis represents the average midpoint of each consecutive interval.

The average rate of descent in deep dives was in the range 1.4–2.2 m/s. Average ascent rates were in the range 1.2–1.8 m/s, but on many dives the rate diminished significantly as the surface was approached (Fig. 6). In 39 ascents from deep dives by female No. 5801 in 1990, the rate of vertical movement (expressed as an average over consecutive periods of 20 s) fell from 1.7 m/s at about 140 m depth to 1.2 m/s in the last complete 20 s interval before surfacing. The maximum rates of descent and ascent over periods of 20 s were 2.55 and 2.35 m/s, respectively.

The estimated period of time available for foraging on the sea bed during each dive was in the range 180–530 s (3–8.8 min), representing 45–60% of the total time submerged. Overall we estimate that about 33–44% of the time is actually spent searching for food on the seabed during bouts of foraging in deep water.

Discussion

The results of this study serve to complement earlier research, for instance by establishing the dynamics of population movement between areas of congregation identified by aerial surveys, and to discredit some existing ideas: for example, that belugas are essentially shallow-water animals. Though satellite telemetry is a powerful tool in its own right, there can be no doubt that direct observation, both on the ground and from the air, greatly improves the interpretation and understanding of information obtained via satellites. This was demonstrated on numerous occasions when the location of our few tagged animals revealed the whereabouts of thousands of whales, which could then be counted and observed from a helicopter or fixed-wing aircraft. The discovery that such a small sample of belugas was sufficient to reveal the broad behaviour of an entire stock has important implications for study design in social species such as this.

Using independent data on variations in the pattern of ice break-up between years, in addition to those from satellite telemetry and aerial surveys, we have gained a reasonably complete understanding of the diving ability and behaviour, movements, distribution, abundance and habitat preference of belugas of this stock in the summer. The picture which emerges is as follows. Belugas of the Canadian high Arctic/Baffin Bay stock follow the receding edge of the fast ice in Lancaster Sound westwards towards Barrow Strait in May and June. As shore leads develop southward into Prince Regent Inlet and along the north coast of Somerset Island, belugas push through them and head into their main areas of summer aggregation, outlets of rivers originating from the interior of Somerset and Prince of Wales Islands. As ice break-up continues, the stock moves in a westward direction, usually having first access to Cunningham Inlet and central Barrow Strait in the middle two weeks of July. Numbers build up in

Cunningham Inlet until the end of July or first week of August, then diminish as most of the animals have by then passed into Peel Sound and are moving south as fast as the ice break-up allows. Large aggregations of belugas occur in freshwater outlets and shallow waters around Somerset Island throughout this period, with up to 2500 in Cunningham Inlet and 4000 in Creswell Bay. Nevertheless, most whales spend time in deep (> 200 m) water on most days. On average, about one third of the animals are offshore at any time, often under heavy ice-cover, and probably feeding on the sea bed.

The ice in Peel Sound continues to break up slowly during the middle part of August. In some years the over-winter ice remains dense south of about 73° 30′N (Wadsworth Island). In others, such as 1991, the whole of Peel Sound becomes almost free of floe ice, allowing the belugas (with many narwhals and harp seals) to penetrate as far south as they wish. In 1991 and 1992 the preferred area of occupation was Franklin Strait, between latitudes of 71° and 72°, but in some years belugas move as far south as the settlement at Spence Bay (69°30′N) where they are hunted.

By the end of August, belugas begin to move out of the summer grounds and by mid September almost the entire stock has moved out of Peel Sound and eastward through Barrow Strait and Lancaster Sound into Baffin Bay. This movement is accomplished rapidly. Having entered Baffin Bay, most of the whales move the short distance into Greenlandic waters, most doing so via southern Jones Sound and Ellesmere Island (Fig. 3), where a hunt is conducted during both the spring and autumn migrations from the village of Grise Fjord.

One remaining area of doubt is whether Creswell Bay on the south-eastern corner of Somerset Island is an early port of call for the bulk of the stock or the final destination for a relatively small proportion. Some 4000 whales occur there in July and much of August. The question could be answered with the deployment of transmitter packs in this area early in the season.

The reasons for whales of this stock travelling long distances to spend such a short period in the heart of the Canadian Arctic archipelago appear to be twofold. Firstly, access to relatively warm freshwater outlets may be beneficial during the period of skin moult (St Aubin, Smith & Geraci 1990; Smith, St Aubin & Hammill 1992). Secondly, there appears to be a food resource on or near the seabed in deep water rich enough to reward energetically costly long dives, and perhaps at least partly justifying the summer migration itself.

The discovery that belugas commonly dive to, and presumably forage on, the seabed in depths as great as 440 m is probably the most surprising outcome of this study. Their unexpected ability to utilize prey at such depths radically alters our understanding of this species' role in the Arctic marine ecosystem. It means that they have access to benthic and pelagic fauna throughout most of the seasonally ice-free parts of the Arctic basin, rather than just the narrow coastal strips previously thought to be within their reach. Exactly what they eat at such depths is unknown. Arctic cod (*Boreogadus saida*) is a common prey of belugas and other marine mammals in coastal waters in the summering area, but there is little evidence of this fish occurring in large quantities on or near the sea bed in deep water (Crawford & Jorgenson 1990). Another possibility is Greenland halibut (*Reinhardtius hyperglossoides*) which occurs on or near the sea bed at considerable depth in this type of habitat (Bigelow & Welsh 1925; Scott 1982; Hudon 1990) and would provide a suitable energetic reward for long dives. The diet might also include invertebrates such as shrimp, crabs and polychaetes which are known to be taken by belugas elsewhere (Tomilin 1957; Brodie 1989). These are probably abundant, and could perhaps be efficiently utilized by belugas because the species has an extraordinarily versatile mouth, which permits it to suck and blow water with great force and directionality.

The constraint on the amount of data that can be sent and received within the ARGOS system proved to be a considerable limitation in this project and will inevitably be so in any study of marine mammals using ARGOS. A maximum of 256 bits of data can be sent in a transmission, and the probability of any such transmission coinciding with a satellite pass, and therefore potentially being received by the satellite (termed an 'uplink'), is small. Even then, many uplinks are corrupted, often by the animal submerging during the propagation of the signal itself. In this experiment only about one third of all uplinks were received free of errors. Add to this the unpredictable surfacing behaviour of the animal and the fact that even when it is at the surface, the transmitter may or may not be exposed long enough for a signal to be triggered, it becomes apparent that successful and efficient use of the technique is not as simple as one might intuitively expect it to be. As transmitters have increased in sophistication, and more sensors have been added, so the data acquisition 'bottleneck' described above has become more restrictive.

A similar, and in some ways related, problem is that of how to make best use of a finite energy store in the transmitter. Because of the need to keep the transmitter package to a minimum size, the volume allocated for the batteries will always be limited. In practice, this means that decisions must be made about how to allocate a finite and inadequate amount of power and, essentially, whether to opt for a relatively large amount of data per day over a short period of time, or a lesser daily amount throughout a longer lifetime of the transmitter. The results presented here mostly reflect our early choice to build up knowledge of the species' behaviour, transmitter performance and longevity, and data-gathering potential during almost continuous functioning of the package with maximum short-term data output. As package retention times improved and performance of the system on belugas became better understood, some of the later packs were designed to extend battery life by switching off for a proportion of each 24-h period. Our eventual goal is to be able to follow and monitor our study animals throughout a full migratory cycle of 12 months, as already achieved in studies of terrestrial mammals and polar bears (*Ursus maritimus*) (Fancy *et al.* 1988). This objective is critically dependent on improving existing techniques for attaching small electronic packages to wet, slippery animals without a collar-holding neck, and must be the next major focus of attention if radio telemetry is to realize its full potential in cetacean research.

Acknowledgements

This project has relied on the generous and expert logistical support of the Polar Continental Shelf Project (PCSP) of Energy, Mines and Resources Canada. We thank Charlie Chambers, Mike Fedak, Colin Hunter, Bernie McConnell, Kevin Nicholas and Jeremy Tomlinson for technical expertise and Gilly Banks, Kathy Frost, Haakon Hop, George Horonowitsch, Amanda Lisle and Gary Sleno for field or logistical help. We are indebted to the Department of Fisheries and Oceans Canada, the National Geographical Society, the Natural Environment Research Council UK and the Worldwide Fund for Nature (UK) for financial support.

References

Bigelow, H. B. & Welsh, W. W. (1925). Fishes of the Gulf of Maine. *Bull. Bur. Fish., Wash.* 40(1): 1–567.

Brodie, P. F. (1989). The white whale – *Delphinapterus leucas* (Pallas, 1776). In *Handbook of marine mammals* 4. *River dolphins and the larger toothed whales*: 119–144. (Eds Ridgway, S. H. & Harrison, R.). Academic Press Limited, London, San Diego, etc.

CLS/Service ARGOS (1989). *Guide to the ARGOS System*. CLS/Service ARGOS, Toulouse, France.

Crawford, R. & Jorgenson, J. (1990). Density distribution of fish in the presence of whales at the Admiralty Inlet landfast ice edge. *Arctic* 43: 215–222.

D.F.O. (1992). *Annual summary of fish and marine mammal harvest data for the Northwest Territories, Vol. 3, 1990–91*. Department of Fisheries and Oceans, Ottawa, Canada.

Fancy, S. G., Pank, L. F., Douglas, D. C., Curby, C. H., Garner, G. W., Amstrup, S. C. & Regelin, W. L. (1988). Satellite telemetry: a new tool for wildlife research and management. *U.S. Fish Wild. Serv. Resour. Publ.* No. 172: 1–54.

Geraci, J. R. & Smith, G. J. D. (1990). Cutaneous response to implants, tags and marks in beluga whales, *Delphinapterus leucas,* and bottlenose dolphins, *Tursiops truncatus. Can. Bull. Fish. aquat. Sci.* 224: 81–95.

Hudon, C. (1990). Distribution of shrimp and fish by-catch assemblages in the Canadian eastern Arctic in relation to water circulation. *Can. J. Fish. aquat. Sci.* **47**: 1710–1723.

Kapel, F. O. (1977). Catch of belugas, narwhals and harbour porpoises in Greenland, 1954–75, by year, month and region. *Rep. int. Whal. Commn* **27**: 507–519.

Keating, K. A., Brewster, W. G. & Key, C. H. (1991). Satellite telemetry: performance of animal-tracking systems. *J. Wildl. Mgmt* **55**: 160–171.

Martin, A. R. & Smith, T. G. (1992). Deep diving in wild, free-ranging beluga whales, *Delphinapterus leucas*. *Can. J. Fish. aquat. Sci.* **49**: 462–466.

Mate, B. R. & Harvey, J. T. (1984). Ocean movements of radio-tagged gray whales. In *The gray whale* Eschrichtius robustus: 577–589. (Eds Jones, M. L., Swartz, S. L. & Leatherwood, S.). Academic Press, Orlando, San Diego, etc.

Mate, B. R., Niekirk, S. L., Kraus, S. D., Mesecar, R. S. & Martin, T. J. (1991). Satellite-monitored movements and dive patterns of radio-tagged North Atlantic right whales *Eubalaena glacialis*. *Abstr. bienn. Conf. Biol. mar. Mammals* **9**. Society for Marine Mammalogy. Unpublished.

McConnell, B. J., Chambers, C., Nicholas, K. S. & Fedak, M. A. (1992). Satellite tracking of grey seals (*Halichoerus grypus*). *J. Zool., Lond.* **226**: 271–282.

Scott, J. S. (1982). Depth, temperature and salinity preferences of common fishes of the Scotian Shelf. *J. northwest Atl. Fish. Sci.* **3**: 29–39.

Smith, T. G., Hammill, M. O., Burrage, D. J. & Sleno, G. A. (1985). Distribution and abundance of belugas, *Delphinapterus leucas*, and narwhals, *Monodon monoceros*, in the Canadian high Arctic. *Can. J. Fish. aquat. Sci.* **42**: 676–684.

Smith, T. G., St Aubin, D. J. & Hammill, M. O. (1992). Rubbing behaviour of belugas, *Delphinapterus leucas*, in a high arctic estuary. *Can J. Zool.* **70**: 2405–2409.

St Aubin, D. J., Smith, T. G. & Geraci, J. R. (1990). Seasonal epidermal molt in beluga whales, *Delphinapterus leucas*. *Can J. Zool.* **68**: 359–367.

Stewart, B. S., Leatherwood, S., Yochem, P. K. & Heide-Jorgensen, M.-P. (1989). Harbor seal tracking and telemetry by satellite. *Mar. Mamm. Sci.* **5**: 361–375.

Tanaka, S. (1987). Satellite radio tracking of bottlenose dolphins *Tursiops truncatus*. *Bull. Jap. Soc. scient. Fish.* **53**: 1327–1338.

Tomilin, A. G. (1957). *Mammals of the USSR and adjacent countries: Cetacea*. Nauk USSR, Moscow. (English Translation, Israel Program for Scientific Translations, Jerusalem).

Würsig, B. & Bastida, R. (1986). Long-range movement and individual associations of two dusky dolphins (*Lagenorhynchus obscurus*) off Argentina. *J. Mammal.* **67**: 773–774.

Symp. zool. Soc. Lond. (1993) No. 66: 211–224

Grey seals off the east coast of Britain: distribution and movements at sea

P. S. HAMMOND,
B. J. McCONNELL and
M. A. FEDAK

Sea Mammal Research Unit
Natural Environment Research
Council
High Cross, Madingley Road
Cambridge CB3 0ET, UK

Synopsis

As part of a programme of research to investigate the interactions between marine mammals and their environment, the SMRU has been studying the distribution and movements of grey seals off the east coast of Britain. Three approaches have been employed: monitoring from Automatic Recording Stations using VHF telemetry; continuous tracking from a yacht using a combination of VHF and underwater ultrasonic telemetry; and satellite-link tracking through Service Argos.

The three different methodological approaches allow investigations from different perspectives. VHF monitoring gives data for up to a year on the presence/absence of animals at sites close to the coast; the use of a number of sites provides information on seal movements between them. Satellite-link tracking gives detailed data for up to several months on the location of seals, wherever they go. Continuous ultrasonic/VHF tracking gives detailed records of movements and behaviour of seals for short periods of up to about ten days. The information on diving behaviour provided by real-time and satellite tracking is presented elsewhere; here we concentrate on the horizontal distribution and movements.

From 1985 to 1991, we successfully monitored over 30 seals via VHF, tracked seven seals real-time, and tracked 12 seals via satellite off the east coast of Britain. Seals were captured at one of three sites: the Humber Estuary, the Farne Islands or the Isle of May, mostly post-moult (March), during the summer (July/August) or post-pupping/breeding (November).

Three important generalizations emerge from the data. Firstly, there is a great deal of variability amongst individual grey seals in their patterns of distribution and movement. This presumably reflects similar variability in foraging habits and perhaps choice of prey. Secondly, long-range movements between haul-out areas play an important role in grey seal behaviour. Why this should be is not clear, but it does confirm that grey seals off the east coast of Britain must be considered as a single population. Thirdly, there is strong evidence of individual specialization in the patterns of movements and behaviour. Older animals tend to be more specialized than younger ones.

ZOOLOGICAL SYMPOSIUM No. 66
ISBN 0–19–854069–8
Copyright © 1993 The Zoological Society of London
All rights of reproduction in any form reserved

Introduction

As part of a programme of research to investigate grey seals at sea around the coast of Britain, the Sea Mammal Research Unit has been studying the distribution, movements and behaviour of free-swimming seals off the coast of eastern England and south-eastern Scotland.

Approximately 8000 of the estimated 85 000 grey seals around Britain breed annually in November/December off the east coast (Hiby, Duck & Thompson 1992). The main breeding sites are at the Isle of May, the Farne Islands and Donna Nook, south of the Humber Estuary, and there is an additional haul-out site at the mouth of the River Tay in St Andrew's Bay (Fig. 1). Further north, grey seals breed on the mainland east coast of Scotland at Helmsdale and also haul out in the Dornoch Firth. The annual moult, when animals spend a high proportion of time ashore, occurs in March.

During the remainder of the year, grey seals spend much of their time at sea. After pupping and breeding, adult animals need to replace the weight they have lost during the preceding weeks. After the moult, animals of breeding age need to build up reserves of energy in the form of blubber for the coming breeding season. Immature animals also need to feed regularly during these periods to ensure survival to breeding age.

We wanted to investigate where grey seals go during these periods; whether they stay in the area in which they bred or moulted or whether they move elsewhere; how much individual animals move around during these periods; and whether there are any identifiable patterns which could be used to characterize movements at sea. Thompson et al. (1991), McConnell, Chambers, Nicholas & Fedak (1992), Hammond, McConnell, Fedak & Nicholas (1992) and Hammond, McConnell, Nicholas et al. (in prep.) have provided detailed descriptions of work which can be used to address these questions.

In this paper, we summarize data presented in these earlier analyses, present a preliminary analysis of some new data and combine them to provide an overall view of the distribution and movements of grey seals at sea off the east coast of Britain.

Methods

The methods for data collection were: (1) conventional VHF telemetry to monitor the broad movements and activity patterns of the seals from a number of sites along the coast; (2) satellite-link UHF telemetry to track individuals over time wherever they went; and (3) a combination of underwater ultrasonic and VHF telemetry from a boat to track individuals continuously and allow real-time observations to be made. Each of these methods has advantages and disadvantages over the others but the combination of them provides a wide perspective on the movements and behaviour of grey seals at sea.

Detailed accounts of the methods have been published in Hammond, McConnell, Fedak *et al.* (1992), Thompson *et al.*. (1991), McConnell, Chambers, Nicholas *et al.* (1992) and Hammond, McConnell, Nicholas *et al.* (in prep.). The following is a summary of the main points.

VHF telemetry

VHF tags were constructed specifically for this study and transmitted in the range 173.2–173.35 MHz. Signals were monitored from up to five Automatic Recording Stations (ARSs—Nicholas, Fedak & Hammond 1992) located along the coast. Approximate areas of sea covered by each ARS, calculated on the assumption that reception occurred within at least a 90° sector and the visible horizon, are shown in Fig. 1.

From the pattern of radio signals received by an ARS, three basic activities could be determined: (1) hauled out of the water, (2) swimming in the vicinity of the ARS, and (3) absent, i.e. out of range of the ARS. There were some periods when an ARS was not functioning and no data were collected. These periods were infrequent and usually short. In addition, there were some periods during which severe radio interference at a particular frequency precluded interpretation of the data for a particular seal.

Ultrasonic/VHF telemetry

The ultrasonic tags transmitted a coded series of clicks giving information on depth (accurate to ±1.5 m) and swimming speed. These data were collected by directional and omnidirectional hydrophones mounted beneath the tracking vessels and stored on hydrophone receivers (Thompson *et al.* 1991; Hammond, McConnell, Nicholas *et al.* in prep.). The direction of the seal at distance was determined from the VHF signals via either a rotatable or a fixed array of three-element Yagi aerials mounted at 90° to each other (Thompson *et al.* 1991; Hammond, McConnell, Nicholas *et al.* in prep.). The VHF data were recorded on a microcomputer.

As a back-up, all data were recorded on magnetic tape together with a voice channel recording visual observations, locations, etc.

Thompson *et al.* (1991) demonstrated that the depth profile of a dive was an indicator of the seal's activity at that time; in particular, time at sea can be determined as either travelling or foraging. This paper is concerned with the distribution and movements at sea and we concentrate on the horizontal movements of the seals. However, we use the available data on dive profiles to indicate whether or not the animal was foraging.

Satellite-link UHF telemetry

Satellite-link transmitter packages have undergone considerable development during the course of this work. Early instruments were in the form of two units containing the battery pack and the transmitter. A subsequent modification had a separate aerial which could be mounted on the top of the head (McConnell, Chambers, Nicholas et al. 1992). The third design introduced a specially designed data-logger to collect and process data on dive depth and swimming velocity (McConnell, Chambers & Fedak 1992). The data-logger and batteries were housed in one unit and the transmitter (including aerial) in another suitable for head mounting (Hammond, McConnell, Nicholas et al. in prep).

A completely new design, composed of a single unit, was introduced at the end of 1990 (McConnell, Chambers & Fedak 1992).

System Argos provides access to information on the location of transmitted signals and their quality, and any other data that have been transmitted. Here, we again concentrate on the locations and tracks but refer to the other data to indicate periods of foraging (Thompson et al. 1991).

Deployment of transmitters

Seals were captured at Donna Nook, the Farne Islands or the Isle of May (Fig. 1) in hand-held nets on land or in tangle nets set around haul-out sites. Each animal was anaesthetized by either a ketamine/diazepam mixture (Baker, Anderson & Fedak 1988) or a tiletamine/zolazepam mixture (Baker, Fedak et al. 1990). Tags were glued to the fur of the seals with epoxy resin (Fedak, Anderson & Curry 1983). VHF tags were attached to the top of the head, ultrasonic tags to the side of the animal about midway from nose to tail, and satellite-link tags to the back, the back and the top of the head, or the back of the neck. Most seals were also measured, weighed and branded for future recognition and an incisor tooth was removed for ageing.

Results

Between 1985 and 1990, 30 animals were monitored by VHF telemetry, seven animals were tracked by ultrasonic/VHF telemetry and seven animals were tracked by satellite-link telemetry.

Post-moult distribution and movements, 1986–1990

Hammond, McConnell, Fedak et al. (1992) investigated the activity patterns of grey seals around the Farne Islands from March to September,

1986–1988. This was a VHF study using a single ARS so that the results provided information on the activity of the seals around the islands during this period, but no information on movements. The main conclusion from the work was that grey seal activity patterns were highly variable both from day to day and from month to month. There was a tendency for immature animals to spend more time around the islands than adults but to haul out less whilst there. Overall, presence around the islands increased from March to June and then declined in July and August. In September, the decline continued for males but immature females spent more time around the islands in this month than any other. Three adult males used the islands as a base throughout the study period but made regular trips away accounting for about 80% of their time.

Thompson *et al.* (1991) tracked three subadult male grey seals from the Farne Islands in August 1986. Two animals were tracked only for short periods in the vicinity of the islands but one animal travelled widely during a nine-day period from the Farne Islands to the Isle of May, up the Firth of Forth and back via the Bass Rock, and into St Andrews Bay. Whilst at the Isle of May it made three foraging trips to the north-east of the island.

Hammond, McConnell, Nicholas *et al.* (in prep.) have described in detail the results of monitoring nine seals tagged at the Isle of May in March 1990 from five VHF ARSs stationed along the coast (Fig. 1). Overall, the data showed a wide range of activity patterns but some general statements could be made. The results were related to the age of the animals. The older seals (three years or more of age) tended to travel widely from March to May and then to show regular behaviour during the summer. The younger seals (one and two years of age) showed the strongest association with the Isle of May or the Farne Islands from March to May but were out of range of the ARSs much of the time during the summer. In October/November, eight seals returned for at least some of the time to the Isle of May.

McConnell, Chambers, Nicholas *et al.* (1992) described the deployment of two satellite-link transmitters on male grey seals at Donna Nook (Fig. 1) in July 1988 and June 1989, which were tracked for 51 and 111 days, respectively. The first was never located more than 10 km from Donna Nook but few positions at sea were expected because the transmitter aerial was on the animal's back. Nevertheless, it did not haul out anywhere else. The second made three 3–4 day journeys between Donna Nook and the Farne Islands. It spent periods of 21, 39, 12 and 32 days associated with these sites during which times it made a number of trips out to sea.

Post-breeding distribution and movements, 1985–1990

The first successful deployment of a satellite-link transmitter on a grey seal was on a post-pupping female at Donna Nook in December 1985

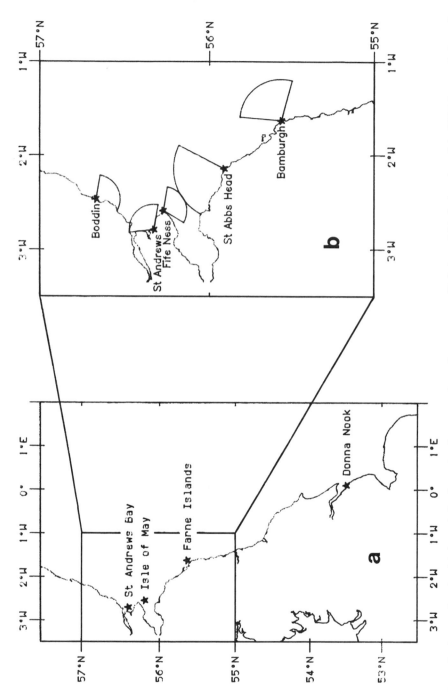

Fig. 1. Study area showing (a) the main haul-out sites for grey seals on the east coast of Britain and (b) the location of the VHF Automatic Recording Stations with their approximate ranges.

(McConnell, Chambers, Nicholas *et al.* 1992). It gave locations at Donna Nook, the Wash and sandbanks off Great Yarmouth over a 29-day period. Locations at sea were neither expected nor received.

Hammond, McConnell, Nicholas *et al.* (in prep.) described the movements of four seals from the Isle of May tracked via a yacht in November and December 1988. One animal spent several weeks moving only between a haulout site on the Isle of May and an area off the north coast of the firth only a few kilometres away where it was assumed to be foraging. The other three animals moved greater distances. One swam almost due north where it was observed following fishing nets set for cod. This seal was later found back at the Isle of May. The second made two trips to the mouth of the River Tay in St Andrews Bay in the space of a few days. The last seal made two foraging trips to an offshore area about 55–75 km east of St Andrews, returning to the Tay Estuary between trips.

Hammond, McConnell, Nicholas *et al.* (in prep.) also described the results from deployment of VHF and satellite-link transmitters at the Isle of May. The patterns of presence and absence for four adult female grey seals tagged with VHF transmitters after pupping on the Isle of May in November 1989 were monitored from four ARSs from St Abbs Head north to Lunan Bay (Fig. 1). Of the three seals which were detected in the vicinity of the Isle of May, one also travelled twice into St Andrews Bay. The fourth seal was only detected from St Abbs Head but no data were available from the Isle of May for this animal.

Four adult female grey seals tagged with satellite-link transmitters in November 1990 were tracked for 19, 4, 4 and 17 days (Hammond, McConnell, Nicholas *et al.* in prep.). One spent most of its time around the Isle of May but moved briefly into St Andrews Bay and also made a trip up to 200 km to the east. Another swam south-east close to the coast, arriving at the Farne Islands less than two days after being tagged. The third stayed close to the Isle of May for the four days for which data were received. The fourth spent about a week at a site deep in the Firth of Forth and then moved east and south, arriving at the Farne Islands ten days after being tagged.

Distribution and movements 1991

In August 1991, single-unit satellite-link transmitters were deployed on four male grey seals (two adult, two subadult) at the Farne Islands. Data from these animals were obtained for 101, 132, 125 and 131 days. A further three transmitters were deployed in November 1991 on post-pupping females but only one provided data, for 41 days.

These five animals spent much (in three cases, all) of their time in the vicinity of the Farne Islands. However, two also made long-distance

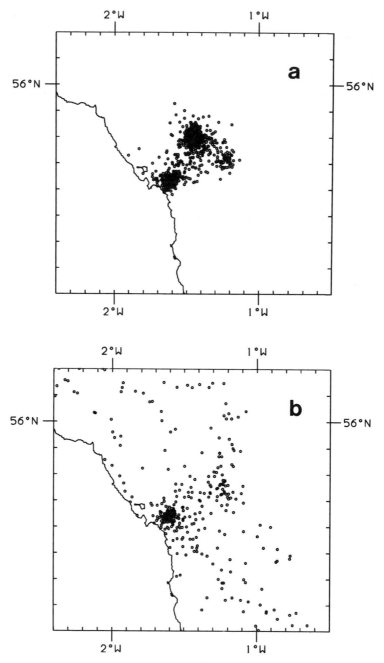

Fig. 2. Locations in the vicinity of the Farne Islands of four grey seals, identified by their transmitter numbers, tracked via satellite in 1991. (a) No. 5810, 184 kg male, tracked 22 August–30 November, (b) No. 5811, 129 kg male, tracked 23 August–31 December,

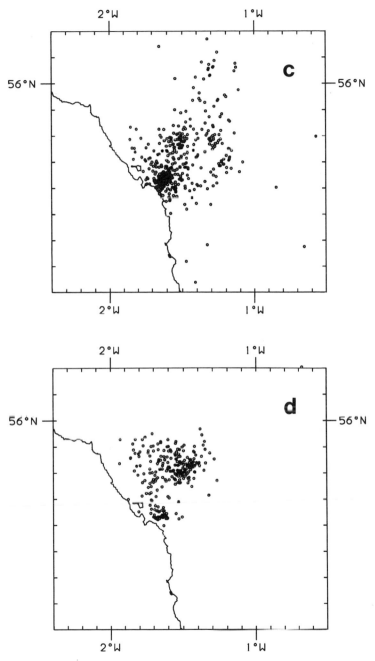

(c) No. 5813, 190 kg male, tracked 24 August–31 December, (d) No. 5815, 162 kg male, tracked 18 November–27 December.

movements which were repeated and regular in one case and wide-ranging in the other. These long-distance movements began in mid to late October and ended in mid December, a period covering the pupping and breeding season for grey seals off the east coast of Britain.

Figure 2 shows the distribution of locations around the Farne Islands for four of these seals. Taking into account that the scatter in the locations is due in part to the error associated with them, it is clear that these seals had favoured sites for foraging (as confirmed by dive profiles) as small as only a few square kilometres in area.

Discussion

Despite the amount of work involved during the last six to seven years to obtain these data, the number of animals fitted with transmitters is still a very small proportion of the total estimated population in the area. Nevertheless, we are now reaching the stage where it is possible to make some general statements and to formulate some hypotheses about the distribution and movements of grey seals off the east coast of Britain.

The most striking result is the variability in movement patterns among individuals. Movements of all the seals tracked via satellite from 1985 to 1991 are plotted in Fig. 3a. They cover a wide area from north of St Andrews Bay to south of the Wash and up to about 300 km offshore. Figure 3b shows a qualitative representation of where grey seals spend their time off the east coast of Britain, based on the tracks in Fig. 3a and the additional data from VHF and ultrasonic/VHF telemetry. There are, not surprisingly, high-density areas around the major haul-out sites of St Andrews Bay, the Isle of May, Farne Islands and Donna Nook, but these areas also include a lot of time spent at sea, as shown in Fig. 2. There is also a high-density area at 54°50′–55°N centred around 0° longitude, but this is a result of only two animals spending time there. The medium-density corridor running along the coast is used for travelling between distant haul-out sites and for foraging. The low-density area further offshore has been drawn to encompass all the available data but is, in fact, determined solely by the offshore tracks of only two seals.

Figure 3b is based on data from a relatively small number of individuals and as description of where grey seals from the east coast of Britain travel to it is somewhat speculative. For example, pups tagged with flipper tags at the Farne Islands have been recovered throughout the North Sea basin as far away as southern England, the Netherlands, Denmark, Norway and Shetland (McConnell, Curry et al. 1984). However, it is the first attempt at describing their distribution at sea based on known movements of individual animals.

Long-distance movements are clearly important to grey seals. This result

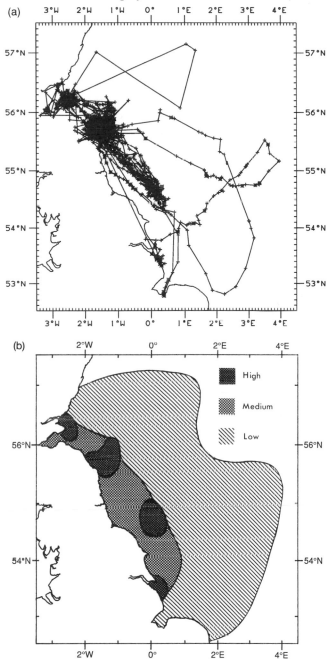

Fig. 3. (a) Movements of 12 grey seals tracked by satellite off the east coast of Britain, 1985–1991. (b) Qualitative representation of where grey seals spend their time off the east coast of Britain, showing areas of high, medium and low density.

confirms that the animals off the east coast of Britain must be considered as a single population unit. These movements may serve different purposes for different animals at different times of the year. The long distances travelled by older animals after the moult, as shown by Hammond, McConnell, Nicholas et al. (in prep.), may be related to searching for good places to feed during the summer in the run-up to the pupping and breeding season. Those in autumn are clearly related to the breeding season. Some animals may use them to visit more than one breeding site, as suggested by McConnell, Chambers, Nicholas et al. (1992), while other subadult animals may use them to move away from breeding sites. Long-distance movements after the breeding season may be migration to a different site at which to moult, as suggested by Hammond, McConnell, Nicholas et al. (in prep.). Those by young animals may be associated with learning to forage effectively.

The results also show strong evidence of individual specialization. The seals monitored via VHF in St Andrews Bay during the summer of 1990 (Hammond, McConnell, Nicholas et al. in prep.) showed very regular patterns of behaviour. One individual female, branded 'DO', was fitted with a transmitter after pupping at the Isle of May in three consecutive years. Her behaviour was similar in each year (Hammond, McConnell, Nicholas et al. in prep.). And the information shown in Fig. 2 shows clearly that individual animals have well-defined patterns of behaviour. Overall, the data suggest that older animals are more specialized; a predictable result, given that grey seal pups have to learn to forage alone from an early age and, therefore, to learn such specialization.

The available information on grey seal diet off the east coast of Britain has been presented in Prime & Hammond (1990), Hammond & Prime (1990) and Hammond, Hall & Prime (in prep.). The diet is very varied but we do not know whether this is a result of each seal having a varied diet or of individual specialization on different prey species. Examination of prey items in individual faecal samples provides evidence of both; some samples contain several species, but many contain only one. The age of the animal which deposited each sample is unknown so there is no information on whether prey composition is a function of age.

Future investigations of the movements and behaviour of grey seals at sea will include the deployment of more transmitters targeted at times of the year and classes of animal (age/sex) for which data are still particularly sparse, in order to obtain a more representative sample. Work is under way to investigate the diet of individual animals so that differences between age-classes and sexes can be assessed. It is also planned to survey one or more of the foraging 'hot spots' shown in Fig. 2 to try to determine why these sites are particularly attractive to grey seals.

Acknowledgements

This work was supported by the Ministry of Agriculture, Fisheries and Food. The National Trust gave permission to work at the Farne Islands and the Nature Conservancy Council, Scotland, allowed access to the Isle of May.

References

Baker, J. R., Anderson, S. S. & Fedak, M. A. (1988). The use of a ketamine-diazepam mixture to immobilise wild grey seals (*Halichoerus grypus*) and southern elephant seals (*Mirounga leonina*). *Vet. Rec.* **123**: 287–289.

Baker, J. R., Fedak, M. A., Anderson, S. S., Arnbom, T. & Baker, R. (1990). Use of tiletamine-zolazepam mixture to immobilise wild grey seals and southern elephant seals. *Vet. Rec.* **126**: 75–77.

Fedak, M. A., Anderson, S. S. & Curry, M. G. (1983). Attachment of a radio tag to the fur of seals. *J. Zool., Lond.* **200**: 298–300.

Hammond, P. S., Hall, A. J. & Prime, J. H. (In preparation). *The diet of grey seals from the Orkney Islands and adjacent sites in northeastern Scotland.*

Hammond, P. S., McConnell, B. J., Fedak, M. A. & Nicholas, K. S. (1992). Grey seal activity patterns around the Farne Islands. In *Wildlife telemetry: remote monitoring and tracking of animals*: 677–686. (Eds Priede, I. G. & Swift, S. M.). Ellis Horwood, Chichester.

Hammond, P. S., McConnell, B. J., Nicholas, K. S., Thompson, D. & Fedak, M. A. (In preparation). *Movements and behaviour of grey seals from the Isle of May.*

Hammond, P. S. & Prime, J. H. (1990). The diet of British grey seals (*Halichoerus grypus*). In *Population biology of sealworm* (Pseudoterranova decipiens) *in relation to its intermediate hosts*. (Ed. Bowen, W. D.). *Can. Bull. Fish. aquat. Sci.* **222**: 243–254.

Hiby, A. R., Duck, C. D. & Thompson, D. (1992). Seal stocks in Great Britain: surveys conducted in 1990 and 1991. *NERC News* **1992** (Jan.): 30–31.

McConnell, B. J., Chambers, C. & Fedak, M. A. (1992). Foraging ecology of southern elephant seals in relation to the bathymetry and productivity of the Southern Ocean. *Antarct. Sci.* **4**: 393–398.

McConnell, B. J., Chambers, C., Nicholas, K. S. & Fedak, M. A. (1992). Satellite tracking of grey seals (*Halichoerus grypus*). *J. Zool., Lond.* **226**: 271–282.

McConnell, B. J., Curry, M. G., Vaughan, R. W. & McConnell, L. C. (1984). Distribution of grey seals outside the breeding season. In *Interactions between grey seals and UK fisheries*: 12–45. Report on research conducted for the Department of Agriculture and Fisheries, Scotland by the Natural Environment Research Council's Sea Mammal Research Unit.

Nicholas, K. S., Fedak, M. A. & Hammond, P. S. (1992). An automatic recording station for detecting and storing radio signals from free ranging animals. In *Wildlife telemetry: remote monitoring and tracking of animals*: 76–78. (Eds Priede, I. G. & Swift, S. M.). Ellis Horwood, Chichester.

Prime, J. H. & Hammond, P. S. (1990). The diet of grey seals from the south-western North Sea assessed from analyses of hard parts found in faeces. *J. appl. Ecol.* **27**: 435–447.

Thompson, D., Hammond, P. S., Nicholas, K. S. & Fedak, M. A. (1991). Movements, diving and foraging behaviour of grey seals (*Halichoerus grypus*). *J. Zool., Lond.* **224**: 223–232.

Symp. zool. Soc. Lond. (1993) No. 66: 225–239

Harbour seal movement patterns

Paul M. THOMPSON

*Aberdeen University
Department of Zoology
Lighthouse Field Station
Cromarty IV11 8YJ, UK*

Synopsis

In many species of phocids, the capacity to store energy as blubber has led to a separation of feeding and breeding activity. Recent marking and telemetric studies have shown that harbour seals may travel extensively and that their breeding activity may also constrain foraging behaviour. This paper assesses the relative importance of variations in food availability, breeding activity and predation pressure in shaping movement patterns. Two broad categories of movements are recognized. The first is those between haul-out sites and the sea. These are primarily for foraging, and appear to occur within 50 km of haul-out sites. The duration, timing and locations of these trips are related to changes in environmental conditions and breeding activity. The second is movements which occur between different haul-out sites. These may involve dispersal, or they can occur seasonally, when seals switch to sites which are more suitable for pupping or are closer to foraging areas. Regional variations exist in the extent of seasonal movements, probably in relation to the relative distribution of foraging areas and suitable haul-out sites.

Introduction

Phocid seals feed at sea, but return ashore during the breeding and moulting seasons. Their ability to carry energy stores as blubber permits them to fast for prolonged periods and in many species this has allowed the separation of breeding and feeding activities (Costa 1991, this volume). The temporal separation of these activities is most obvious in species such as the grey seal (*Halichoerus grypus*) which remains ashore throughout the lactation and mating period (Anderson & Fedak 1987). Wide spatial separation has also taken place in species such as harp (*Phoca groenlandica*) and hooded seals (*Cystophora cristata*) which undertake long migrations between suitable feeding and breeding habitats (Reeves & Ling 1981).

Harbour seals (*Phoca vitulina*) are often seen in the same geographical areas throughout the year, and their more aquatic behaviour during the breeding season provides the opportunity for seals to feed during the

ZOOLOGICAL SYMPOSIUM No. 66
ISBN 0–19–854069–8

Copyright © 1993 The Zoological Society of London
All rights of reproduction in any form reserved

pupping and mating periods. Until recently, this led to the belief that the species is rather sedentary, with little spatial or temporal separation of breeding and feeding activities (Scheffer & Slipp 1944; Fisher 1952; Bigg 1969). Over the last 15 years, the development of techniques for marking and tracking individual seals has provided more detailed data on harbour seal movements. The collection of such data has been driven primarily by fisheries management questions which require information on the distribution of foraging seals or the degree of mixing between adjacent populations (Harwood & Croxall 1988). More recently, however, interest in population genetics and wildlife epidemiology following the 1988 phocine distemper virus outbreak (Heide-Jorgensen *et al.* 1992) has highlighted the importance of information about movement patterns to this and a range of other applied issues. More detailed data on movements can also help elucidate current questions on pinniped social structure and mating patterns (Le Boeuf 1991).

The primary reason for an individual's movement is likely to be to obtain food, but movements will also be influenced by the need to come ashore to breed or rest, to find a mate and to avoid predators. One would therefore expect individual behaviour to vary considerably, depending both upon local environmental conditions and on the age, sex or breeding status of the individual concerned. Harbour seals are an ideal species for comparative studies of the factors causing these variations in behaviour because of the wide range of environmental conditions they inhabit. Breeding groups are found over an extensive geographical range along the shores of the North Pacific and North Atlantic, on haul-out habitats ranging from intertidal sandbanks to rocky shores and glacial ice flows (Bigg 1981). Predation pressure also varies widely. In reserve areas in the Dutch Wadden Sea, predation is probably insignificant while in other parts of the range it may either be by terrestrial predators including humans (Bonner 1989) and coyotes, *Canis latrans*, (Steiger *et al.* 1989) or aquatic predators such as killer whales, *Orcinus orca*, (Jefferson, Stacey & Baird 1991) and white sharks, *Carcharodon carcharias* (Stewart & Yochem 1985). Food availability and habitat productivity are also likely to differ considerably, although there are currently few quantitative data on the nature and extent of these differences.

The first section of this paper describes the techniques which have been used to assess the movements of harbour seals. This provides background on the methodologies used in work reviewed later in the paper, and outlines the relative advantages and limitations of the techniques currently available. The second section reviews the available information on harbour seal movements. This draws heavily on work carried out around the UK but, where possible, comparisons are made between different populations and between seasons, age or sex classes from the same population. In particular, these comparisons aim to assess the relative importance of changes in food availability, breeding activity and predation pressure in producing observed variations in movement patterns. Because current data are limited in extent, an additional aim is to provide a framework for further comparative studies by categorizing movement types and highlighting areas of interest for future research.

Methodology

Early information on harbour seal movements was based on occasional observations of seals at sea (Spalding 1964: Wahl 1977), and on changes in the abundance of seals at their haul-out sites (Fisher 1952). However, both types of data have their limitations. Sightings at sea tend to be biased towards areas where there is heavy marine traffic or fishing activity; changes in haul-out abundance could result either from seals moving between sites or from variations in the amount of time that individual seals spend ashore.

More detailed studies of movements require individual seals to be recognizable, ideally for many years. It may be possible for pelage patterns (Yochem, Stewart, Mina et al. 1990) or unique scars (Davis & Renouf 1987) to be used to identify some individual harbour seals, but this technique has only been used extensively in a study of the more distinctively marked Kuril seal. *P.v.stenegerii* (Niizuma 1986). The most common form of long-term marking has been to apply numbered cattle or sheep ear-tags through the webbing of the seals' rear flippers. Because harbour seals do not usually remain ashore when approached, the most extensive tagging schemes have concentrated upon marking the more easily caught pups or juveniles (e.g. Bonner & Witthames 1974; van Haaften 1981; Wiig & Oien 1988). In some areas, haul-out groups can be approached closely enough to read tags on live seals (P. M. Thompson 1989) but, in most studies, information on the movements of tagged individuals is available only from dead or moribund seals. As a result, only a low percentage of tags are usually recovered and it can be difficult to interpret results, particularly when seals have been washed ashore after death.

Where more detailed observational studies have been carried out, a variety of more visible marks have been applied. These include marine paint (Davis & Renouf 1987) or epoxy-based numbers or marks (P. M. Thompson 1989), and streamers or neoprene tags which are glued to the hair (Jeffries 1986; Allen 1988). All of these marks may last until the following moult and can provide useful information on the behaviour or movements of animals in the short term. However, many marked seals are re-sighted infrequently, or not at all. There is therefore the danger that, if variations in behaviour exist, such studies are biased towards the behaviour of more sedentary individuals. Longer-term marks from either hot-iron (W. T. Stobo & B. Beck pers. comm.) or freeze-brands (T. Härkönen pers. comm.) have also been applied, and at many sites these remain the only methods available for carrying out long-term observational studies of known individuals.

The above marking techniques are useful for identifying movements between haul-out sites, but they provide little information on the distribution of seals at sea. In order to collect data on the characteristics of foraging trips, transmitters or data recorders have been attached to seals. The first radio-tracking studies of harbour seals were carried out in the late 1970s when VHF radios were applied to seals from the coasts of Oregon (Brown & Mate 1983) and Alaska (Pitcher & McAllister 1981). However, these tags were attached to ankle-bracelets and, because VHF signals do not transmit through sea water, provided data only on the location of seals while at haul-out sites. The development of techniques to glue transmitters and other instruments to the pelage of seals (Fedak, Anderson & Curry 1983) subsequently permitted researchers to attach VHF tags to the back or the head of seals, making it possible to locate seals whilst they were at the surface between dives. Nevertheless, the limited range of VHF radios (5–20 km at sea level) has meant that harbour seals have often moved too far offshore to be located from the coast on a regular basis (Harvey 1987; Allen 1988; P. M. Thompson 1989). Consequently, data collection has been biased towards terrestrial sites and information on at-sea locations has frequently been collected opportunistically. This has made analyses of data and comparisons between studies difficult, and does not permit an unbiased assessment of key feeding sites. In a few areas, the characteristics of a study area have permitted more systematic collection of data on foraging locations by triangulation from coastal vantage points (e.g. P. M. Thompson & Miller 1990). In other studies, automatic logging devices at haul-out sites have been used to collect data routinely on the presence or absence of seals in inshore areas (Yochem, Stewart, DeLong et al. 1987; Allen 1988; P. M. Thompson, Fedak et al. 1989). Changes in the pattern of VHF signals when seals dive also permit these data to be used to assess the haul-out behaviour and surfacing patterns of seals.

Although VHF tags have been used most widely, other techniques have also been applied to studies of harbour seal movements. The limited range of VHF tags, together with the problems of systematic data collection, make satellite telemetry an attractive alternative for producing unbiased estimates of locations at sea. Current satellite transmitters are rather large for harbour seals, especially in view of the need to mount packages on the head if locations are to be obtained during relatively short inter-dive periods (Stewart et al. 1989). Nevertheless, recent reductions in the size of satellite transmitters mean that the technology should soon be suitable for smaller pinnipeds. Where detailed information is required on diving behaviour, acoustic transmitters have successfully been deployed on harbour seals. The short range of these tags (<1 km) requires researchers to follow individual seals in a hydrophone-equipped boat (D. Thompson *et al.* 1991). This may often be logistically difficult and expensive, but the technique has provided important information on the heart-rate patterns (Fedak, Pullen & Kanwisher 1988), swimming depths and speeds, and movements of foraging harbour seals (Sea Mammal Research Unit (SMRU) unpubl. data). Finally, micro-processor controlled time-depth recorders (TDRs) have been used to collect detailed data on the diving patterns and characteristics of foraging trips (Stewart *et al.* 1989; D. J. Boness, W. D. Bowen & O. T. Oftedal pers. comm.). However, the problems inherent in recapturing instrumented harbour seals at most sites prevent the widespread use of these instruments.

Patterns of movement

In this section, harbour seal movements are described under two broad categories. The first includes movements between a favoured terrestrial haul-out site and the sea. The second concerns movements between different haul-out sites.

Movements between haul-out sites and the sea

When seals return regularly to offshore areas over periods of several weeks or months, it seems reasonable to assume that these movements are primarily for foraging. Systematic data on at-sea locations of instrumented seals have been collected in only a few areas but, in all cases, harbour seals do not appear to forage more than 50 km from their haul-out sites (Stewart *et al.* 1989; P. M. Thompson & Miller 1990; P. M. Thompson, Pierce *et al.* 1991; SMRU unpubl. data; D. Oxman & J. T. Harvey pers. comm.). In those cases where seals have travelled further than 50 km, they have also switched to alternative haul-out sites closer to foraging areas (unpubl. data). In the Moray Firth in north-east Scotland, seals caught from one haul-out area in the Dornoch Firth travelled regularly to the same area during the summer (Fig. 1). Adult males and females showed no differences in their overall pattern of summer locations (unpubl. data), but there were individual differences in locations within the broad area used by all seals (P. M. Thompson, Miller *et al.* in press). In general, the seabed in the Moray Firth is fairly uniform, and locations appeared to cluster around relatively scarce submarine features such as rocky reefs or offshore banks (P. M. Thompson & Miller 1990; unpubl. data). Where inshore habitats are more diverse, a wider range of foraging locations might be expected and comparative work in different habitats would be of interest.

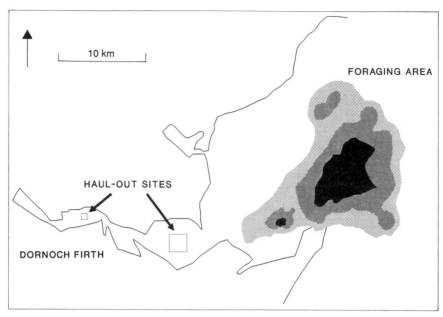

Fig. 1. Foraging area used by seven adult female harbour seals which were radio-tracked during summer. Data are presented as 25, 50 and 75% harmonic mean isoclines around all daily locations which were > 2 km from a haul-out site. (Data from P. M. Thompson, Miller *et al.* in press.)

Seasonal changes in at-sea locations indicate that both breeding activities and changes in food availability influence foraging movements. During the summer, breeding females from the Moray Firth showed a marked restriction in their range size during the early part of lactation. At this time, they remained in inshore haul-out areas and came ashore on almost every low tide (P. M. Thompson, Miller *et al.* in press). These studies, together with data from TDRs deployed on Sable Island (D. J. Boness, W. D. Bowen & O. T. Oftedal pers. comm.), indicate that female foraging movements are severely constrained during early lactation, although longer feeding trips do resume later in lactation. The extent to which males also restrict their foraging movements during the breeding season remains unclear. Much of the evidence remains indirect and anecdotal, but it appears that harbour seals are serially polygynous, with males defending areas of water in which they display visually (Sullivan 1981; Godsell 1988; P. Thompson 1988) and acoustically (D. Thompson pers. comm.) to attract females. It therefore seems likely that successful males also restrict their foraging activity in order to maximize the time spent in display areas. Decreases in male body condition over the summer months (Pitcher 1986) support this suggestion, but further work is required to understand the influence of breeding activity on male movements and activity during the summer.

Seasonal and between-year changes in the local distribution of prey can also lead to variations in foraging range or location. Summer locations of Moray Firth harbour seals indicated that most foraging occurred 20–40 km from the haul-out areas (P. M. Thompson & Miller 1990; Fig. 1). In contrast, seals followed during the winter of 1988/89 reduced their foraging range to 5–10 km when large numbers of herring and sprat moved into inshore areas to overwinter (P. M. Thompson, Pierce et al. 1991). The following year, few overwintering herring and sprat were recorded in the area and seals were regularly located further offshore (unpubl. data).

Studies in the UK have shown that trips to sea often last several days and are interspersed with periods of a day or two spent in the haul-out area (P. M. Thompson, Fedak et al. 1989; P. M. Thompson, Miller et al. in press). In some cases, these trips are remarkably regular and are similar in pattern to the foraging trips of lactating otariids (Gentry & Kooyman 1986). Harbour seals using intertidal sites must return to the water over high tide, even when they have returned inshore to rest. In the Moray Firth, the frequency distribution of the duration of periods spent in the water is therefore highly skewed (P. M. Thompson & Miller 1990), with most being less than 12 h. However, when seals have been located during these short periods at sea, they have generally been close to haul-out sites. Seals travel further from haul-out sites when in the water for longer periods, and the trips of over 12 h account for at least 60% of a seal's time in the water (P. M. Thompson & Miller 1990). One cannot rule out the possibility that seals forage opportunistically when in the water around haul-out sites. Nevertheless, it seems reasonable to assume that most foraging in this area occurs on these longer trips to sea and it is probably more realistic to exclude those of < 12 h when comparing trip characteristics.

Although seals may be at sea for several days, it seems unlikely that they are foraging actively throughout this whole period. If trip durations are to be used as a measure of foraging effort, especially for comparative work, it is important to understand how activity varies during these trips. If longer trips do include periods of rest, as seems likely, information is required on the factors which influence whether a seal remains at sea or returns ashore when not feeding. Although the distance between foraging areas and haul-out areas might be expected to be the key factor, the limited evidence available suggests that other factors are also important. Recent studies in Monterey Bay, California, have shown that juvenile harbour seals returned to haul-out sites each day, making overnight feeding trips along the continental shelf (D. Oxman & J. T. Harvey pers. comm.). It might be assumed that these seals were feeding closer to haul-out sites than those in our UK studies. However, when an adult male from the Moray Firth foraged just 8 km from his haul-out site he regularly remained at sea for periods of several days (P. M. Thompson, Pierce et al. 1991). Differences between these results could be caused by thermoregulatory factors, either because of differences in environmental temperature, or because of the age and condition of the seals under study. Alternatively, the presence of aquatic predators along the Pacific coast could favour a return to land whenever seals are not actively foraging. Further investigations are required on the effect of body size and condition on haul-out and foraging activity within a single population before these differences can be explained fully. In the meantime, one cannot assume that longer foraging trips necessarily indicate that seals are having to travel further to feed.

The timing of trips to sea in relation to tidal and diel cycles varies considerably both within and between studies. Seals using sites which are available throughout the tidal cycle generally have activity patterns which are dominated by the diel cycle, but the tidal cycle may have a significant, but smaller, influence (Stewart 1984; Calambokidis et al. 1987; P. M. Thompson, Fedak et al. 1989). At estuarine sites, where sandbank haul-out sites are only available for a few hours over low tide, the tidal cycle inevitably has the more dominant effect but diel patterns may also be present (Schneider & Payne 1983; Allen, Ainley et al. 1984; P. M. Thompson & Miller 1990). The influence of the tidal cycle appears to act primarily through its effect on haul-out site availability. Nevertheless, increases in the time that seals spend at sea on rising as opposed to falling tides (P. M. Thompson, Fedak et al. 1989; P. M. Thompson & Harwood 1990) suggest that the tidal cycle may sometimes have an indirect effect on behaviour through changes in food availability. In most studies the predominant diel pattern has been for seals to spend most time at sea at night. Such behaviour is believed to result from nocturnal changes in the behaviour of many of the prey known to be taken by harbour seals. However, other patterns have also been identified. Seals in the Moray Firth fed more often during the day when feeding upon wintering clupeoids, indicating that they preferred foraging when these prey were in tight schools near the sea bed (P. M. Thompson, Pierce et al. 1991). On California's San Miguel Island, Yochem, Stewart, Delong et al. (1987) found individual differences in the behaviour of 17 harbour seals which were radio-tagged during the autumn. Six seals came ashore most often during the day, seven individuals hauled out most frequently at night, and the remainder showed no preference. If haul-out patterns at these sites are also influenced largely by feeding behaviour, these data indicate that there may be individual differences in foraging strategy within the population. However, haul-out patterns will not necessarily reflect periods of feeding activity if seals on longer trips rest at sea between active feeding bouts. In Orkney, for example, diurnal haul-out patterns disappeared when two individuals started making longer trips away from the haul-out areas during winter (P. M. Thompson, Fedak et al. 1989).

Movements between haul-out sites

Harbour seals travel between haul-out sites for a variety of reasons. Some movements result from animals dispersing to new social groups, while others may occur because the physical characteristics of a site make it suitable for particular activities such as pupping. Alternatively, seals may switch to a new haul-out site because it is closer to their current foraging area. In general, one would expect dispersal movements to predominate at one or a few stages of the life-cycle, while movements based on site characteristics should occur on a seasonal basis in line with the annual cycle. If seasonal changes in food availability are predictable, movements to sites near feeding areas may also be similar from year to year, but between-year differences in site-use could also occur.

Pup movements

Contrary to the belief that harbour seals are sedentary, tagging experiments have shown that individuals may travel long distances. In the UK, the most extensive movements have been recorded from pups in their first year of life. Pups tagged in East Anglia were recovered as far away as the Dutch Wadden Sea (> 300 km) (Bonner & Witthames 1974) and pups from Orkney travelled to adjacent populations in Shetland (> 150 km), the Moray Firth (> 175 km) and Norway (> 475 km) (P. M. Thompson, Kovacs & McConnell in press). Similarly, pups from Sable Island have travelled widely, being found along the Nova Scotia coast (200 km) and as far south as Cape Cod (> 800 km) (W. D. Bowen pers. comm.). Nevertheless, although long-distance movements in excess of 200 km have been recorded in several studies (Bonner & Witthames 1974; Wigg & Oien 1988; P. M. Thompson, Kovacs et al. in press), the frequency distribution of distances travelled can be highly skewed. Although several pups from Orkney travelled over 100 km, 80% of recoveries were < 50 km from their tagging site (P. M. Thompson, Kovacs et al. in press). Similarly, most of the weaned pups freeze-branded in the Skaggerak have remained within 20 km of their capture site until at least three to four years old (T. Härkönen pers. comm.). These data indicate that although young pups may disperse widely, resulting in mixing between populations, most remain in their natal area. Although several studies have tagged large numbers of pups, the low recovery rates do not yet permit an analysis of sex-related differences in dispersal rates or of the proximate factors which cause individuals to disperse.

Adult movements

During the breeding season, older seals have generally remained faithful to a single site or small group of local sites (Jeffries 1986; Allen 1988). Groups of seals at different sites within a local haul-out area may also show

consistent differences in sex or age structure (Knudtson 1977; Slater &
Markowitz 1983; Allen, Ribic & Kjelmyr 1988; P. M. Thompson 1989;
Kovacs, Jonas & Welke 1990), indicating that an individual's status will
affect its choice of site, or at least its success in gaining access to a site.

On the other hand, seasonal movements between haul-out sites have been
reported from many areas, involving distances of just 10–20 km (P. M.
Thompson 1989) to over 200 km (Allen 1988). Return movements have
often been recorded and there appears to be relatively little dispersal of
adults. Nevertheless, some studies have found that immatures travel more
extensively than adults (e.g. Herder 1986; H. R. Huber & S. J. Jeffries pers.
comm.) and it is possible that some of these are dispersal movements.

Seasonal changes in site-use often appear to be related to the characteristics of the haul-out site. The most frequently recorded pattern is for
groups of females with pups to predominate at certain sites during the
pupping season (Jeffries 1986; Kovacs et al. 1990). These sites are often in
relatively sheltered or isolated areas and may be chosen because the lack of
disturbance from heavy seas or terrestrial predators makes them suitable for
suckling. The consistent use of these sites suggests that individual females
return each year to give birth at favoured sites. Long-term data are
unavailable from most studies but, on Sable Island, observations of 35
marked females over the period 1989–91 showed that 73% of between-year
movements between birth sites were of less than 2 km (W. D. Bowen &
D. J. Boness pers. comm.). Other sites may be used predominantly during
the moulting period, when male seals haul out for a high proportion of their
time (Niizuma 1986; P. M. Thompson, Fedak et al. 1989). These sites are
often on higher sand bars, which allow seals to haul out for more of the tidal
cycle, or on steep beaches which minimize the movement required to remain
close to the water's edge (Jeffries 1986; P. M. Thompson 1989). The use of
such sites allows seals to remain ashore for longer during the moult.
However, capturing seals at these sites is usually easier than elsewhere,
which indicates that seals may also be more vulnerable to terrestrial
predators whilst there. This could explain why these sites are not used at
other times. In general, where alternative sites are within a few kilometres of
each other, movements between them probably result only from differences
in the characteristics of these sites. In one such group of sites in Orkney,
marked animals moved between sites in line with changes in abundance, but
radio-tracking studies confirmed that seals did not necessarily use sites
which were nearest to their feeding areas (P. M. Thompson 1989).

In other studies, changes in haul-out distribution apparently result from
changes in foraging grounds. In several estuaries along the Pacific coast of
North America there are seasonal increases in abundance at haul-out sites
during major runs of prey species (Brown & Mate 1983; Roffe & Mate
1984; Jeffries 1986). Similarly, seals in the Moray Firth hauled out in

greater numbers at a site on the outer part of the estuary during one winter when radio-tagged seals foraged closer to this site (unpubl. data). However, there appear to be differences in the nature and extent of these seasonal movements. Around the UK, only one of the adults which have been radio-tagged is known to have moved more than 100 km (unpubl. data; SMRU unpubl. data), and there appears to be little mixing of older seals from breeding groups on the east coast of Britain. In contrast, several comparable studies of Pacific harbour seals have recorded movements in excess of 200 km (Jeffries 1986; Herder 1986; Allen 1988), suggesting that local environmental conditions may lead to more widespread movements in this population. Differences also appear to exist in the pattern of seasonal movements of Pacific and north-west Atlantic harbour seals. In the Pacific, seals moved away from capture sites to alternative sites to both the north and south (Jeffries 1986; Allen 1988). Few data are available from marked seals in the north-west Atlantic, but trends in abundance suggest that here there is a general southwards movement in winter, from areas such as the Bay of Fundy to the New England coast (Schneider & Payne 1983; Rosenfeld, George & Terhune 1988).

The data available therefore indicate that both the characteristics of haul-out sites and the proximity to foraging areas can be important in determining whether seals switch sites. In addition, mating patterns may influence the extent to which adult males and females move between sites (Greenwood 1980, 1983). Only limited data exist on sex differences in the distribution of harbour seals during the non-breeding season. Nonetheless, studies of individually recognizable Kuril seals indicate that adult males continued to use haul-out sites in the breeding area throughout the year (Niizuma 1986). In contrast, known females came ashore at breeding sites only rarely during the winter. Such a difference might result from changes in haul-out behaviour, but by-catches in the salmon net fishery elsewhere on the Japanese coast were biased towards females during the winter, which suggests that females are more widely distributed than males at this time (Hayama et al. 1986). Similarly, studies along the Pacific coast indicate that adult females sometimes travel further during the winter (Herder 1986; Allen 1988), although sample sizes have been too small for statistical analyses of these trends. These data are preliminary, but they do suggest that males are more likely to remain in the breeding areas where they subsequently defend display areas.

Conclusions

Harbour seals do not have the discrete feeding and breeding periods, or areas, seen in many other phocids. Nevertheless, breeding activities clearly constrain foraging, and may influence movement patterns both during and outside the breeding season. Currently, the effects of predation are less clear, perhaps being most likely where females move to undisturbed sites to pup. However, more detailed studies may uncover more subtle differences in the behaviour of seals using areas under different predation threats. For example, variations in foraging trip characteristics may exist as a result of differences in the relative risks of resting on land and in the water. Overall, variations in the distribution and abundance of prey resources probably have a more profound effect on movement patterns, leading in particular to longer-distance seasonal movements between haul-out sites and feeding areas. Although poorly documented, there may be major differences in the nature of seasonal movements in different harbour seal populations. These are also likely to be a consequence of differences in resource distribution. However, before these can be confirmed, more extensive data are required on the extent of age and sex differences in behaviour within these populations. Recent advances in telemetric and data logging systems, in combination with more traditional techniques, now provide excellent opportunities for further comparative studies of this kind.

Acknowledgements

I would like to thank everyone who allowed me to discuss their unpublished results and I. L. Boyd, A. D. Hawkins, P. A. Racey and D. Tollit for their constructive criticism of earlier drafts of this paper. Studies in the Moray Firth were carried out under a contract from the Scottish Office Agriculture and Fisheries Department to Professor P. A. Racey and myself.

References

Allen, S. G. (1988). *Movement and activity patterns of harbor seals at the Point Reyes Peninsula, California.* MSc diss.: University of California at Berkeley.
Allen, S. G., Ainley, D. G., Page, G. W. & Ribic, C. A. (1984). The effect of disturbance on harbor seal haul out patterns at Bolinas Lagoon, California. *Fish. Bull. U.S. Fish Wildl. Serv.* **82**: 493–500.
Allen, S. G., Ribic, C. A. & Kjelmyr, J. E. (1988). Herd segregation in harbor seals at Point Reyes, California. *Calif. Fish Game* **74**: 55–59.
Anderson, S. S. & Fedak, M. A. (1987). The energetics of sexual success of grey seals and comparison with the costs of reproduction in other pinnipeds. *Symp. zool. Soc. Lond.* No. **57**: 319–341.
Bigg, M. A. (1969). The harbour seal in British Columbia. *Bull. Fish. Res. Bd Can.* No. **172**: 1–33.
Bigg, M. A. (1981). Harbour seal, *Phoca vitulina* and *P. largha*. In *Handbook of marine mammals* **2**: *Seals*: 1–28. (Eds Ridgway, S. H. & Harrison, R. J.). Academic Press, London, New York etc.
Bonner, W. N. (1989). Seals and man—a changing relationship. *Biol. J. Linn. Soc.* **38**: 53–60.

Bonner, W. N. & Witthames, S. R. (1974). Dispersal of common seals (*Phoca vitulina*), tagged in the Wash, East Anglia. *J. Zool., Lond.* **174**: 528–531.

Brown, R. F. & Mate, B. R. (1983). Abundance, movements, and feeding habits of harbor seals, *Phoca vitulina*, at Netarts and Tillamook Bays, Oregon. *Fish. Bull. U.S. Fish Wildl. Serv.* **81**: 291–301.

Calambokidis, J., Taylor, B. L., Carter, S. D., Steiger, G. H., Dawson, P. K. & Antrim, L. D. (1987). Distribution and haul-out behaviour of harbor seals in Glacier Bay, Alaska. *Can J. Zool.* **65**: 1391–1396.

Costa, D. P. (1991). Reproductive and foraging energetics of pinnipeds: implications for life history patterns. In *The behaviour of pinnipeds*: 300–344. (Ed. Renouf, D.). Chapman & Hall, London etc.

Davis, M. B. & Renouf, D. (1987). Social behaviour of harbour seals, *Phoca vitulina*, on haulout grounds at Miquelon. *Can. Fld Nat.* **101**: 1–5.

Fedak, M. A., Anderson, S. S. & Curry, M. G. (1983). Attachment of a radio tag to the fur of seals. *J. Zool., Lond.* **200**: 298–300.

Fedak, M. A., Pullen, M. R. & Kanwisher, J. (1988). Circulatory responses of seals to periodic breathing: heart-rate and breathing during exercise and diving in the laboratory and open sea. *Can J. Zool.* **66**: 53–60.

Fisher, H. D. (1952). The status of the harbour seal in British Columbia, with particular reference to the Skeena River. *Bull. Fish. Res. Bd Can.* No. 93: 1–58.

Gentry, R. L. & Kooyman, G. L. (Eds) (1986). *Fur seals: maternal strategies on land and at sea*. Princeton University Press, Princeton, New Jersey.

Godsell, J. (1988). Herd formation and haul-out behaviour in harbour seals (*Phoca vitulina*). *J. Zool., Lond.* **215**: 83–98.

Greenwood, P. J. (1980). Mating systems, philopatry and dispersal in birds and mammals. *Anim. Behav.* **28**: 1140–1162.

Greenwood, P. J. (1983). Mating systems and the evolutionary consequences of dispersal. In *The ecology of animal movement*: 116–131. (Eds Swingland, I. R. & Greenwood, P. J.). Oxford University Press, Oxford.

Harvey, J. T. (1987). *Population dynamics, annual food consumption, movements and dive behaviours of harbor seals*, Phoca vitulina richardsi, *in Oregon*. PhD diss.: Oregon State University.

Harwood, J. & Croxall, J. P. (1988). The assessment of competition between seals and commercial fisheries in the North Sea and the Antarctic. *Mar. Mamm. Sci.* **4**: 13–33.

Hayama, S., Wada, T., Nakaoka, T. & Uno, H. (1986). On the migration model of Kuril seals. In *Proceedings of the symposium on the ecology and protection of the Kuril seal*: 140–157. (Eds Wada, K., Itoo, T., Niizuma, A., Hayama, S. & Suzuki, M.). Tokai University Press, Tokyo. [In Japanese with English summary.]

Heide-Jorgensen, M.-P., Härkönen, T., Dietz, R. & Thompson, P. M. (1992). Retrospective of the 1988 European seal epizootic. *Dis. aquat. Organ.* **13**: 37–62.

Herder, M. J. (1986). *Seasonal movements and hauling site fidelity of harbor seals*, Phoca vitulina richardsii, *tagged at the Klamath River, California*. MA diss.: Humboldt State University.

Jefferson, T. A., Stacey, P. J. & Baird, R. W. (1991). A review of killer whale interactions with other marine mammals: predation to co-existence. *Mammal Rev.* **21**: 151–180.

Jeffries, S. J. (1986). *Seasonal movements and population trends of harbor seals* (Phoca vitulina richardsi) *in the Columbia River and adjacent waters of Washington and Oregon: 1976–1982.* Report to the US Marine Mammal Commission, Contract MM2079357–5.

Knudtson, P. M. (1977). Observations on the breeding behavior of the harbor seal in Humboldt Bay, California. *Calif. Fish Game* **63**: 66–70.

Kovacs, K. M., Jonas, K. M. & Welke, S. E. (1990). Sex and age segregation by *Phoca vitulina concolor* at haul-out sites during the breeding season in the Passamaquoddy Bay region, New Brunswick. *Mar. Mamm. Sci.* **6**: 204–214.

Le Boeuf, B. J. (1991). Pinniped mating systems on land, ice and in the water: emphasis on the Phocidae. In *The behaviour of pinnipeds*: 45–65. (Ed. Renouf, D.). Chapman & Hall, London etc.

Niizuma, A. (1986). Socio-ecology and reproductive strategy of the Kuril seal. In *Proceedings of the symposium on the ecology and protection of the Kuril seal*: 59–102. (Eds Wada, K., Itoo, T., Niizuma, A., Hayama, S. & Suzuki, M.). Tokai University Press, Tokyo. [In Japanese with English summary.]

Pitcher, K. W. (1986). Variation in blubber thickness of harbor seals in southern Alaska. *J. Wildl. Mgmt* **50**: 463–466.

Pitcher, K. W. & McAllister, D. C. (1981). Movements and haulout behavior of radio-tagged harbor seals, *Phoca vitulina*. *Can. Fld Nat.* **95**: 292–297.

Reeves, R. R. & Ling, J. K. (1981). Hooded seal, *Cystophora cristata*. In *Handbook of marine mammals* **2**: *Seals*: 171–194. (Eds Ridgway, S. H. & Harrison, R. J.). Academic Press, London, New York etc.

Roffe, T. J. & Mate, B. R. (1984). Abundances and feeding habits of pinnipeds in the Rogue river, Oregon. *J. Wildl. Mgmt* **48**: 1262–1274.

Rosenfeld, M., George, M. & Terhune, J. M. (1988). Evidence of autumnal harbour seal, *Phoca vitulina*, movement from Canada to the United States. *Can. Fld Nat.* **102**: 527–529.

Scheffer, V. B. & Slipp, J. W. (1944). The harbour seal in Washington State. *Am. Midl. Nat.* **32**: 373–416.

Schneider, D. C. & Payne, P. M. (1983). Factors affecting haul-out of harbour seals at a site in southeastern Massachusetts. *J. Mammal.* **64**: 518–520.

Slater, L. M. & Markowitz, H. (1983). Spring population trends in *Phoca vitulina richardsi* in two central California coastal areas. *Calif. Fish Game* **69**: 217–226.

Spalding, D. J. (1964). Comparative feeding habits of the fur seal, sea lion and harbour seal on the British Columbia coast. *Bull. Fish. Res. Bd Can.* No. 146: 1–52.

Steiger, G. H., Calambokidis, J., Cubbage, J. C., Skilling, D. E., Smith, A. V. & Gribble, D. H. (1989). Mortality of harbor seal pups at different sites in the inland waters of Washington. *J. Wild. Dis.* **25**: 319–328.

Stewart, B. S. (1984). Diurnal hauling patterns of harbor seals at San Miguel Island, California. *J. Wildl. Mgmt* **48**: 1459–1461.

Stewart, B. S., Leatherwood, S., Yochem, P. K. & Heide-Jorgensen, M.-P. (1989). Harbor seal tracking and telemetry by satellite. *Mar. Mamm. Sci.* **5**: 361–375.

Stewart, B. S. & Yochem, P. K. (1985). Radio-tagged harbor seal, *Phoca vitulina richardsi*, eaten by white shark, *Carcharodon carcharias*, in the southern California Bight. *Calif. Fish Game* **71**: 113–115.

Sullivan, R. M. (1981). Aquatic displays and interactions in harbor seals, *Phoca vitulina*, with comments on mating systems. *J. Mammal.* **62**: 825–831.

Thompson, D., Hammond, P. S., Nicholas, K. S. & Fedak, M. A. (1991). Movements, diving and foraging behaviour of grey seals (*Halichoerus grypus*). *J. Zool., Lond.* **224**: 223–232.

Thompson, P. (1988). Timing of mating in the common seal (*Phoca vitulina*). *Mammal Rev.* **18**: 105–112.

Thompson, P. M. (1989). Seasonal changes in the distribution and composition of common seal (*Phoca vitulina*) haul-out groups. *J. Zool., Lond.* **217**: 281–294.

Thompson, P. M., Fedak, M. A., McConnell, B. J. & Nicholas, K. S. (1989). Seasonal and sex-related variation in the activity patterns of common seals (*Phoca vitulina*). *J. appl. Ecol.* **26**: 521–535.

Thompson, P. M. & Harwood, J. (1990). Methods for estimating the population size of common seals, *Phoca vitulina. J. appl. Ecol.* **27**: 924–938.

Thompson, P. M., Kovacs, K. M. & McConnell, B. J. (In press). Natal dispersal of harbour seal (*Phoca vitulina*) from breeding sites in Orkney, Scotland. *J. Zool., Lond.*

Thompson, P. M. & Miller, D. (1990). Summer foraging activity and movements of radio-tagged common seals (*Phoca vitulina* L.) in the Moray Firth, Scotland. *J. appl. Ecol.* **27**: 492–501.

Thompson, P. M., Miller, D., Cooper, R. & Hammond, P. S. (In press). Changes in the distribution and activity of female harbour seals during the breeding season: implications for their lactation strategy and mating patterns. *J. Anim. Ecol.*

Thompson, P. M., Pierce, G. J., Hislop, J. R. G., Miller, D. & Diack, J. S. W. (1991). Winter foraging by common seals (*Phoca vitulina*) in relation to food availability in the inner Moray Firth, N. E. Scotland. *J. Anim. Ecol.* **60**: 283–294.

van Haaften, J. L. (1981). The common or harbour seal. In *Marine mammals of the Waddon Sea*: 15–32. (Eds Reijnders, P. J. H. & Wolff, W. J.). A. A. Balkema, Rotterdam.

Wahl, T. R. (1977). Sight records of some marine mammals offshore from Westport, Washington. *Murrelet* **58**: 21–23.

Wiig, O. & Oien, N. (1988). Recoveries of common seals *Phoca vitulina* L. tagged along the Norwegian coast. *Fauna norv. Ser. A* **9**: 51–52.

Yochem, P., Stewart, B. S., DeLong, R. L. & DeMaster, D. P. (1987). Diel haul-out patterns and site fidelity of harbor seals on San Miguel Island, California, in autumn. *Mar. Mamm. Sci.* **3**: 323–333.

Yochem, P. K., Stewart, B. S., Mina, M., Zorin, A., Sadovov, V. & Yablokov, A. (1990). Non-metrical analyses of pelage patterns in demographic studies of harbor seals. *Rep. int. Whal. Commn spec. Issue* No. 12: 87–90.

Symp. zool. Soc. Lond. (1993) No. 66: 241–261

Recent advances in diet analysis of marine mammals

G. J. PIERCE, P. R. BOYLE, J. WATT and M. GRISLEY

Department of Zoology
University of Aberdeen
Tillydrone Avenue
Aberdeen AB9 2TN, UK

Synopsis

Analysis of marine mammal diets has traditionally been based on identification of fish otoliths recovered from stomachs and faeces. However, not all prey species have otoliths which appear in stomachs or faeces. One promising new approach is the identification of fish muscle protein residues using specific antisera. This methodology has been validated for salmonid and sandeel proteins and has been successfully applied to field studies of common seal and grey seal diets. Further development is required for routine use on a wider scale. Fish prey have also been identified from other hard remains, although there are few guides or keys. Discriminant analysis of bone morphometrics may assist identification of prey species. The sizes of fish eaten are normally estimated from measurements of otoliths or bones, but a major difficulty is the differential degradation of these hard parts. Degradation of otoliths and bones *in vitro* has helped to quantify reduction in size due to digestion.

Introduction

Analysis of marine mammal diets has traditionally relied on identification of hard remains of prey in stomach contents and faeces. In the case of piscivorous species, diets are primarily assessed by identifying and counting fish otoliths. Prey importance may be further quantified by calculating fish weight from measurements of otoliths. Similar methods may be applied to species feeding on cephalopods, the beaks of which are resistant to digestion. For species feeding primarily on other invertebrates, prey may be identified from remains of exoskeletons or shells.

In the present paper we focus on some specific aspects of methodology for diet analysis of piscivorous species: the development of serological methods for identifying fish proteins, the extension of conventional methods of identification by using fish bones, and estimating prey size from degraded

ZOOLOGICAL SYMPOSIUM No. 66
ISBN 0–19–854069–8

Copyright © 1993 The Zoological Society of London
All rights of reproduction in any form reserved

remains (for a more general treatment of the subject, see Pierce & Boyle 1991).

Serological methods and salmon in diets of UK seals

Background

There is strong evidence that conventional methods are not adequate for quantifying the importance of salmon in seal diets. Captive-feeding experiments have shown that, while grey seals will eat salmon heads, the recovery rate of ingested salmon otoliths in faeces may be as low as 2% in optimum conditions. Salmonid bones are almost totally digested (Boyle, Pierce & Diack 1990) and salmon otoliths are rarely found in faecal samples from harbour seals *Phoca vitulina* and are not recorded in faeces of grey seals *Halichoerus grypus* in the UK (SMRU 1984, 1988; Hammond & Harwood 1985; Pierce, Boyle & Thompson 1990; Prime & Hammond 1990; Pierce, Thompson *et al.* 1991; Pierce, Miller *et al.* 1991; Thompson *et al.* 1991).

Species-specific differences in fish muscle protein composition have been demonstrated by means of isoelectric focusing (Fig. 1, see also Laird, Mackie & Ritchie 1982) or liquid chromatography (Osman, Ashoor & Marsh 1987). However, the banding patterns are sufficiently complex that it would be almost impossible to identify to species in this way from stomach contents of an animal which had fed on a mixed diet of several different fish species. Added to this there is the complication of degradation of proteins during digestion, but it has been possible to raise antisera to species-specific proteins resistant to digestion.

Serological methods for prey identification have previously been successfully applied to studies of trophic relationships in aquatic ecosystems, primarily on invertebrates (Feller *et al.* 1979; Feller & Gallagher 1982; Calver 1984; Grisley & Boyle 1985, 1988; Boyle, Grisley & Robertson 1986), but also on seabirds (Walter, O'Neill & Kirby 1986).

Methodology

Although proteins obviously are degraded by mammalian digestive processes, some survive lengthy periods of acid digestion, as demonstrated by isoelectric focusing of *in vitro* digestates of fish muscle (Fig. 2).

We raised antisera to salmon proteins to evaluate the applicability of serological methods to salmon in seal diets (Boyle, Pierce *et al.* 1990; Pierce, Diack & Boyle 1990; Pierce, Boyle & Diack 1991a: see Fig. 3). The basic timing of the protocol reflects what is known about the time course of immune responses, but other details must be adjusted to suit particular needs (Hudson & Hay 1980). Anti-salmon antisera were then tested on

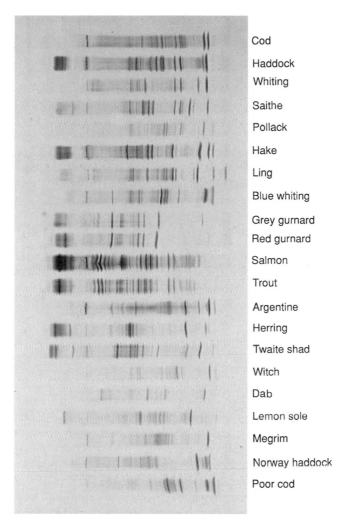

Cod

Haddock

Whiting

Saithe

Pollack

Hake

Ling

Blue whiting

Grey gurnard

Red gurnard

Salmon

Trout

Argentine

Herring

Twaite shad

Witch

Dab

Lemon sole

Megrim

Norway haddock

Poor cod

Fig. 1. Species differences in muscle proteins as revealed by isoelectric focusing banding patterns. From top row: cod *Gadus morhua*, haddock *Melanogrammus aeglifinus*, whiting *Merlangius merlangus*, saithe *Pollachius virens*, pollack *Pollachius pollachius*, hake *Merluccius merluccius*, ling *Molva molva*, blue whiting *Micromesistius poutassou*, grey gurnard *Eutrigla gurnardus*, red gurnard *Aspitrigla cuculus*, salmon *Salmo salar*, trout *Salmo trutta*, argentine *Argentina sphyraena*, herring *Clupea harengus*, twaite shad *Alosa fallax*, witch *Glyptocephalus cynoglossus*, dab *Limanda limanda*, lemon sole *Microstomus kitt*, megrim *Lepidorhombus whiffiagonis*, Norway haddock *Sebastes viviparus*, poor cod *Trisopterus minutus*.

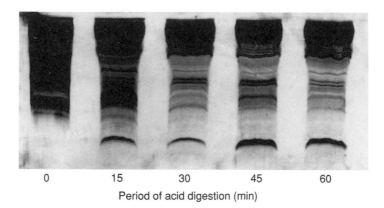

0   15   30   45   60

Period of acid digestion (min)

Fig. 2. IEF of *in-vitro* digestates of fish (sandeel) muscle.

protein extracts from muscles of a range of fish species, fish muscle digested *in vitro*, faeces of captive seals fed on salmon, seal digestive tracts which contained salmonid hard remains, and stomach contents of captive dolphins *Tursiops truncatus* fed on salmon.

Antisera raised to muscle protein extracts were the most satisfactory in terms of titre (strength) and specificity. The antisera were polyspecific, reacting with several different proteins in salmon muscle. There were individual differences in antisera from different rabbits, although these were mostly quantitative (Fig. 4).

By using fused rocket immuno-electrophoresis, it was shown that although the antisera precipitated proteins from muscle extracts of other fish species, the only 'reaction of identity' was with trout *Salmo trutta* proteins. Reactions of identity are distinguished on the gel by linkage of precipitin peaks of the unknown sample to those of a standard extract from the target fish species (Fig. 5), indicating that the same or extremely similar proteins were present in standard and unknown samples. Thus the antisera could be used to recognize salmonid flesh residues. In contrast, antisera raised to gadids, e.g. whiting *Merlangius merlangus*, were less specific: they reacted strongly with muscle proteins of other gadids and less closely related species (Fig. 6), and in this case linkages could not be used to detect the target prey in unknown samples.

With the salmon antisera, reactions of identity were also obtained for protein extracts from *in vitro* digests of salmon, salmonid remains in digestive tract contents of seals and dolphins, and faeces from captive seals fed on salmon (Fig. 7). Antisera to salmon were applied in a small-scale study of seal stomach contents on the east coast of Scotland (Pierce, Boyle &

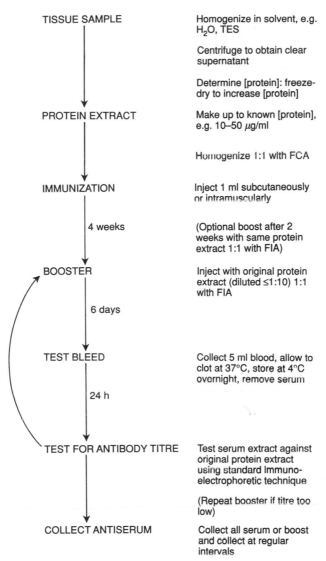

Fig. 3. Aspects of immunization protocol which may affect antiserum quality. TES = N-tris (hydroxymethyl) methyl-2-aminoethane sulphonic acid; FCA = Freunds complete adjuvant; FIA = Freunds incomplete adjuvant.

Diack 1991a). Salmonid proteins were identified in four digestive tracts although three of these also contained recognizable salmonid bones.

While it was possible to detect salmon protein residues in seal faeces with our antiserum, the reaction was relatively weak. Of antisera raised to muscle protein extracts of other fish species, the most successful in terms of

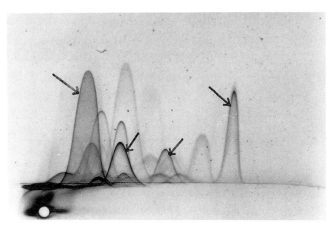

Fig. 4. Crossed immuno-electrophoresis gels showing individual differences in antisera to salmon from two different rabbits. Four reference peaks occurring in both antisera are arrowed.

strength of specific reaction to faecal protein residues was that raised to sandeel *Ammodytes marinus*. This antiserum was successfully applied to faecal samples from the Moray Firth common seal population (Pierce, Boyle, Diack & Clark 1990).

The potential for using serological methods to distinguish common and grey seal faeces has also been investigated (G. J. Pierce, J. S. W. Diack & P. R. Boyle, unpubl. data). The lining (mucosa) of seal digestive tracts is highly antigenic, but the resultant antisera could not be used to distinguish common seal and grey seal faeces, because a high proportion of proteins were similar in both species.

Antisera may be stored frozen (−20 °C) for prolonged periods without

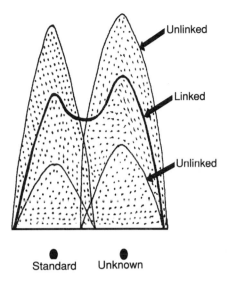

Fig. 5. Diagram of fused rocket gel showing linkage of precipitin peaks from the unknown sample to those from a standard extract (reaction of identity).

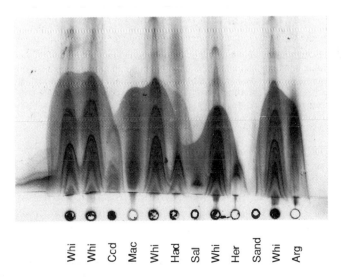

Fig. 6. Fused rocket gel indicating low specificity of antisera raised to gadid (whiting) muscle. Rockets from left to right are: Whiting, whiting, cod, mackerel *Scomber scombrus*, whiting, haddock, salmon, whiting, herring, sandeel, whiting, argentine.

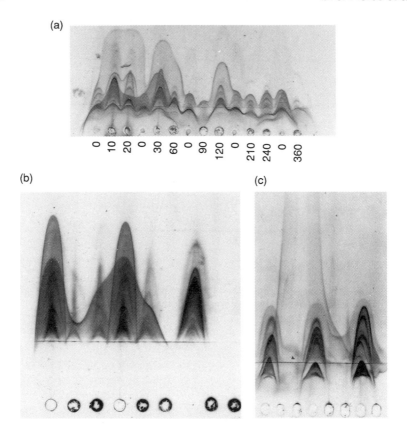

Fig. 7. Reactions of identity obtained for protein extracts from (a) *in vitro* digests of salmon, from left to right: 0, 10, 20, 0, 30, 60, 0, 90, 120, 0, 210, 240, 0, 360 min; (b) salmonid remains in a seal digestive tract; from left to right: standard salmon, oesophagus, stomach, salmon, stomach, intestine, salmon, intestine, colon; (c) faeces from captive grey seals fed on salmon; from left to right: salmon, faeces, faeces, salmon, faeces, faeces, salmon, faeces.

any obvious loss of activity. However, fish muscle proteins deteriorate during frozen storage (Laird, Mackie & Hattula 1980), although Osman *et al.* (1987) report that water-soluble protein extracts could be stored frozen for six months without any apparent deterioration. Thus, it is probably advisable to make standard protein extracts from fresh fish when needed.

This method is potentially applicable on a wider scale and probably offers the best hope for reliable quantification of the incidence of salmon in seal diets. However, there are disadvantages to using serological methods: (1) separate antisera must be raised to all species of interest, which is expensive, time-consuming, and requires use of live animals in which to raise antisera, and (2) in its present form, the method is essentially qualitative. Although there are quantitative methods for using antisera (e.g. ELISA), completely species-specific antisera and extensive calibration using captive-feeding experiments would be required.

Fish eye lenses frequently survive passage through seal digestive tracts and the use of lens proteins for species identification needs to be investigated. If the serological approach is to be used routinely, it would be preferable to isolate species-specific proteins to which antisera may be raised. Monoclonal antibody techniques appropriate for this purpose have the additional advantage that antiserum production may ultimately be carried out *in vitro* (Hudson & Hay 1980; Campbell 1984).

Identification of fish bones

Background

Studies on seal diets have been largely dependent on identification and measurement of fish otoliths, and various keys and guides to otoliths are available, notably Härkönen (1986). Cartilaginous fish, however, lack large paired morphologically distinct otoliths, and some teleosts also have small or fragile otoliths which are degraded or lost during passage through the gut. Other hard remains of fish, e.g. skeletal elements, scales and denticles, have been identified in some seal studies, but information on bone morphology is widely scattered in the literature (see Pierce & Boyle 1991), and identifications have been based on comparisons with reference collections of bones.

Studies of coastal otter diet have relied heavily on bones found in spraints for prey identification (Fairley 1972; Watson 1978; Mason & MacDonald 1980; Gormally & Fairley 1982; Herfst 1984; Murphy & Fairley 1985; Watt 1991). Most bones pass through otter digestive tracts intact, facilitating identification. These studies have also relied principally on reference collections for bone identification.

There are relatively few relevant guides to bones. Watson (1978) presented a preliminary key to the caudal vertebrae of inshore fish species eaten by otters, but its geographical and species coverage are limited. Izquierdo's (1988) guide to dentaries and articulars of fish species in Spanish waters covers a wide range of species but omits many which have been shown to be important in the diets of both seals and otters in the UK. There is also information on the identification of fish bones in the archaeological literature: see Casteel (1976) and Wheeler & Jones (1989).

Identification of bones in addition to otoliths increased the range of species detected, and the frequency with which most prey-types were detected, in seal faeces (Pierce, Boyle & Diack 1991b). For species with very small or fragile otoliths, such as salmonids and lumpsuckers (*Cyclopterus lumpus*), use of bones for identification markedly increased their apparent frequency of occurrence in the diet (Pierce, Diack & Boyle 1989; Pierce, Boyle & Diack 1991b).

Bones may be used to discriminate between species which are difficult to identify from their otoliths alone, e.g. between saithe, pollack and haddock, and between species of the Pleuronectidae. The use of bones in addition to otoliths may thus increase the precision of identification, and this is particularly important for groups of species with very different otolith size–body weight regressions or of widely differing commercial importance. Since many of the identifiable bones, e.g. jaw bones, are paired, they offer potential for estimating minimum numbers of prey consumed.

Development of guides

At the University of Aberdeen we have established, with support from two research grants from the Scottish Office Agriculture and Fisheries Department, a reference collection of skeletons of approximately 80 North Sea fish species, including most species of commercial importance.

Procedures have been evolved for the production of clean, disarticulated skeletons: fish are cooked in a microwave oven and the bulk of flesh is detached from the skeleton by hand. The microwave is preferred to boiling since no distortion of bone shape is caused. Bones are then left to soak in a solution of BIOTEX, for a few hours to a few days depending on bone size, washed with tap water, oven-dried, identified, and stored in airtight plastic vials. For some species, e.g. clupeids, it may also be necessary to apply a fat solvent.

Bones of potential value for species identification have been selected on the basis of diagnostic value, ease of recognition, and observed survival in faeces. Dentaries and articulars of a range of species are covered by Izquierdo (1988). Ongoing work at the University of Aberdeen is using premaxillae and vertebrae (Watt, Pierce & Boyle in prep.).

Sufficient adult specimens of fish species commonly found in seal diets have been obtained to establish regressions of fish size on otolith and bone size. Selected measurements of jaw bones and vertebrae have been shown to be good predictors of fish length (Wise 1980; Pierce & Boyle 1991). By selecting from a range of measurements it should be possible to estimate fish size from damaged as well as intact bones.

Discriminant analyses on measurements taken on caudal vertebrae and jaw bones have been used to demonstrate the potential for using numerical taxonomic methods to identify bones to species or family (Heba 1988; Izquierdo 1988).

Heba (1988) made a series of measurements on caudal vertebrae (Fig. 8)

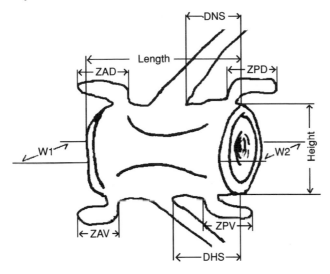

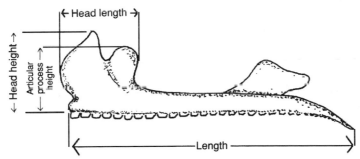

Fig. 8. Measurements made on premaxillae and vertebrae. W1 = anterior width, W2 = posterior width, L = length, H = height, DNS = distance from centrum posterior to neural spine, DHS = distance from centrum posterior to haemal spine, ZAD = length of anterior dorsal zygapophysis, ZAV = length of anterior ventral zygapophysis, ZPD = length of posterior dorsal zygapophysis, ZPV = length of posterior ventral zygapophysis.

of a range of North Sea fish and these data are reanalysed here. When dealing with isolated vertebrae it is not possible to correct for fish size and, in any case, caudal vertebrae decrease in size posteriorly. Therefore, the analysis used ratios between different measurements. Classification to family level using these data was not 100% successful (Table 1), but could no doubt be improved by increasing the number of variables. In particular, waisting and sculpturing of the centrum were not considered. However, it may be difficult to orientate individual vertebrae unless thoracic vertebrae

Table 1. Discriminant analysis on caudal vertebrae[a]

(a) Families. Characters used: DHS/W1, W2/Height, ZAV/Height, DNS/W2, ZAV/Length; see Fig. 8 for definitions

Group	Correct	Number of cases classified into group						
	(%)	Gadid	Angler	Pleuro-nectid	Gurnard	Clupeid	Mackerel	Salmonid
Gadid	91.1	102	1	5	0	3	0	1
Angler	80.0	0	4	0	0	0	0	1
Pleuronectid	80.0	1	1	8	0	0	0	0
Gurnard	100.0	0	0	0	5	0	0	0
Clupeid	60.0	1	0	0	1	6	0	2
Mackerel	80.0	0	0	0	1	0	4	0
Salmonid	90.0	0	0	1	0	0	0	9

(b) Haddock, pollack, saithe. Characters used: H/Length, DHS/W1, ZAV/DHS, ZPD/W1; see Fig. 8 for definitions

Species	Correct	Number of cases classified into species		
	(%)	Haddock	Pollack	Saithe
Haddock	60.0	9	2	4
Pollack	66.7	1	6	2
Saithe	90.0	0	2	18

[a] Vertebrae used (number of specimens in parentheses if > 1): Gadids: blue whiting, cod (2), four-bearded rockling *Rhinonemus cimbrius*, greater fork beard *Phycis blennoides*, haddock (3), hake (2), ling, pollack (2), saithe (4), three-bearded rockling *Gaidropsaurus vulgaris* (2), torsk *Brosme brosme*, poor cod, Norway pout *Trisopterus esmarkii*, whiting. Angler *Lophius piscatorius*. Pleuronectids: dab, plaice *Pleuronectes platessa*. Gurnard: grey gurnard. Clupeids: herring, twaite shad. Mackerel. Salmonids: salmon, trout.

are also available for comparison. The limited data also suggest that discrimination between related species with similar otoliths is possible, e.g. between haddock, pollack and saithe. Obviously sample sizes were very small and a much larger-scale project is needed to test these results.

Measurements on premaxillae (Fig. 8) may also be used to assist in species identification (G. J. Pierce & J. Watt unpubl. data; Table 2). Again, misclassifications could be reduced by using additional characters, e.g. tooth sockets.

Numerical taxonomic methods are arguably more difficult to use than guides and keys based on qualitative features. However, by using less obvious aspects of morphology, it may thus be possible to distinguish between species that have bones of similar appearance. Also, if features used in guides are coded numerically and included in discriminant analyses, the reliability of these features for identification can thus be quantified.

Table 2. Discriminant analysis on premaxillae[a]

(a) Families. Characters used: HH/HL, HL/L, HH/L, APH/L, APH/HH, APH/HL; character names abbreviated from Fig. 8

Group	Correct	Number of cases classified into group					
	(%)	Gadid	Pleuro-nectid	Bothid	Clupeid	Salmonid	Sandeel
Gadid	93.0	267	0	19	1	0	0
Pleuronectid	75.3	3	113	34	0	0	0
Bothid	96.4	1	0	27	0	0	0
Clupeid	83.1	0	0	0	54	11	0
Salmonid	88.9	0	0	0	1	8	0
Sandeel	100.0	0	0	0	0	0	12

(b) Haddock, pollack, saithe. Characters used: HH/L, APH/HH; character names abbreviated from Fig. 8

Species	Correct	Number of cases classified into species		
	(%)	Haddock	Pollack	Saithe
Haddock	86.8	46	1	6
Pollack	53.3	0	8	7
Saithe	77.8	0	4	14

(c) Poor cod, Norway pout, bib. Characters used: APH/HL, HH/L, APH/HH, HL/L; character names abbreviated from Fig. 8

Species	Correct	Number of cases classified into species		
	(%)	Poor cod	Norway pout	Bib
Poor cod	89.5	17	2	0
Norway pout	91.3	2	21	0
Bib	100.0	0	0	4

[a] Species used: Gadids: blue whiting, three-bearded rockling, four-bearded rockling, five-bearded rockling *Ciliata mustela*, cod, haddock, hake, ling, pollack, saithe, silvery pout *Gadiculus argenteus*, torsk, poor cod, Norway pout, bib *Trisopterus luscus*, whiting. Pleuronectids: dab, flounder *Platichthys flesus*, halibut *Hippoglossus hippoglossus*, lemon sole, long rough dab *Hippoglossoides platessoides*, plaice, witch. Bothids: brill *Scophthalmus rhombus*, megrim, scaldfish *Arnoglossus laterna*, turbot *Psetta maxima*. Clupeids: herring, sprat *Sprattus sprattus*. Salmon. Sandeel: *Ammodytes marinus*.

Quantification of diet composition

Background

A distinct source of error when measurements on otoliths are used to estimate the size of fish eaten is that otoliths are degraded by digestive processes. Previous work using captive seals and *in vitro* digestion has produced results on the relative digestibility of otoliths from different fish species and, in some cases, estimates of the degree of size reduction during passage through the gut (SMRU 1984; Härkönen 1986; Jobling & Breiby 1986; Harvey 1989). Otoliths are formed mainly of calcium carbonate, which is susceptible to acid digestion, and might thus be expected to be more digestible than bone, which is formed of less digestible calcium phosphate. It is not known if all otolith measurements decline at an equal rate, how this affects back-calculated fish weights, and whether similar size reduction is found in other fish hard parts.

The other class of prey organisms for which size can be estimated from hard parts is the cephalopods. Cephalopod beaks are reported not to be reduced in size during passage through marine mammal guts (Harvey 1989) and size is normally estimated from lower beak hood lengths—regressions for many species appear in Clarke (1986)—although upper beak hood length was used by Kashiwada, Reckseik & Karpov (1979). Regressions relating body size and upper and lower beak hood lengths for two common UK cephalopods, the octopus *Eledone cirrhosa* (G. J. Pierce unpubl. data), and the squid *Loligo forbesi* (Hastie 1990) are given in Table 3.

A preliminary study of otolith and bone digestion
Methods
In order to compare the resistance of bones and otoliths to mammalian digestive processes and to compare the utility of different measurements, we carried out *in vitro* acid digestion on otoliths and premaxillae of cod.

Thirty-six cod *Gadus morhua* of a range of sizes were prepared as described above. Total length and wet weight were recorded for complete specimens. Four measurements were made on the left otoliths: weight, length, breadth and thickness. Thickness was measured at five points along the otolith and the maximum value was used in analyses. Four measurements were made on the left premaxilla: weight, length, head height and head length. Each set of measurements was regressed against fish weight. Regression coefficients are summarized in Table 3. Fifteen premaxillae and 13 otoliths were digested *in vitro* by placing each bone/otolith in 100 ml of pH 2 HCl at 37 °C in a water bath for up to 15 h. The procedure was also attempted at pH 0.8 for a small number of otoliths. The original recipe included pepsin (Pierce, Diack & Boyle 1990) but this had no apparent effect on the time-course of digestion of hard parts. Measurements were made at 0, 30 and 60 min, and then at least every 2 h thereafter. Bones and otoliths were blotted prior to weighing. All measurements were converted to estimated fish weights by using the previously established regressions (Table 3).

Table 3. Regression of body weight on hard part measurements

(a) Cod body weight

Variable	Regression equation	n	r^2
Otolith weight	Log BW = 9.0981 + 1.5122 Log OW	21	0.987
Otolith length	Log BW = −5.3261 + 4.5412 Log OL	23	0.976
Otolith breadth	Log BW = −1.0471 + 4.2829 Log OB	23	0.974
Otolith thickness	Log BW = 1.5700 + 2.3700 OT	23	0.914
PMX weight	Log BW = 7.9714 + 0.9524 Log PW	23	0.992
PMX length	Log BW = −4.2818 + 3.3008 Log PL	24	0.971
PMX head height	Log BW = 0.2269 + 2.9212 Log PHH	24	0.962
PMX head length	Log BW = −0.0925 + 3.1374 Log PHL	24	0.971

(b) Cephalopod body weight & beaks

Variable	Regression equation	n	r^2
Loligo UH	Log BW = −2.3384 + 3.1576 Log UH	112	0.891
Loligo LH	Log BW = 1.2883 + 2.8745 Log LH	112	0.764
Loligo beaks	Log LH = −0.7863 + 0.9149 Log UH	112	0.806
Eledone UH	Log BW = 2.1103 + 2.3374 Log UH	31	0.742
Eledone LH	Log BW = 3.7158 + 1.7060 Log LB	31	0.557
Eledone beaks	Log LH = −0.3440 + 1.0557 Log UH	31	0.858

BW = body weight, Log = natural logarithm. PMX = premaxilla, LH = lower beak hood length, UH = upper beak hood length. The regression equations given are those with the highest r^2 value. In all cases, simple linear regression on untransformed data gave lower r^2 values. The *Loligo* regressions are based on approximately equal numbers of males and females but only female *Eledone* were available. For *Eledone* the regression LH = −0.2954 + 0.83421 UH has r^2 = 0.854; c.f. Pierce, Miller *et al.* (1991).

Results

Weights of individual otoliths declined at fairly constant rates although reduction in other measurements was less regular. Absolute weight loss after 6 h digestion increased with otolith size, although percentage weight loss was lower for bigger otoliths (Fig. 9). The rate of otolith digestion was obviously higher at the lower pH value (Fig. 10). When fish weight was back-calculated from all measurements at intervals over the time-course of digestion, fish size declined more rapidly when estimated from otolith measurements than when estimated from measurements on premaxillae (Fig. 11). After 6 h digestion, absolute apparent body weight loss was greater for larger fish. The percentage reduction, however, was smaller for larger fish except when based on otolith thickness (Fig. 12). This latter result presumably reflects the nature of the fish weight–otolith thickness relationship: otolith thickness increases more slowly as fish weight increases.

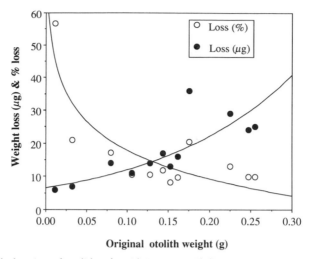

Fig. 9. Weight loss in cod otoliths after 6 h *in vitro* acid digestion.

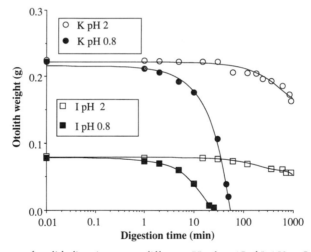

Fig. 10. The rate of otolith digestion at two different pH values (Cod I, 153 g, Cod K, 1243 g).

Discussion

All measurements of otoliths and jaw bones declined during acid digestion, but otolith size was reduced more rapidly than bone size and this is reflected in a greater drop in back-calculated fish weight. Otolith thickness appears to be a particularly unsuitable measurement: not only is it the most difficult measurement to make but the negative allometric relationship between

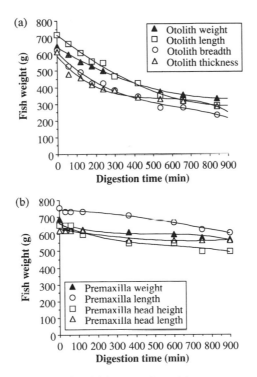

Fig. 11. Reductions in back-calculated fish size as derived from measurements on (a) otoliths and (b) premaxillae of the same fish (Cod B).

otolith thickness and fish weight means that fish weights are likely to be seriously underestimated for larger fish. Gadid otoliths are among the most resistant to digestion (Härkönen 1986), and size reduction is thus likely to be more marked in otoliths of other species.

From this small-scale study it is clear that the methodology could be used for comparisons between species and for other bones. However, despite standardized conditions, reduction in linear dimensions was somewhat irregular, perhaps influenced by local weaknesses in the structure of individual otoliths and bones, and cracks appeared in some otoliths. *In vivo*, hard parts may not be exposed immediately to the acidic conditions in the stomach, since they are initially encased in flesh and (in the case of otoliths) other bones. Even though premaxilla length is unlikely to be a useful measurement, owing to mechanical damage to the shaft of the bone, the head of the premaxilla is more robust. The *in vitro* experiments also do not consider the possible effect of digestion in the intestine and of changes in pH during digestion in the stomach.

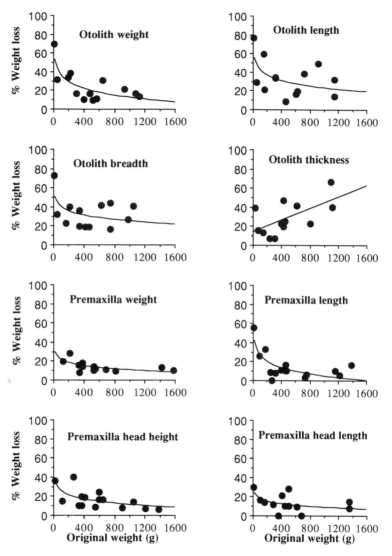

Fig. 12. Changes in back-calculated cod body weight after 6 h digestion, as a function of original fish size.

Acknowledgements

We thank Jane Diack, Ishbel Clark, Sara Frears, Lee Hastie, Hassan Heba, Linda Key, Andy Lucas, Al Miller and Helen Reilly for technical assistance. Fish specimens were obtained from SOAFD research cruises. We thank Andy Bryant and Dominic Tollit for comments on the manuscript. Much of the work described was funded by the Scottish Office Agriculture and Fisheries Department.

References

Boyle, P. R., Grisley, M. S. & Robertson, G. (1986). Crustacea in the diet of *Eledone cirrhosa* (Mollusca: Cephalopoda) determined by serological methods. *J. mar. biol. Ass. U.K.* **66**: 867–879.

Boyle, P. R., Pierce, G. J. & Diack, J. S. W. (1990). Sources of evidence for salmon in the diets of seals. *Fish. Res., Amst.* **10**: 137–150.

Calver, M. C. (1984). A review of ecological applications of immunological techniques for diet analysis. *Aust. J. Ecol.* **9**: 19–25.

Campbell, A. M. (1984). *Monoclonal antibody technology: the production and characterization of rodent and human hybridomas.* Elsevier, Amsterdam. (*Lab. Techn. Biochem. Molec. Biol.* **13**).

Casteel, R. W. (1976). *Fish remains in archaeology and paleo-environmental studies.* Academic Press, London, New York & San Francisco.

Clarke, M. R. (Ed.) (1986). *A handbook for the identification of cephalopod beaks.* Clarendon Press, Oxford.

Fairley, J. S. (1972). Food of otters (*Lutra lutra*) from Co. Galway, Ireland, and notes on aspects of their biology. *J. Zool., Lond.* **166**: 469–474.

Feller, R. J. & Gallagher, E. D. (1982). Antigenic similarities among estuarine soft bottom benthic taxa. *Oecologia* **52**: 305–310.

Feller, R. J., Taghon, G. L., Gallagher, E. D., Kenny, G. E. & Jumars, P. A. (1979). Immunological methods for food web analysis in a soft-bottom benthic community. *Mar. Biol., Berl.* **54**: 61–74.

Gormally, M. J. & Fairley, J. S. (1982). Food of otters *Lutra lutra* in a freshwater lough and an adjacent brackish lough in the West of Ireland. *J. Zool., Lond.* **197**. 313–321.

Grisley, M. S. & Boyle, P. R. (1985). A new application of serological techniques to gut content analysis. *J. exp. mar. Biol. Ecol.* **90**: 1–9.

Grisley, M. S. & Boyle, P. R. (1988). Recognition of food in *Octopus* digestive tract. *J. exp. mar. Biol. Ecol.* **118**: 7–32.

Hammond, P. S. & Harwood, J. (Eds) (1985). *The impact of grey seals on North Sea resources.* Sea Mammal Research Unit, Natural Environment Research Council, Cambridge.

Härkönen, T. (1986). *Guide to the otoliths of the bony fishes of the northeast Atlantic.* Danbiu ApS, Hellerup, Denmark.

Harvey, J. T. (1989). Assessment of errors associated with harbour seal (*Phoca vitulina*) faecal sampling. *J. Zool., Lond.* **219**: 101–111.

Hastie, L. (1990). *A preliminary study of geographic variation in the squid* Loligo forbesi *Steenstrup in Scottish coastal waters.* Unpubl. MSc thesis: University of Aberdeen.

Heba, H. M. A. (1988). *Fish skeletal structures as a tool for dietary analysis: use of caudal vertebrae.* Unpubl. MSc thesis: University of Aberdeen.

Herfst, M. S. (1984). Habitat and food of the otter *Lutra lutra* in Shetland. *Lutra* **27**: 57–70.

Hudson, L. & Hay, F. C. (1980). *Practical immunology*. Blackwell Scientific Publications, Oxford.

Izquierdo, E. R. (1988). *Contribución al atlas osteológico de los teleósteos Ibericos* 1. *Dentario y articular.* Ediciones de la Universidad Autonoma de Madrid, Madrid.

Jobling, M. & Breiby, A. (1986). The use and abuse of fish otoliths in studies of feeding habits of marine piscivores. *Sarsia* 71: 265–274.

Kashiwada, J., Reckseik, C. W. & Karpov, K. A., (1979). Beaks of the market squid, *Loligo opalescens*, as tools for predator studies. *Rep. Calif. coop. oceanic Fish. Invest.* 20: 65–69.

Laird, W. M., Mackie, I. M. & Hattula, T. (1980). Studies of the changes in the proteins of cod-frame minces during frozen storage at -15 °C. In *Advances in fish science and technology*: 428–434. (Ed. Connell, J. J.). Fishing News Books Ltd, Farnham, Surrey.

Laird, W. M., Mackie, I. M. & Ritchie, A. H. (1982). Differentiation of species of fish by isoelectric focussing on agarose and polyacrylamide gels—a comparison. *J. Ass. publ. Anal.* 20: 125–135.

Mason, C. F. & MacDonald, S. M. (1980). The winter diet of otters (*Lutra lutra*) on a Scottish sea loch. *J. Zool., Lond.* 192: 558–561.

Murphy, K. P. & Fairley, J. S. (1985). Food and sprainting places of otters on the west coast of Ireland. *Ir. Nat. J.* 21: 477–479.

Osman, M. A., Ashoor, S. H. & Marsh, P. C. (1987). Liquid chromatographic identification of common fish species. *J. Ass. off. anal. Chem.* 70: 618–625.

Pierce, G. J. & Boyle, P. R. (1991). A review of methods for diet analysis in piscivorous marine mammals. *Oceanogr. mar. Biol.* 29: 409–486.

Pierce, G. J., Boyle, P. R. & Diack, J. S. W. (1991a). Digestive tract contents of seals in Scottish waters: comparison of samples from salmon nets and elsewhere. *J. Zool., Lond.* 225: 670–676.

Pierce, G. J., Boyle, P. R. & Diack, J. S. W. (1991b). Identification of fish otoliths and bones in faeces and digestive tracts of seals. *J. Zool., Lond.* 224: 320–328.

Pierce, G. J., Boyle, P. R., Diack, J. S. W. & Clark, I. (1990). Sandeels in the diets of seals: application of novel and conventional methods of analysis to faeces from seals in the Moray Firth area of Scotland. *J. mar. biol. Ass. U.K.* 70: 829–840.

Pierce, G. J., Boyle, P. R. & Thompson, P. M. (1990). Diet selection by seals. In *Trophic relationships in the marine environment. Proceedings of the 24th European marine biology symposium*: 222–238. (Eds Barnes, M. & Gibson, R. N.). Aberdeen University Press, Aberdeen.

Pierce, G. J., Diack, J. S. W. & Boyle, P. R. (1989). Digestive tract contents of seals in the Moray Firth area of Scotland. *J. Fish Biol.* 35 (Suppl. A): 341–343.

Pierce, G. J., Diack, J. S. W. & Boyle, P. R. (1990). Application of serological methods to identification of fish prey in diets of seals and dolphins. *J. exp. mar. Biol. Ecol.* 137: 123–140.

Pierce, G. J., Miller, A., Thompson, P. M. & Hislop, J. R. G. (1991). Prey remains in grey seal (*Halichoerus grypus*) faeces from the Moray Firth, north-east Scotland. *J. Zool., Lond.* 224: 337–341.

Pierce, G. J., Thompson, P. M., Miller, A., Diack, J. S. W., Miller, D. & Boyle, P. R.

(1991). Seasonal variation in the diet of common seals (*Phoca vitulina*) in the Moray Firth area of Scotland. *J. Zool., Lond.* **223**: 641–652.

Prime, J. H. & Hammond, P. S. (1990). The diet of grey seals from the south-western North Sea assessed from analyses of hard parts found in faeces. *J. appl. Ecol.* **27**: 435–447.

SMRU (1984). *Interactions between grey seals and UK fisheries. A report on research conducted for the Department of Agriculture and Fisheries Scotland by the Natural Environment Research Council's Sea Mammal Research Unit 1980 to 1983.* Sea Mammal Research Unit, Natural Environment Research Council, Cambridge.

SMRU (1988). *Multispecies fishery assessment in the North Sea: estimation of mortality caused by marine mammals.* Final Report on Contract with DGXIV–B–1 of the Commission of the European Communities, Sea Mammal Research Unit, Natural Environment Research Council, Cambridge.

Thompson, P. M., Pierce, G. J., Hislop, J. R. G., Miller, D. & Diack, J. S. W. (1991). Winter foraging activity by common seals (*Phoca vitulina*) in relation to food availability in the inner Moray Firth, N.E. Scotland. *J. Anim. Ecol.* **60**: 283–294.

Walter, C. B., O'Neill, E. & Kirby, R. (1986). 'ELISA' as an aid in the identification of fish and molluscan prey of birds in marine ecosystems. *J. exp. mar. Biol. Ecol.* **96**: 97–102.

Watson, H. (1978). *Coastal otters in Shetland.* Vincent Wildlife Trust, London.

Watt, J. (1991). *Prey selection by coastal otters* (Lutra lutra L). Unpubl. PhD thesis: University of Aberdeen.

Wheeler, A. & Jones, A. K. G. (1989). *Fishes.* Cambridge Manuals in Archaeology. Cambridge University Press, Cambridge.

Wise, M. H. (1980). The use of fish vertebrae in scats for estimating prey size of otters and mink. *J. Zool., Lond.* **192**: 25–31.

Symp. zool. Soc. Lond. (1993) No. 66: 263–291

Milk secretion in marine mammals in relation to foraging: can milk fatty acids predict diet?

Sara J. IVERSON

*Canadian Institute of Fisheries
   Technology
Technical University of Nova Scotia
1360 Barrington Street, PO Box 1000
Halifax, Nova Scotia, B3J 2X4
Canada*

Synopsis

Patterns of milk composition and delivery differ greatly among mammals. However, despite these species-specific differences, lactation is very expensive in all species and thus diet and nutrient reserves of individual females must play a critical role in lactation performance. Pinnipeds, as well as mysticete whales, exhibit extreme adaptations to the constraints imposed by the separation of maternal feeding from milk transfer. This paper will consider three main questions: (1) how are milk secretion processes adapted to the temporal separation of foraging and milk secretion; (2) how are changes in diet or nutrient reserves likely to affect milk composition or yield; and (3) can specific milk constituents be used to indicate foraging behaviour or diet of individual animals? Substantial quantities of nutrients and metabolites are required by the mammary gland for the secretion of milk constituents. Nutrient partitioning and milk secretion are physiological processes which are both highly regulated and biochemically constrained. The general principles of these processes appear to be shared among all mammals, including marine mammals. It is concluded that neither the levels nor the types of most milk constituents are likely to be affected by maternal diet in marine mammals. However, milk yield may be reduced during a low plane of nutrition. Unlike other constituents, such as protein or carbohydrate, dietary fatty acids essentially remain intact through the digestion process (in carnivorous mammals) and many of these are secreted in milk or deposited in adipose tissue with no or minimal modification. Recent studies on species such as the California sea lion, Antarctic fur seal, hooded seal and harbour seal suggest a strong potential for determining prey items and diet of marine mammals through fatty acid signatures in the milk, particularly given the complex array of fatty acids which exist in marine organisms.

ZOOLOGICAL SYMPOSIUM No. 66
ISBN 0–19–854069–8

Copyright © 1993 The Zoological Society of London
All rights of reproduction in any form reserved

Introduction

Milk is the complex lacteal secretion of the mammary gland which is responsible for the provision of nutrients and energy to the growing neonate. Among mammals, there are large differences in the composition of milk, as well as in milk output, frequency of nursing and duration of lactation. However, despite species-specific differences in milk secretion patterns, lactation represents the greatest energetic cost of reproduction in female mammals (Blaxter 1962; Millar 1977; Oftedal 1985), requiring large amounts of nutrient transfer and elevated maintenance costs. Thus maternal diet and nutrition must play a critical role in the ability of individual mothers to meet the nutrient requirements of milk secretion.

Most mammals are able to mobilize body reserves, especially fat, to partially compensate for energy deficits during lactation (Young 1976; Bauman & Elliot 1983). This ability has been critical in the evolution of lactation strategies of many species, but in particular to those which must cope with long-term spatial and temporal separation of maternal feeding and milk secretion. Many marine mammals (e.g. pinnipeds and mysticete whales) have evolved unusual patterns in both milk composition and output which reflect this separation of foraging and lactation. Thus, understanding the influence of maternal diet and nutrient reserves on milk secretion in individuals is important to interpretations of reproductive patterns, foraging ecology and life history strategies in these species.

The interaction between maternal feeding and lactation in marine mammals may be considered by posing three questions: (1) how are milk secretion processes adapted to the constraints imposed by the temporal separation of foraging and milk secretion; (2) how are changes in diet or nutrient reserves likely to affect milk composition or yield; and (3) can specific milk constituents indicate foraging behaviour or diet of individual animals? Each of these questions involves understanding the relationship between the supply of nutrients and metabolic substrates to the secretion of milk products at the level of the mammary gland. Although direct data are often lacking for marine mammals, in some cases observations from other species provide reasonable clues; pinnipeds are emphasized in this paper, since most quantitative work has been done on this marine mammal group. Finally, because milk fat represents the primary and most interesting constituent of marine mammal milks and because lipid metabolism plays such a large role in the lactation process in these species, most attention is given here to lipids and their secretion in milk.

The overall objective of this paper is twofold: the first aim is to provide an overview of the physiological and biochemical processes involved in secretion of the major organic constituents of milk, which is necessary to interpret data on milk composition and production. On the basis of understanding these processes, the second aim is to explore the use of milk fatty acids as a means of determining marine mammal diets.

Foraging and lactation in marine mammals: adaptation vs. physiological response

In pinnipeds, two general lactation strategies have evolved to cope with the constraints imposed on the mother which must feed at sea but suckle her pup on land (Bonner 1984; Oftedal, Boness & Tedman 1987). Otariids lactate during an initial perinatal fast of about 5–9 days, then begin alternating regular foraging trips with suckling periods of 1–2 days. Thus milk is initially synthesized from substrates derived from maternal tissues and subsequently from nutrients acquired while foraging during lactation. Lactation generally lasts one year or more in these species. By contrast, phocids, in particular large phocids, are able to store greater energy reserves as blubber prior to parturition, and fast during a brief intense lactation period of 4–50 days. Hence all substrates for milk synthesis are derived from maternal tissues. Smaller phocid species, such as the harbour seal (*Phoca vitulina*), may not be able to store sufficient energy reserves to support lactation, and may initiate foraging during the nursing period (Bowen, Oftedal & Boness 1992).

In mysticete whales, their larger body size (3000–150 000 kg; Macdonald 1984: 214–235) permits far greater energy reserves than phocids possess. Most mysticetes undergo seasonal migrations and give birth and begin the first months of the 6–10 month lactation period in tropical areas of low food availability (Harrison 1969; Rice & Wolman 1971; Gaskin 1982). Hence during the first half of lactation milk must be synthesized primarily from stored nutrients.

Thus, pinnipeds and mysticetes have evolved lactation patterns which may reflect such factors as the need to conserve maternal water and to reduce lactation length during fasting. For instance, milks of these species tend to be concentrated and especially high in fat (30–60%; Gaskin 1982; Oftedal, Boness & Tedman 1987), and are often associated with large daily milk energy outputs. But to evaluate the effects of diet, nutrition, and foraging strategies on milk composition or production, it is important to distinguish between evolutionary adaptations and physiological responses on an ecological time-scale which might affect an individual animal during lactation. For example, milk fat content is correlated with average duration of maternal foraging trips when compared across various species of otariids (Trillmich & Lechner 1986). However, this does not mean that if an individual female otariid changes her foraging trip duration she will be able to 'adjust' the composition of her milk.

Diet, physiology, or even body size have been suggested as factors which might affect the proximate milk composition of an individual (e.g., Kretzmann *et al.* 1991). However, given the physiological mechanisms responsible for milk secretion, it is unlikely that such factors would affect milk composition. For instance, in no mammal does the proximate milk composition reflect the composition of its diet. Although the patterns observed in the proximate composition of milk across species may have evolved in response to nutritional and environmental constraints, they are clearly not a function of the diet itself. Rabbits (order Lagomorpha) and black bears (*Ursus americanus*) produce relatively high-fat milks (15% and 28% fat, respectively; Cowie 1969; Iverson & Oftedal 1992), even though they feed primarily on succulent and leafy vegetation. Pinnipeds uniformly produce high-fat milks regardless of whether species typically feed on high-fat fish or low-fat invertebrates or whether they fast entirely. Within species, the fat concentration of milk is equally unlikely to be directed by diet. The proximate composition of milk within a species is also unlikely to be influenced by body size. For example, although there exist hundreds of breeds of the domestic dog (*Canis lupus*), which vary tremendously in body size (e.g. dachshunds to Saint Bernards), the proximate composition of dog milks exhibits little variability and no correlation to body size (Russe 1961; Oftedal 1984a).

Finally, the factors which might influence the composition of milk are different than those which influence the yield of milk. Therefore, attempts to link ecological factors to the proximate milk composition of individuals should first have a physiological basis. It is for this reason that we must understand milk secretion and nutrient partitioning to the mammary gland as physiological processes, which are both highly regulated and biochemically constrained. Certain aspects of these processes may be highly influenced by diet and nutrition, but most patterns are genetically and evolutionarily determined among differing taxa.

Milk secretion

The concept of nutrient partitioning

Substantial quantities of nutrients and metabolites are required by the mammary gland for the synthesis and secretion of milk. In fact, the demand for nutrients by the mammary gland in all mammals is so great that the general body metabolism of the mother must be altered and organized in such a way that the appropriate nutrients are efficiently partitioned to the mammary glands for milk synthesis and secretion. Thus, milk secretion takes place at the expense of other biological processes (Patton & Jensen 1976; Bauman & Currie 1980; Williamson 1980; Bauman & Elliot 1983).

A number of physical and metabolic changes in tissues occur during lactation to support milk secretion. The hormonal regulation of this process can be viewed as a combination of several types of controls, namely, homeostasis (maintaining physiological equilibrium of the whole body) and homeorhesis ('orchestrated changes for the priorities of a physiological state') (Bauman & Currie 1980). The primary objectives of homeorhesis are to increase the availability of fatty acids, amino acids, glucose and metabolites to the mammary gland. This often begins with increased food intake and digestion, accompanied by hypertrophy of the digestive tract and increased absorptive capacity (Cripps & Williams 1975; Millar 1979), and continues with changes in tissue metabolism, mobilization of body reserves, increased mammary blood flow and a priority of nutrient use by the mammary gland (Hanwell & Peaker 1977; Mepham 1983). A number of hormones may be involved in homeorhetic control, which are not yet fully understood, but these probably include prolactin, progesterone and growth hormone (Bauman & Elliot 1983).

The nutrient requirements of the mammary gland are essentially species-specific or characteristic of a phylogenetic group. Within species, changes in nutrient requirements often occur with stage of lactation (i.e. with changes in milk composition or yield). These changes are hormonally and biochemically directed (Cowie, Forsyth & Hart 1980; Jenness 1986). The regulation of nutrient utilization during lactation in all species allows milk secretion by the mammary gland to be robust to many fluctuations in dietary intake or composition. Thus, during periods of feeding, nutrients from dietary intake are partitioned first to the mammary glands, but during fasting or when nutrient intake is insufficient to meet demands, body nutrient stores are mobilized for milk secretion. In other words, homeorhetic regulatory mechanisms alter body metabolism to make available what the mammary gland requires (Bauman & Currie 1980; Williamson 1980). Although there are numerous examples of homeorhesis in nutrient partitioning, some of the best examples occur in female phocids, mysticete whales and winter-dormant bears, which fast throughout all or a large portion of lactation and produce milk derived entirely from the mobilization of body stores.

Dietary lipid and nutrient partitioning

Of the major physical and metabolic tissue changes which occur in mammals during lactation, those involving lipid metabolism and adipose tissue stores (and thus energy requirements) are probably of greatest importance and magnitude (Bauman & Elliot 1983). In marine mammals this is particularly the case since a considerable component in the reproductive strategy involves alternating between deposition and mobilization of fat reserves for milk synthesis.

Fatty acids are essentially the building blocks of lipids, being the largest constituent of neutral lipid compounds such as triglycerides (glycerol esterified to three fatty acids) and phospholipids. The most common fatty acids in plants and animals are composed of 14–22 carbon atoms in even-numbered straight chains, containing a terminal methyl end and a terminal carboxyl end. The chain may be saturated with no double bonds or unsaturated, containing from one to six double bonds[1]. Organisms are able to biosynthesize, modify chain-length and introduce double bonds to fatty acids, but can only do so subject to biochemical limitations and differences in these processes depending on the phylogenetic group.

In monogastrics, ingested lipid is hydrolysed in the stomach and small intestine to monoglyceride, glycerol and free fatty acids. These products are transported across the mucosal lining of the small intestine during which they are re-esterified into triglycerides and incorporated into circulating lipoproteins (i.e. the chylomicrons) for transport to tissues (Borgstrom 1977; Patton 1981). Thus, with the exception of ruminants, fatty acids essentially remain intact during digestion and those found in chylomicrons reflect the original dietary composition. Circulating very low density lipoproteins (VLDL) carry endogenously derived triglycerides, for instance from the mobilization of adipose tissue. The circulating chylomicrons along with VLDL are the only lipoprotein classes which have been found to contribute triglyceride fatty acids to the mammary gland (Annison 1983; Hamosh & Hamosh 1985).

Lipoprotein lipase (LPL) is the key enzyme involved in channelling circulating triglycerides to various organs. LPL is a tissue-bound enzyme responsible for hydrolysing circulating triglyceride carried in chylomicrons and VLDL, thus facilitating the uptake of fatty acids by most tissues, including adipose tissue and the mammary gland (Hamosh & Hamosh 1985). During fasting or starvation, activity of LPL in adipose tissue decreases dramatically, which stops the uptake of fatty acids, while adipocyte hormone-sensitive lipase becomes active, which catalyses the release of fatty acids from adipose tissue into the circulation (Allen 1976).

The homeorhetic control of LPL activity, studied primarily in humans, rodents and ruminants, is also one of the more striking examples of nutrient partitioning during lactation. Shortly before parturition, LPL activity in adipose tissue decreases sharply, while LPL activity in the mammary gland (previously non-existent) appears (Hamosh et al. 1970; Bauman & Elliot 1983; Hamosh & Hamosh 1985). Although numerous factors may be involved in this change, prolactin is believed to be primarily responsible for LPL regulation during lactation (Zinder et al. 1974; Bauman & Elliot 1983). LPL activity in adipose tissue remains almost completely depressed

---

[1] According to IUPAC nomenclature, fatty acids are herein designated as carbon chain length:number of double bonds and n-x, if present, denotes the position (x) of the last double bond relative to the terminal methyl carbon.

throughout most of lactation because of the continued secretion of prolactin, particularly in response to suckling (Hamosh & Hamosh 1985), while LPL activity in the mammary gland rises sharply after parturition (McBride & Korn 1963; Robinson 1963; Hamosh 1989) and may continue to rise during lactation (Mehta, Jones & Hamosh 1982). Thus most dietary fatty acids as well as those mobilized from body stores are directed to the mammary gland during lactation. In humans, evidence suggests that the increase in milk fat content from colostrum to mature milk is accompanied by an increase in milk LPL levels (Mehta et al. 1982; Hamosh 1989). In guinea pigs and rats, LPL activity is correlated with the degree of uptake of lipid by the mammary gland (McBride & Korn 1964; Mendelson & Scow 1972; Scow 1977).

Recent evidence suggests that a similar pattern may occur in phocids. In harbour seal and grey seal (*Halichoerus grypus*) females at parturition, general activity levels of LPL (released into the circulation after heparin administration) are low or non-existent, consistent with depression in adipose tissue as well as perhaps a delay in its rise in the mammary gland. However, during mid to late lactation, LPL levels in both species are up to tenfold higher than levels at parturition and sixfold higher than those in humans (Iverson, Hamosh & Bowen 1991, and unpubl. data). During late lactation, both the concentration of milk fat and the amount secreted are at maximum levels in fasting grey seal mothers (Iverson, Bowen et al. 1993). Thus, this dramatic rise in LPL activity is probably due to mammary gland activity, since adipose tissue will only be releasing fatty acids. The pattern of rapidly increasing lipid content of milk during early lactation in some phocids (e.g. Peaker & Goode 1978; Riedman & Ortiz 1979; Iverson, Bowen et al. 1993), might be in part explained by rising LPL levels at the onset of lactation, although this has not yet been verified. This hypothesis is supported by evidence that prolactin (which regulates these changes in mammary LPL) is essential for the maintenance of lactation, at least in otariids (Boyd 1991). Given that milk fat content increases over lactation in these phocid species while females fast, the pattern of change is clearly not related to diet.

The milk secretory process and pinnipeds

Despite the extreme variation in milk composition and output among mammals, there appear to be only a few mechanisms responsible for the secretion of all major organic constituents of milk and these mechanisms appear to be identical among all mammals including pinnipeds (Tedman & Bryden 1981; Mather & Keenan 1983; Tedman 1983). The morphology and ultrastructure of the mammary gland, at the level of ducts and alveoli, is also relatively homogeneous among all species studied (Patton & Jensen 1976; Mepham 1983; Oftedal, Boness & Tedman 1987). In the developed mammary gland, milk is synthesized in secretory (epithelial) cells which line the central alveolar cavities (lumina). Milk is then secreted from the cells, each of which produces the complete milk product, into the alveolar lumina which drain into the system of ducts towards the body surface and teat (Jenness 1974; Mepham 1983). Thus, although relatively little work has been done on pinnipeds, it appears that the underlying processes of milk secretion are similar to those of other mammals. The unusual patterns observed in milk delivery in many pinnipeds suggest that these species may have some physiological adaptations uncommon in other mammals. Is there evidence for such adaptations?

The secretion of milk fat globules in all mammalian species occurs by apocrine action (the extrusion of lipid droplets from the cell, taking with them some of the surrounding cell membrane) (Mather & Keenan 1983; Tedman 1983). Since milk fat content of most mammals studied ranges from about 4 to 8%, we might expect differences in the total amount of membrane required and in turnover of membrane components in pinnipeds secreting milk containing 50–60% fat. However, pinnipeds may partially alleviate membrane utilization by secreting larger diameter globules which require less membrane per unit fat secreted (Tedman 1983).

Given the magnitude of daily nutrient output by lactating phocids, we might expect the mammary glands in these species to be unusually large compared to those of other mammals. Mammary glands isolated from a single lactating Weddell seal and a single hooded seal (Oftedal, Boness & Tedman 1987) were larger than predicted by the allometric equation for terrestrial mammals (Hanwell & Peaker 1977). However, data from lactating harbour seals ($n = 13$) indicate that mammary gland weight (1.9 kg; Bowen et al. 1992) is reasonably predicted (1.8 kg) from the allometric equation. If mammary glands are not larger than expected in most phocids, we might expect a greater synthetic capacity of secretory cells or a greater efficiency and degree of the homeorhetic regulation of nutrient partitioning to the secretory cells. Phocids may be particularly efficient in the provisioning (from blubber through LPL regulation as discussed above) and transport (through increased mammary blood flow; e.g. Linzell 1974, Hanwell & Peaker 1977) of nutrients, metabolites and energy to the sites of synthesis.

Since the evacuation of milk from the alveolar lumina is of critical importance in sustaining and maximizing the secretory process in other mammals (Mepham 1983), we might expect that the phocid neonate is more proficient at milk removal and is able to digest greater loads of nutrients than are the young of other mammals. Pinniped neonates are extremely efficient at lipid digestion compared to other species (Iverson, Sampugna & Oftedal 1992) but we do not know whether it is the mother or pup that controls milk delivery (e.g. see Iverson, Bowen et al. 1993).

In otariids, the secretion mechanisms which enable females to maintain lactation despite long intervals of sometimes many days between suckling are not understood. However, if we assume that a constant rate of milk secretion into alveoli occurs during both suckling and foraging trips, the maximum volume of milk stored in mammary glands during the foraging trip of 2.5–6 days in the California sea lion (*Zalophus californianus*) or northern fur seal (*Callorhinus ursinus*), would be 60 and 170 $g/kg^{0.75}$, respectively (calculated from Costa & Gentry 1986, and Oftedal, Iverson & Boness 1987). Assuming that mammary glands in otariids scale to body mass as in other species, this maximum is similar to the range of the total daily secretion volumes (45–158 $g/kg^{0.75}$; Oftedal 1984b) of a number of terrestrial carnivores and rodents, as well as the rabbit which suckles only once per day. Furthermore, milk secretion rates in otariids are likely to be somewhat reduced during foraging trips in any case owing to end-product inhibition and increased hydrostatic pressure within the mammary gland (e.g., Mepham 1983). However, the extraordinary length (greater than 20 days) of foraging trips in some Juan Fernandez fur seals (*Arctocephalus philippii*) (J. Francis & D. J. Boness pers. comm.) cannot easily be explained by the above argument.

Do diet or nutrient reserves affect milk composition or yield in marine mammals?

Dietary constituents and milk products

The principal constituents of all milks are water, lipids secreted in the form of globules, proteins comprising caseins and whey, carbohydrates and major inorganic salts (Jenness 1974; Johnson 1974; Davies, Holt & Christie 1983; Tedman 1983). The major organic constituents of diets in general are proteins, carbohydrates and lipids. In monogastrics, the primary digestion products of these dietary components are amino acids, simple monosaccharides (e.g. glucose) and fatty acids, respectively.

The major milk proteins of both casein and whey are synthesized within the mammary gland from individual amino acids present in the secretory cells; these amino acids are in turn derived either from direct uptake (all essential and some non-essential amino acids) or from biosynthesis (non-essential amino acids) within the mammary gland (Mepham 1971; Mather & Keenan 1983). The synthesis and secretion rates of these milk proteins are directed by a specific group of genes which become hormonally expressed during pregnancy (Mather & Keenan 1983; Mercier & Gaye 1983).

The primary milk carbohydrates of most mammals are synthesized within the mammary gland from simple monosaccharides taken up by the gland.

Carbohydrate concentration and composition tend to be specific to species and lactation stage (Messer & Green 1979; Oftedal 1984b), and are regulated by the presence of specific enzymes and proteins in the mammary gland under hormonal influence (Brew, Vanaman & Hill 1968; Brew 1970; Kuhn 1983). In most pinniped milks, carbohydrate is absent or less than 1% by weight throughout lactation (Oftedal, Boness & Tedman 1987). This is probably true also of many cetaceans and hence is of minor importance in these species. Thus, by analogy to other species, neither the nature nor the amount of protein and carbohydrate in the milks of marine mammals will be directed by the composition of diet.

However, the relationship between diet and the nature of milk fat differs from that of protein or carbohydrate. Milk lipid is composed of 98–99% triglyceride in all mammals studied, including otariids and phocids (Iverson, Sampugna *et al.* 1992). However, striking differences occur among species in the composition of the triglyceride fatty acids. Milk fatty acids originate from two sources, namely (1) the direct uptake of circulating fatty acids and (2) *de novo* synthesis within the mammary gland from metabolites such as acetate and NADPH which provide sources of carbon and energy. Thus, the variation in milk fatty acid composition among species must arise from differences in the composition of circulating fatty acids (from diet or mobilization) and from species-specific differences in types and rates of *de novo* synthesis.

The extent to which *de novo* synthesis occurs in the mammary glands of carnivores or pinnipeds is not known. *De novo* synthesis of fatty acids is easily detectable when short- (4:0 and 6:0) or medium-chain (8:0–12:0) fatty acids are present in milk, since the enzymes which produce these fatty acids are found only in the mammary glands of some species (Strong & Dils 1972; Dils, Clark & Knudsen 1977; Dils 1983). The fatty acids usually produced in other tissues (e.g. liver, adipose and some species' mammary tissue) are 16:0 or 18:0. The short- and medium-chain fatty acids are found in relatively specific proportions primarily in the milks of ruminants and some monogastrics (e.g. primates, rodents, rabbits and elephants), but not in carnivores and marine mammals. Thus, milk fatty acids in carnivores are thought to be derived primarily from direct uptake from the circulation. Given the nature of dietary lipid digestion and absorption in monogastric species (see pp. 267–268), milk fatty acid composition will be strongly directed by dietary fatty acid composition.

The relationship of diet and nutrient reserves to the proximate composition and yield of milk in marine mammals

Given that lactation is essential for the survival of mammalian offspring, we might expect high priority to be given to milk secretion and to the constancy of nutrient ratios to which the young are adapted. The effects of diet on proximate milk composition and production have been studied primarily in dairy breeds, other domestic animals, and humans. Although no direct studies have been conducted on marine mammals, given the similarity in the milk secretion processes in all mammals studied to date, observations from non-dairy species may be informative. The protein sequencing genes of the mammary gland require a specific array of amino acids (Bauman & Elliot 1983) and only if important component amino acids are unavailable are rates of synthesis likely to be limited; however, the product is unlikely to be altered. Considerable variations in diet, including a poor-quality diet, do not affect the nature of milk proteins and appear to have only a small effect on milk protein content in humans (Lonnerdal *et al.* 1976; Jenness 1979). During starvation in some species, there may be differences in the precursors used for milk fat synthesis (Annison 1983), but in non-dairy species there are virtually no reliable data which indicate that milk fat concentration will be altered by changes in diet (e.g. National Research Council 1991). Numerous studies on the effects of diet composition, quality, quantity or malnutrition on milk fat in humans have concluded that milk fat concentration is little if at all changed by such variables (World Health Organization 1985: 3–22; Jensen 1989). In species adapted to fasting during lactation, total milk fat must be independent of dietary fat. The fact that fasting phocids produce milk containing a relatively constant protein content, up to 60% fat, and increasing milk energy output during the fasting period (e.g. Iverson, Bowen *et al.* 1993), serves as a good example of the control of nutrient partitioning and the unresponsiveness of proximate milk composition to diet.

Despite regulation mechanisms which 'place a priority' on milk secretion, lactation cannot continue without sufficient metabolic substrates for synthesis. When nutrients or reserves become severely limited, production cannot be sustained. However, in non-dairy species there is little evidence that the proximate composition of milk will become significantly altered, unless disease states or metabolic disorders occur. Instead, milk yield will decrease or stop altogether. This has been demonstrated in studies of both penned and pasture-fed ungulates (reviewed in Oftedal 1985). There is also evidence that this occurred in the California sea lion during the reduced food availability associated with the 1983–84 El Niño (Iverson, Oftedal & Boness 1991). Milk composition was not altered, but milk output was reduced. Thus, on a low plane of nutrition, milk output rather than milk composition appears to be altered. Milk yield might also be influenced by body size if body size is an indicator of total nutrient availability, as in phocids adapted to using only body stores to support lactation. In grey seals, smaller females produce less milk than larger females, but milk composition follows the normal postpartum increase in fat content with lactation (Iverson, Bowen *et al.* 1993).

Analysis and interpretation

To assess the relationship between diet and proximate composition of milk in any species, it is important to account for species-specific changes in composition which occur as a function of the lactation process and its regulation, e.g., with lactation stage (Oftedal 1984b; Oftedal, Boness & Tedman 1987; Jensen 1989). Existing data for most marine mammal species are often characterized by small sample sizes and little or imprecise information on lactation stage (Oftedal, Boness & Tedman 1987). Thus it is important to evaluate sampling schemes.

Given the predominance of lipid over all other components in marine mammal milk, accurate assays for milk fat are critical. Firstly, gastric samples cannot be used to assess fat content of milk, not even on a relative basis. Gastric lipase is present in neonates of both phocids and otariids and may result in up to 56% hydrolysis of milk triglyceride in the stomach (Iverson, Sampugna et al. 1992). Hydrolysis is associated with more rapid passage of fat than other components of dry matter and both are a function of time since ingestion (Iverson, Kirk et al. 1991; Iverson 1988). Thus estimated fat concentration (on both a fresh and a dry weight basis) will also vary with time since ingestion.

The use of extraction methods appropriate to the condition of the sample is also essential for total isolation and accurate quantitation of lipid. Milks, including those of pinnipeds, possess endogenous lipases which hydrolyse triglyceride to free fatty acids and partial glycerides (Hernell & Olivecrona 1974; Freed et al. 1986). The Roese-Gottlieb method for analysis of total milk fat is widely recommended for appropriate samples (Horwitz et al. 1975; Jensen et al. 1985). However, it may significantly underestimate fat content (by as much as 25%) for samples of milk which have undergone extensive hydrolysis, since this method does not extract free fatty acids (Iverson 1988). This has important implications for the analysis of milks that have been stored in various preservatives, such as sodium azide, which do not inhibit hydrolysis. In these cases, extractions using chloroform/methanol, such as the Folch or Bligh and Dyer procedures (reviewed in Christie 1982), should be used. In all cases, gravimetric methods which isolate the fat are clearly the most accurate methods of analysis (Horwitz et al. 1975; Jensen 1989). Volumetric methods, such as the Gerber method or creamotocrit, are designed to measure milk fat of the cow, and if used on other species must be calibrated to the specific gravity of the milk fat of that species under the conditions measured (Jenness & Patton 1959).

Lastly, although gastric samples from pups cannot be used to indicate total milk fat content, they can be used to assess maternal milk fatty acids. The fatty acid composition of milk and gastric samples from five harp seal (*Phoca groenlandica*) and five hooded seal (*Cystophora cristata*) mother-pup pairs was identical, even after extensive hydrolysis (Iverson 1988). Gastric hydrolysis products (except for short- and medium-chain free fatty acids) remain inside the intact milk fat globule until the intestine (Patton *et al.* 1982). Thus, in marine mammals, although the original milk may be both diluted and hydrolysed, any given globule will still contain the original fatty acid composition.

Fatty acids as indicators of diet composition in marine mammals

Fatty acids in the marine food web

Perhaps the most intriguing aspect of the relationship between maternal diet and milk secretion is the question of what the milk lipid composition can tell us about the foraging ecology of a species or individual. Marine mammal diets are high in unusual fatty acids which often occur in specific individual arrangements depending on the prey and geographical location. Thus, differences in the fatty acid composition of marine mammal milks can lead to the identification of individual prey types and perhaps to establishing the species composition of diets (Iverson 1988).

An important consequence of the differing restrictions to fatty acid biosynthesis or modification within animals, plants and bacteria is that individual isomers (fatty acids with the same chain length and number of double bonds, but with different configurations or positions of double bonds) as well as 'families' of fatty acids arise in the food chain which can be attributed to specific origins (Cook 1985). Since many animals deposit these dietary fatty acids in body tissues with no or minimal modification, it is possible to distinguish between fatty acids that could be biosynthesized by the animal and those that could only come from the diet.

Marine lipids are well-known for their characteristically high levels of long-chain and polyunsaturated fatty acids (PUFA), which originate from various unicellular phytoplankton and seaweed (Ackman 1980). However, just as noteworthy are the occurrences in some marine lipids of unusual or novel fatty acids (e.g. those that are odd- or branched-chained or non-methylene interrupted, or that contain functional groups such as cyclopropane rings) which may be attributed to a single phylogenetic group or even to a single species from a specific ecological community. For instance, jellyfish were indicated in the diet of ocean sunfish, *Mola mola* (Hooper, Paradis & Ackman 1973) on the basis of the presence of a single unusual fatty acid, *trans*-6-hexadecenoic acid, originally discovered in the giant leatherback turtle, *Dermochelys coriacea coriacea* (Ackman, Hooper & Sipos 1972). Rare, individual isomers are also found among various species. For instance 16:2n-4 and 16:4n-1 are metabolically inert in animals and arise from certain algae (Ackman & McLachlan 1977). Thus, their presence in the depot fats of some fish (e.g. sturgeon, *Acipenser oxyrhynchus*) can be indicative of feeding habits (Ackman, Eaton & Linke 1975). The unique monounsaturate 22:1n-11 originates, together with very large concentrations of 20:1n-9, from the fatty alcohols (wax esters) of some copepods (Pascal & Ackman 1976; Ackman, Sebedio & Kovacs 1980), and appears as a marker in omnivorous fish such as the Atlantic cod, *Gadus morhua*, and herring, *Clupea harengus harengus* (Ackman 1980). Specific odd-chain fatty acids with unique double-bond positions have been found to vary with season and geographical location in the mullet, *Mugil cephalus* (Deng *et al.* 1976) and in the smelt, *Osmerus mordax* (Paradis & Ackman 1976), apparently as a result of consuming specific amphipods. Thus, previous evidence indicates that individual fatty acids or their positional isomers in many marine animals must or most likely arise from dietary intake of specific phyla, classes or even species. The presence of intact wax esters and glycerol ethers (fatty acid-containing lipids) can also be used in tracing trophic relations (Lewis 1967; Ackman 1980).

Given the nature of lipid digestion in carnivores where dietary fatty acids remain largely intact (see pp. 267–268), the concept of fatty acids as trophodynamic tracers readily extends to marine mammals. For example, some of the first examples of such tracers included the presence of the unusual methyl-branched fatty acids in lipids of the ocean sunfish and blubber of the sperm whale, *Physeter macrocephalus* (Pascal & Ackman 1975), and high levels of the long-chain fatty acids 20:1 and 22:1 distinguishing oils of the North Atlantic from those of the Antarctic finwhale, *Balaenoptera physalus* (Ackman & Eaton 1966).

Milk fatty acids as indicators of prey type and diet in marine mammals: evidence

There are many difficulties inherent in determining the diet of pinnipeds posed by collecting data from free-ranging animals at sea (Harwood & Croxall 1988). Information on the diets of whales is even less accessible. The use of fatty acid composition in milk, in combination with information from more traditional methods (i.e., recovered hard parts from stomach contents and faeces), may provide a better understanding of marine mammal diets both nearshore and offshore. Over the past few decades, growing sophistication in methodology and equipment (e.g. gas liquid chromatography—GLC—and mass spectroscopy) has enabled the increasingly complex and rapid analysis of lipid classes and their fatty acids. Although the 'basic' composition of any marine oil may be summarized by about 14 fatty acids, 50 to 70 fatty acids and isomers can routinely be identified and quantified in most marine lipids (Ackman, Ratnayake & Olsson 1988; Iverson 1988). This allows extremely detailed examination of any sample.

In the following discussion, all isomers have been identified and quantified; however, to simplify presentation only a summary of fatty acids is shown in most figures.

The relationship between foraging and milk fatty acid composition can be considered in two ways. The first is the use of fatty acids as tracers, i.e. the presence of a single unusual component which can be traced to a specific prey species, as described above. The second is to consider the array of fatty acids present; i.e. to match the pattern (components and relative levels of components) of all fatty acids in prey to the pattern in milk. This signature method may potentially be more useful because, in addition to identification of prey types, we may be able to evaluate the relative contribution of various prey to the overall diet.

The use of fatty acid signatures is possible because the array of fatty acids can differ substantially among various prey species. To use these differences to infer diet from milk, fatty acids must be generally grouped into three categories: (1) those components which could be readily biosynthesized by the animal or mammary gland (primarily 16:0, 16:1, 18:0 and 18:1); (2) those components that could be biosynthesized but at the levels found are probably mostly of dietary origin (e.g. 14:0, 20:1, some 22:1); and (3) those components which could originate only from the diet (e.g. 22:1n-11, all long chain PUFA with n-3 and n-6 double bond positions, and unusual fatty acids). One long-chain PUFA, 22:5n-3, cannot be used to evaluate diet because it is believed to be an intermediate of 20:5n-3 and 22:6n-3 (Ackman, Ratnayake *et al.* 1988). However, since it is a relatively minor component (usually < 1% in prey and 1–4% in marine mammal lipids. Iverson 1988) it does not interfere with prey identification. Generally, categories 2 and 3 represent 'indicator' fatty acids (dappled areas, Fig. 1) which contain information about the types of food eaten. The potential differences among prey in the indicator fatty acids 20:1, 22:1, 20:5n-3 and 22:6n-3 are clearly illustrated in Fig. 1. Thus, the extent to which marine mammals can biosynthesize fatty acids should be taken into account when evaluating fatty acid patterns as indicators of diet.

Fatty acids in marine mammal blubber can also be used to infer diet. However, blubber and milk may tell us different things about the diet. The fatty acids in the blubber of a pinniped or cetacean represent an integration of the dietary history of the individual. In an animal which undergoes seasonal periods of fasting and extensive depletion of fat stores (e.g. during breeding or moulting), followed by intensive blubber deposition (e.g. prior to the subsequent breeding season), blubber lipids may reflect dietary intake over the period of several months. Thus, milk lipids secreted while the female is fasting (e.g. in a large phocid) may directly reflect blubber lipids and thus dietary history. However, in species which feed during lactation (e.g. an otariid), milk lipids will reflect the most recent dietary intake, since

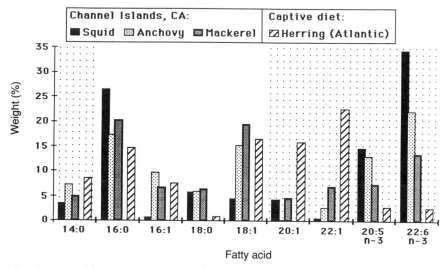

Fig. 1. Fatty acid composition of selected prey items of the California sea lion. Dappled areas represent typical 'indicator' fatty acids (see text). References for prey fatty acid values: Ackman (1980); Ackman (1982); U.S.D.A. (1987).

dietary fatty acids will be directed to the mammary gland by LPL. If dietary intake is insufficient to meet demands of milk lipid secretion, fatty acids may reflect a combination of both diet and blubber mobilization.

Does the fatty acid composition of marine mammal blubber reflect that of diet? Unfortunately little experimental work has been done to date. Grahl-Nielsen & Mjaavatten (1991) studied incorporation of dietary fatty acids into the blubber of adult grey and harbour seals. However, in their study a period of fasting did not precede the diet trial and animals appear to have been kept at maintenance rations during the trial, so that no new dietary fatty acids would have been deposited. Thus, their study was poorly designed to evaluate the use of fatty acid signatures and their conclusion that fatty acids could not be used in this matter must be doubted. Although there are few data available for adults, lipid deposition processes are likely to be similar to those in the young for which there are data. Hooded seal pups are born with a blubber layer which represents fatty acid biosynthesis by the foetus or transfer across the placental membrane, but includes none of the typical indicator (dietary) fatty acids of mothers' milk (Fig. 2; Iverson, Oftedal, Bowen et al. in prep.). However, at weaning, the fatty acid composition of pup blubber mirrors the fatty acids which have been transferred over the four-day lactation period (Fig. 2). Thus, initial deposition from diet is unmodified in pups, and this may be true also for adults. Ackman, Epstein & Eaton (1971) suggest that some modification of fatty acids may occur after deposition; however, modification of fatty acids in

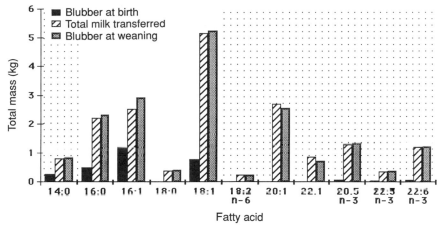

Fig. 2. Total deposition of dietary fatty acids over lactation in suckling hooded seal pups in comparison to the fatty acid content of milk delivered. Dappled areas represent typical 'indicator' fatty acids (see text). From Iverson, Oftedal, Bowen *et al.* (in prep.).

adipose tissue is unlikely to occur at significant rates during lactation (e.g. Bauman & Elliot 1983; Turner, Anderson & Blintz 1989).

To evaluate the incorporation of blubber fatty acids into milk during fasting, milk and blubber must be taken from the same individuals. The hooded seal produces a milk which is 61% fat and secretes about 5 kg milk lipid per day entirely from body reserves (Oftedal, Boness & Bowen 1988). Evidence from this species indicates that the fatty acids in milk are derived mostly from blubber stores (Fig. 3), although minor differences occur in

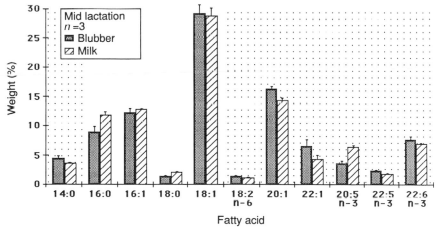

Fig. 3. Fatty acids of maternal blubber in comparison to those secreted in milk in lactating hooded seals at 2 days postpartum. Dappled areas represent typical 'indicator' fatty acids (see text). From Iverson, Oftedal, Bowen *et al.* (in prep.).

some components at different lactation stages (Iverson, Oftedal, Bowen *et al.* in prep.). Lipid transfer from diet to deposition and subsequently to milk is extraordinarily efficient and probably little modification occurs during the process. Thus even if there is some *de novo* synthesis of 16:0–18:1 and some differences or selectivity in mobilized fatty acids, the array of fatty acids secreted in milk while fasting is still highly similar to the array in maternal blubber throughout lactation (Fig. 3). These limited results suggest that both milk and blubber may be used to infer information about dietary history. To the extent that blubber can be used, then the diets of males or juveniles can also be investigated.

If the pattern of milk fatty acids resembles that of blubber when the female is fasting, does the pattern change when feeding resumes? During the perinatal fast in otariids, milk fatty acid secretion must be derived from the mobilization of body stores as described above for a phocid. During this fast, milk lipid presumably reflects blubber and thus dietary history prior to parturition. However, after the onset of foraging trips, lipids from dietary intake will be channelled to the mammary gland by LPL, and thus milk lipid should reflect recent intake. If the diet prior to arrival on the rookery is different from the diet consumed during short foraging trips, the pattern in milk fatty acids should reflect this change, as illustrated by the following examples.

During the breeding season at South Georgia, Antarctic fur seals (*Arctocephalus gazella*) feed almost exclusively on krill (Croxall & Pilcher 1984); however, prior to parturition the fur seal diet is unknown. The fatty acid composition of milk samples obtained from lactating females during the perinatal fast is substantially different than that of milk obtained at the return from foraging trips (Fig. 4a). Although actual prey items have not been analysed, krill (*Euphausia superba*) from the Southern Ocean, which feed upon a specific diatom, have a characteristic fatty acid composition (Ackman & Eaton 1966) which is remarkably similar to the pattern of milk fatty acids during foraging trips (Fig. 4a).

Similarly, milk fatty acid composition of California sea lions also changes dramatically between the perinatal period and intervals between foraging trips (Fig. 4b). In 1983, sea lions in the Channel Islands consumed a mixture of mackerel, *Trachurus symmetricus* (22% mass, 5.3% fat), squid, *Loligo opalescens* (45% mass, 1.0% fat), anchovy, *Engraulis mordax* (14% mass, 2.4% fat) and whiting, *Merluccius productus* (19% mass, 1.2% fat) (Costa, Antonelis & DeLong 1991). Fatty acid contribution of each prey species to the diet can be estimated by using the reported mass contribution corrected for fat content and literature values for the fatty acid compositions of these species (e.g. Fig. 1). Even with this rough estimate of dietary intake, milk fatty acids secreted by females during foraging intervals in 1983 reflect levels of indicator fatty acids of prey species (e.g. 14:0, 20:1–22:6; Fig. 1,

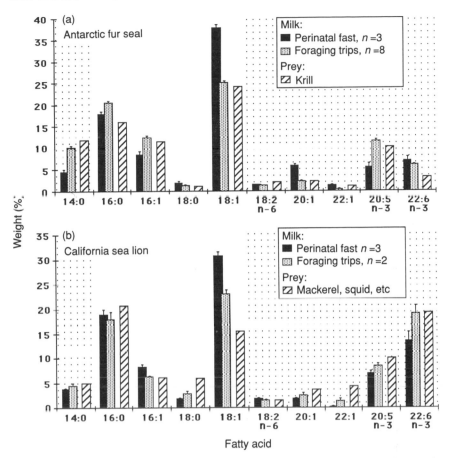

Fig. 4. Changes in milk fatty acid composition from the perinatal fast to feeding in the Antarctic fur seal (a) and the California sea lion (b). The diet during foraging trips in the fur seal (a) was assumed to be krill and in the sea lion (b) to be a mixture of mackerel, squid, anchovy and whiting (see text and Fig. 1). Dappled areas represent typical 'indicator' fatty acids (see text). Most data from Iverson (1988).

Fig. 4b). Additionally, although category 1 fatty acids may originate in part from *de novo* synthesis, in both species of otariid, changes in the levels of 16:1, 18:0 and 18:1 during feeding reflect the differences in the respective prey eaten (Fig. 4a, b).

The use of fatty acid signatures extends beyond marine ecosystems to other species with unique dietary lipid compositions. A lactation pattern comparable to that of otariids is found in the black bear, which secretes milk both during a period of fasting in winter dormancy and during spring and summer foraging. The indicator fatty acid isomers of primary

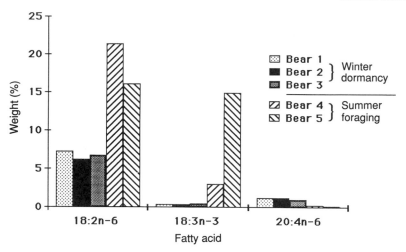

Fig. 5. Important milk fatty acids of dietary origin as indicators of individual foraging habits in the black bear: comparison of milk secreted during and after the fast of winter dormancy. Adapted from data of Iverson & Oftedal (1992).

importance in this case are 18:2n-6 and 18:3n-3, which in marine mammals and their prey usually account for less than 2% and 1% of total fatty acids, respectively. During the period of winter dormancy, levels of 18:2n-6, 18:3n-3 and 20:4n-6 in black bear milk ($n = 9$) exhibit extremely little variation (Iverson & Oftedal 1992). However, during summer foraging, levels of 18:2n-6 and 18:3n-3 in milks of individual females are variable and extraordinarily high (15–22%), while 20:4n-6 disappears (Fig. 5). Unfortunately, information about the specific dietary items of these individual females is unavailable. However, these patterns clearly reflect individual differences in foraging habits and diets that are high in fruits and leafy vegetation, lipids of which may contain 5–50% of either 18:2n-6 or 18:3n-3, depending on the item, and 0% of 20:4n-6 (Kamel & Kakuda 1992; Peng 1992). Such high levels of 18:2n-6 and especially 18:3n-3 are found only in the milks of other species which feed solely on vegetation, e.g. the koala, *Phascolarctos cinereus*, and the horse, *Equus caballus* (Parodi 1982), red and mantled howler monkeys, *Alouatta* sp. (S. J. Iverson & O. T. Oftedal unpubl. data), and human vegetarians (Jensen 1989).

If milk fatty acids resemble those of recent diet, do they change with a switch in diet? Although there are no controlled studies, a comparison of captive animals with their wild counterparts suggests that fatty acid signatures do indeed reflect changes in diet. The California sea lion in the Channel Islands consumes a mixed diet of mackerel, squid, anchovy and whiting during foraging trips (Costa *et al.* 1991; e.g. Fig. 4b). The fatty acid composition of milk obtained from a captive California sea lion fed

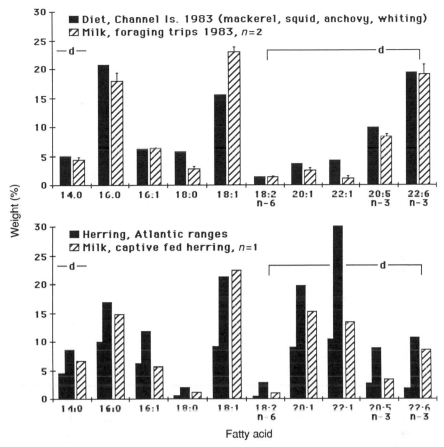

Fig. 6. Milk fatty acid composition of wild (Channel Islands, CA) and captive California sea lions in relation to estimated diets (see text and Fig. 1). Areas marked with '-d-' represent important dietary (or 'indicator') fatty acids (see text). Most data from Iverson (1988).

exclusively on Atlantic herring was substantially different from that in the wild population and the pattern in both milks reflected that estimated for their prey (Fig. 6). Herring are notable for their high levels of isomers of fatty acids 20:1 and 22:1 (Fig. 1), and levels of these components found in the captive sea lion milk have not been encountered in milks from wild sea lions (Iverson 1988). Because the fatty acid composition of lots of herring may vary substantially with factors such as geographical origin (Ackman 1980), a range of values from the literature is shown (Fig. 6). Despite the lack of a controlled study, it is apparent that the changes in diet drive the changes in milk fatty acid patterns.

Additional evidence comes from data on wild and captive harbour seals (S. J. Iverson, W. D. Bowen & D. J. Boness unpubl.). During mid to late

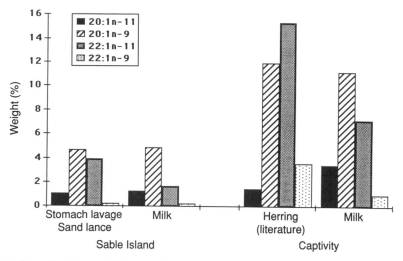

Fig. 7. Long-chain fatty acid isomers of 20:1 and 22:1 in milk from a wild (Sable Island, Nova Scotia) and a captive harbour seal in comparison to wild and captive diets. Wild diet (sand lance) was assessed from fatty acid analysis of stomach contents; captive diet (herring) was derived from literature. From S. J. Iverson, W. D. Bowen & D. J. Boness (unpubl.)

lactation, harbour seals on Sable Island, Nova Scotia, begin regular foraging trips (Bowen *et al.* 1992). Stomach lavage data reveal that the primary prey item at this time is sand lance (*Ammodytes dubius*). Levels of indicator isomers of 20:1 and 22:1 in milk from a harbour seal female that had begun foraging trips are comparable to levels found in lavage samples analysed (Fig. 7). However, the fatty acid pattern of milk from a captive harbour seal, fed solely on herring, reflects the large increases primarily in levels of 20:1n-9 and 22:1n-11 found in herring. Again, the composition of actual herring consumed is unavailable and must be derived from literature values, which probably accounts for discrepancies in some components (Fig. 7). Similarly high levels in these fatty acids are observed in oils from finwhales feeding on the krill *Meganyctiphanes norvegica* (high in 20:1 and 22:1) in Nova Scotia waters as compared with oils from finwhales feeding on the krill *E. superba* (low in 20:1 and 22:1) in the Southern Ocean (Ackman & Eaton 1966).

In conclusion, the potential for use of fatty acid patterns in milks of marine mammals to identify and perhaps quantify prey species appears to be great. During ongoing feeding, fatty acids are partitioned to the mammary gland and thus resemble dietary intake. Unfortunately, information on trophic relationships involving marine mammals through fatty acid signatures is primarily based upon literature values for expected prey items and therefore precludes reliable conclusions. Furthermore, much of the descriptive data from previous marine mammal and prey studies are difficult or impossible to interpret owing to incomplete reports, apparent misidentification of fatty acid peaks and/or failure to separate components with similar retention times during GLC analysis (reviewed in Iverson 1988). Most reports are also based on the analysis of a single lot of fish, seal or whale oil comprising many individuals and often reports list only the analysis of 'edible parts' for human consumers. Thus, detailed and systematic studies of marine mammals and their prey must be conducted in order to exploit fully the potential of using fatty acids to determine trophic interactions.

Acknowledgements

I would like to thank my colleague Dr O. T. Oftedal for the fruitful discussions which have contributed much to the direction of this paper, particularly in the area of the physiology of lactation and milk secretion. Frequent discussions and collaborative research with Dr W. D. Bowen have contributed substantially to my ideas on the relationships between pinniped foraging and fatty acids. I am especially grateful to W. D. Bowen for critically reviewing earlier drafts of this manuscript and I thank Dr R. G. Ackman and Dr D. J. Boness for helpful comments. I thank Dr D. P. Costa for kindly providing the milk samples from the Antarctic fur seal and captive California sea lion. Unpublished research reported herein was supported in part by grants from the American Heart Foundation, Georgetown University Medical Center and the Smithsonian Institution, Washington D.C., and by the Department of Fisheries and Oceans, Canada.

References

Ackman, R. G. (1980). Fish lipids, part 1. In *Advances in fish science and technology*: 86–103. (Ed. Connell, J. J.). Fishing News Books Ltd, Surrey, UK.

Ackman, R. G. (1982). Fatty acid composition of fish oils. In *Nutritional evaluation of long-chain fatty acids in fish oil*: 25–88. (Eds Barlow, S. M. & Stansby, M. E.). Academic Press, London & New York.

Ackman, R. G. & Eaton, C. A. (1966). Lipids of the fin whale (*Balaenoptera physalus*) from North Atlantic waters. III. Occurrence of eicosenoic and docosenoic fatty acids in the zooplankton *Meganyctiphanes norvegica* (M. Sars) and their effect on whale oil composition. *Can. J. Biochem.* 44: 1561–1566.

Ackman, R. G., Eaton, C. A. & Linke, B. A. (1975). Differentiation of freshwater characteristics of fatty acids in marine specimens of the Atlantic sturgeon, *Acipenser oxyrhynchus*. *Fish. Bull. US. Fish Wildl. Serv.* 73: 838–845.

Ackman, R. G., Epstein, S. & Eaton, C. A. (1971). Differences in the fatty acid compositions of blubber fats from northwestern Atlantic finwhales (*Balaenoptera physalus*) and harp seals (*Pagophilus groenlandica*). *Comp. Biochem. Physiol.* **40B**: 683–697.

Ackman, R. G., Hooper, S. N. & Sipos, J. C. (1972). Distribution of *trans*-6-hexadecenoic and other fatty acids in tissues and organs of the Atlantic leatherback turtle, *Dermochelys coriacea coriacea* L. *Int. J. Biochem.* **3**: 171–179.

Ackman, R. G. & McLachlan, J. (1977). Fatty acids in some Nova Scotian marine seaweeds: a survey for octadecapentaenoic and other biochemically novel fatty acids. *Proc. Nova Scotian Inst. Sci.* **28**: 47–64.

Ackman, R. G., Ratnayake, W. M. N. & Olsson, B. (1988). The 'basic' fatty acid composition of Atlantic fish oils: potential similarities useful for enrichment of polyunsaturated fatty acids by urea complexation. *J. Am. Oil Chem. Soc.* **65**: 136–138.

Ackman, R. G., Sebedio, J.-L. & Kovacs, M. I. P. (1980). Role of eicosenoic and docosenoic fatty acids in freshwater and marine lipids. *Mar. Chem.* **9**: 157–164.

Allen, W. V. (1976). Biochemical aspects of lipid storage and utilization in animals. *Am. Zool.* **16**: 631–647.

Annison, E. F. (1983). Metabolite utilization by the ruminant mammary gland. In *Biochemistry of lactation*: 399–435. (Ed. Mepham, T. B.). Elsevier Science Publishers, Amsterdam.

Bauman, D. E. & Currie, W. B. (1980). Partitioning of nutrients during pregnancy and lactation: a review of mechanisms involving homeostasis and homeorhesis. *J. Dairy Sci.* **63**: 1514–1529.

Bauman, D. E. & Elliot, J. M. (1983). Control of nutrient partitioning in lactating ruminants. In *Biochemistry of lactation*: 437–468. (Ed. Mepham, T. B.). Elsevier Science Publishers, Amsterdam.

Blaxter, K. L. (1962). *The energy metabolism of ruminants.* Hutchinson, London.

Bonner, W. N. (1984). Lactation strategies in pinnipeds: problems for a marine mammalian group. *Symp. zool Soc. Lond.* No. 51: 253–272.

Borgstrom, B. (1977). Digestion and absorption of lipids. *Int. Rev. Physiol.* **12**: 305–323.

Bowen, W. D., Oftedal, O. T. & Boness, D. J. (1992). Mass and energy transfer during lactation in a small phocid, the harbor seal (*Phoca vitulina*). *Physiol. Zool.* **65**: 844–866.

Boyd, I. L. (1991). Changes in plasma progesterone and prolactin concentrations during the annual cycle and the role of prolactin in the maintenance of lactation and luteal development in the Antarctic fur seal (*Arctocephalus gazella*). *J. Reprod. Fert.* **91**: 637–647.

Brew, K. (1970). Lactose synthetase: evolutionary origins, structure and control. *Essays Biochem.* **6**: 93–118.

Brew, K., Vanaman, T. C. & Hill, R. L. (1968). The role of α-lactalbumin and the A protein in lactose synthetase: a unique mechanism for the control of a biological reaction. *Proc. natn. Acad. Sci. U.S.A.* **59**: 491–497.

Christie, W. W. (1982). *Lipid analysis.* Pergamon Press, New York.

Cook, H. W. (1985). Fatty acid desaturation and chain elongation in eucaryotes. In *Biochemistry of lipids and membranes*: 181–211. (Eds Vance, D. E. & Vance, J. E.). The Benjamin/Cummings Publishing Co., Inc., Menlo Park, CA.

Costa, D. P., Antonelis, G. A. & DeLong, R. L. (1991) The effects of El Niño on the foraging energetics of the California sea lion. In *Pinnipeds and El Niño: responses to environmental stress*: 156–165. (Eds Trillmich, F. & Ono, K. A.). Springer-Verlag, Berlin, Heidelberg.

Costa, D. P. & Gentry, R. L. (1986). Free-ranging energetics of northern fur seals. In *Fur seals: maternal strategies on land and at sea*: 79–101. (Eds Gentry, R. L. & Kooyman, G. L.). Princeton University Press, Princeton.

Cowie, A. T. (1969). Variations in the yield and composition of the milk during lactation in the rabbit and the galactopoietic effect of prolactin. *J. Endocr.* **44**: 437 450.

Cowie, A. T., Forsyth, I. A. & Hart, I. C. (1980). *Hormonal control of lactation*. Springer-Verlag, Berlin, Heidelberg.

Cripps, A. W. & Williams, V. J. (1975). The effect of pregnancy and lactation on food intake, gastrointestinal anatomy and the absorptive capacity of the small intestine in the albino rat. *Br. J. Nutr.* **33**: 17–32.

Croxall, J. P. & Pilcher, M. N. (1984). Characteristics of krill *Euphausia superba* eaten by Antarctic fur seals *Arctocephalus gazella* at South Georgia. *Bull. Br. Antarct. Surv.* No. 63: 117–125.

Davies, D. T., Holt, C. & Christie, W. W. (1983). The composition of milk. In *Biochemistry of lactation*: 71–117. (Ed. Mepham, T. B.). Elsevier Science Publishers, Amsterdam.

Deng, J.-C., Orthoefer, F. T., Dennison, R. A. & Watson, M. (1976). Lipids and fatty acids in the mullet (*Mugil cephalus*): seasonal and locational variations. *J. Food Sci.* **41**: 1479–1483.

Dils, R. R. (1983). Milk fat synthesis. In *Biochemistry of lactation*: 141–157. (Ed. Mepham, T. B.). Elsevier Science Publishers, Amsterdam.

Dils, R. R., Clark, S. & Knudsen, J. (1977). Comparative aspects of milk fat synthesis. *Symp. zool. Soc. Lond.* No. 41: 43–55.

Freed, L. M., York, C. M., Hamosh, M., Sturman, J. T., Oftedal, O. T. & Hamosh, P. (1986). Bile salt stimulated lipase; the enzyme is present in non-primate milk. In *Human lactation* 2. *Maternal-environmental factors*: 595–602. (Eds Hamosh, M. & Goldman, A. S.). Plenum Press, New York.

Gaskin, D. E. (1982). *The ecology of whales and dolphins*. Heinemann Educational Books, London.

Grahl-Nielsen, O. & Mjaavatten O. (1991). Dietary influence on fatty acid composition of blubber fat of seals as determined by biopsy: a multivariate approach. *Mar. Biol., Berlin* **110**: 59–64.

Hamosh, M. (1989). Enzymes in human milk: their role in nutrient digestion, gastrointestinal function, and nutrient delivery to the newborn infant. In *Textbook of gastroenterology and nutrition in infancy*: 121–134. (Ed. Lebenthal, E.). Raven Press, New York.

Hamosh, M., Clary, T. R., Chernick, S. S. & Scow, R. O. (1970). Lipoprotein lipase activity of adipose and mammary tissue and plasma triglyceride in pregnant and lactating rats. *Biochim. Biophys. Acta* **210**: 473–482.

Hamosh, M. & Hamosh, P. (1985). Lipoproteins and lipoprotein lipase. In *Handbook of physiology—the respiratory system* 1: 387–418. (Eds Fishman, A. P. & Fisher, A. B.). The American Physiological Society, Washington D.C.

Hanwell, A. & Peaker, M. (1977). Physiological effects of lactation on the mother. *Symp. zool. Soc. Lond.* No. 41: 297–312.

Harrison, R. J. (1969). Reproduction and reproductive organs. In *The biology of marine mammals*: 253–348. (Ed. Andersen, H. T.). Academic Press, New York & London.

Harwood, J. & Croxall, J. P. (1988). The assessment of competition between seals and commercial fisheries in the North Sea and the Antarctic. *Mar. Mammal Sci.* **4**: 13–33.

Hernell, O. & Olivecrona, T. (1974). Human milk lipases. I. Bile salt stimulated lipase. *J. Lipid Res.* **15**: 367–372.

Hooper, S. N., Paradis, M. & Ackman, R. G. (1973). Distribution of *trans*-6-hexadecenoic acid, 7-methyl-7-hexadecenoic acid and common fatty acids in lipids of the ocean sunfish *Mola mola*. *Lipids* **8**: 509–516.

Horwitz, W., Senzel, A., Reynolds, H. & Park, D. L. (Eds) (1975). *Official methods of analysis*. Association of Official Analytical Chemists, Washington D.C.

Iverson, S. J. (1988). *Composition, intake and gastric digestion of milk lipids in pinnipeds*. PhD thesis: Univ. of Maryland, College Park.

Iverson, S. J., Bowen, W. D., Boness, D. J. & Oftedal, O. T. (1993). The effect of maternal size and milk energy output, and its effect on pup growth in grey seals (*Halichoerus grypus*). *Physiol. Zool.* **66**: 61–88.

Iverson, S. J., Hamosh, M. & Bowen, W. D. (1991). Lipid clearing enzymes during lactation in species adapted to high milk fat output and intake. *Abstr. pediatr. Res.* **29**: 44A.

Iverson, S. J., Kirk, C. L., Hamosh, M. & Newsome, J. (1991). Milk lipid digestion in the neonatal dog: the combined actions of gastric and bile salt stimulated lipases. *Biochim. biophys. Acta* **1083**: 109–119.

Iverson, S. J. & Oftedal, O. T. (1992). Fatty acid composition of black bear (*Ursus americanus*) milk during and after the period of winter dormancy. *Lipids* **27**: 940–943.

Iverson, S. J., Oftedal, O. T. & Boness, D. J. (1991). The effect of El Niño on pup development in the California sea lion (*Zalophus californianus*). II. Milk intake. In *Pinnipeds and El Niño: responses to environmental stress*: 180–184. (Eds Trillmich, F. & Ono, K. A.). Springer-Verlag, Berlin, Heidelberg.

Iverson, S. J., Oftedal, O. T., Bowen, W. D., Boness, D. J. & Sampugna, J. (In preparation). *The transfer of lipids from mother to pup in the hooded seal* (Cystophora cristata): *comparison of maternal depot fatty acids to milk fatty acids and subsequent incorporation in the pup*.

Iverson, S. J., Sampugna, J. & Oftedal, O. T. (1992). Positional specificity of gastric hydrolysis of long-chain n-3 polyunsaturated fatty acids of seal milk triglycerides. *Lipids* **27**: 870–878.

Jenness, R. (1974). Biosynthesis and composition of milk. *J. invest. Derm.* **63**: 109–118.

Jenness, R. (1979). The composition of human milk. *Semin. Perinatol.* **3**: 225–239.

Jenness, R. (1986). Lactational performance of various mammalian species. *J. Dairy Sci.* **69**: 869–885.

Jenness, R. & Patton, S. (1959). *Principles of dairy chemistry*. John Wiley & Sons, New York.

Jensen, R. G. (1989). Lipids in human milk—composition and fat-soluble vitamins. In *Textbook of gastroenterology and nutrition in infancy*: 157–208. (Ed. Lebenthal, E.). Raven Press, New York.

Jensen, R. G., Clark, R. M., Bitman, J., Wood, D. L. & Clandinin, M. T. (1985). Methods for the sampling and analysis of human milk lipids. In *Human lactation: milk components and methodologies*: 97–112. (Eds Jensen, R. G. & Neville, M. C.). Plenum Press, New York.

Johnson, A. II. (1974). The composition of milk. In *Fundamentals of dairy chemistry*: 1–57. (Eds Webb, B. H., Johnson, A. H. & Alford, J. A.). AVI Publishing, Westport, Conn.

Kamel, B. S. & Kakuda, Y. (1992). Fatty acids in fruits and fruit products. In *Fatty acids in foods and their health implications*: 263–295. (Ed. Chow, C. K.). Marcel Dekker, Inc., New York.

Kretzmann, M. B., Costa, D. P., Higgins, L. V. & Needham, D. J. (1991). Milk composition of Australian sea lions, *Neophoca cinerea*: variability in lipid content. *Can. J. Zool.* **69**: 2556–2561.

Kuhn, N. J. (1983). The biosynthesis of lactose. In *Biochemistry of lactation*: 159–176. (Ed. Mepham, T. B.). Elsevier Science Publishers, Amsterdam.

Lewis, R. W. (1967). Fatty acid composition of some marine animals from various depths. *J. Fish. Res. Bd Can.* **24**: 1101–1115.

Linzell, J. L. (1974). Mammary blood flow and methods of identifying and measuring precursors of milk. In *Lactation* **1**: 143–225. (Eds Larson, B. L. & Smith, V. R.). Academic Press, New York & London.

Lonnerdal, B., Forsum, E., Gebre-Medhin, M. & Hambraeus, L. (1976). Breast milk composition in Ethiopian and Swedish mothers. II. Lactose, nitrogen and protein contents. *Am. J. clin. Nutr.* **29**: 1134–1141.

Macdonald, D. (Ed.) (1984). *The encyclopedia of mammals* **1**. George Allen & Unwin, London; Facts on File Publications, New York.

Mather, I. H. & Keenan, T. W. (1983). Function of endomembranes and the cell surface in the secretion of organic milk constituents. In *Biochemistry of lactation*: 231–283. (Ed. Mepham, T. B.). Elsevier Science Publishers, Amsterdam.

McBride, O. W. & Korn, E. D. (1963). The lipoprotein lipase of mammary gland and the correlation of its activity to lactation. *J. Lipid Res.* **4**: 17–20.

McBride, O. W. & Korn, E. D. (1964). Uptake of free fatty acids and chylomicron glycerides by guinea pig mammary gland in pregnancy and lactation. *J. Lipid Res.* **5**: 453–458.

Mehta, N. R., Jones, J. B. & Hamosh, M. (1982). Lipases in preterm human milk: ontogeny and physiologic significance. *J. pediatr. Gastroenterol Nutr.* **1**: 317–326.

Mendelson, C. R. & Scow, R. O. (1972). Uptake of chylomicron-triglyceride by perfused mammary tissue of lactating rats. *Am. J. Physiol.* **223**: 1418–1423.

Mepham, T. B. (1971). Amino acid utilization by the lactating mammary gland. In *Lactation*: 297–315. (Ed. Falconer, I. R.). Butterworths, London.

Mepham, T. B. (1983). Physiological aspects of lactation. In *Biochemistry of lactation*: 3–28. (Ed. Mepham, T. B.). Elsevier Science Publishers, Amsterdam.

Mercier, J.-C. & Gaye, P. (1983). Milk protein synthesis. In *Biochemistry of lactation*: 177–227. (Ed. Mepham, T. B.). Elsevier Science Publishers, Amsterdam.

Messer, M. & Green, B. (1979). Milk carbohydrates of marsupials. II. Quantitative and qualitative changes in milk carbohydrates during lactation in the tammar wallaby (*Macropus eugenii*). *Aust. J. biol. Sci.* **32**: 519–531.

Millar, J. S. (1977). Adaptive features of mammalian reproduction. *Evolution, Lawrence, Kans.* **31**: 370–386.

Millar, J. S. (1979). Energetics of lactation in *Peromyscus maniculatus*. *Can. J. Zool.* **57**: 1015–1019.

National Research Council (1991). *Nutrition during lactation*. Institute of Medicine, Subcommittee on Nutrition during Lactation. National Academy Press, Washington D.C.

Oftedal, O. T. (1984a). Lactation in the dog: milk composition and intake by puppies. *J. Nutr.* **114**: 803–812.

Oftedal, O. T. (1984b). Milk composition, milk yield and energy output at peak lactation: a comparative review. *Symp. zool. Soc. Lond* No. 51: 33–85.

Oftedal, O. T. (1985). Pregnancy and lactation. In *Bioenergetics of wild herbivores*: 215–238. (Eds Hudson, R. J. & White, R. G.). CRC Press, Boca Raton, Florida.

Oftedal, O. T., Boness, D. J. & Bowen, W. D. (1988). The composition of hooded seal (*Cystophora cristata*) milk: an adaptation to postnatal fattening. *Can. J. Zool.* **66**: 318–322.

Oftedal, O. T., Boness, D. J. & Tedman, R. A. (1987). The behavior, physiology, and anatomy of lactation in the Pinnipedia. *Curr. Mammal.* **1**: 175–245.

Oftedal, O. T., Iverson, S. J. & Boness, D. J. (1987). Milk and energy intakes of suckling California sea lion *Zalophus californianus* pups in relation to sex, growth, and predicted maintenance requirements. *Physiol. Zool.* **60**: 560–575.

Paradis, M. & Ackman, R. G. (1976). Localization of a source of marine odd chain-length fatty acids. I. The amphipod *Pontoporeia femorata* (Kröyer). *Lipids* **11**: 863–870.

Parodi, P. W. (1982). Positional distribution of fatty acids in triglycerides from milk of several species of mammals. *Lipids* **17**: 437–442.

Pascal, J. C. & Ackman, R. G. (1975). Occurrence of 7-methyl-7-hexadecenoic acid, the corresponding alcohol, 7-methyl-6-hexadecenoic acid, and 5-methyl-4-tetradecenoic acid in sperm whale oils. *Lipids* **10**: 478–482.

Pascal, J. C. & Ackman, R. G. (1976). Long chain monoethylenic alcohol and acid isomers in lipids of copepods and capelin. *Chem. Phys. Lipids* **16**: 219–223.

Patton, J. S. (1981). Gastrointestinal lipid digestion. In *Physiology of the gastrointestinal tract*: 1123–1146. (Ed. Johnson, L. R.). Raven Press, New York.

Patton, S. & Jensen, R. G. (1976). *Biomedical aspects of lactation with special reference to lipid metabolism and membrane functions of the mammary gland*. Pergamon Press, Oxford.

Patton, J. S., Rigler, M. W., Liao, T. H., Hamosh, P. & Hamosh, M. (1982). Hydrolysis of triacylglycerol emulsions by lingual lipase; a microscopic study. *Biochim. biophys. Acta* **712**: 400–407.

Peaker, M. & Goode, J. A. (1978). The milk of the fur-seal, *Arctocephalus tropicalis gazella**; in particular the composition of the aqueous phase. *J. Zool.,*

Lond. **185**: 469–476. [*Species incorrectly identified; species is actually the southern elephant seal, *Mirounga leonina.*]

Peng, A. C. (1992). Fatty acids in vegetables and vegetable products. In *Fatty acids in foods and their health implications*: 185–236. (Ed. Chow, C. K.). Marcel Dekker, Inc., New York.

Rice, D. W. & Wolman, A. A. (1971). The life history and ecology of the grey whale (*Eschrichtius robustus*). *Spec. Publ. Am. Soc. Mammal.* No. 3: 1–142.

Riedman, M. & Ortiz, C. L. (1979). Changes in milk composition during lactation in the northern elephant seal. *Physiol. Zool.* **52**: 240–249.

Robinson, D. S. (1963). Changes in the lipolytic activity of the guinea pig mammary gland at parturition. *J. Lipid Res.* **4**: 21–23.

Russe, I. (1961). Die Laktation der Hunden. *Zentbl. VetMed.* **8**: 252–281.

Scow, R. O. (1977). Metabolism of chylomicrons in perfused adipose and mammary tissue of the rat. *Fed. Proc.* **36**: 182–185.

Strong, C. R. & Dils, R. R. (1972). The fatty acid synthetase complex of lactating guinea-pig mammary gland. *Int. J. Biochem.* **3**: 369–377.

Tedman, R. A. (1983). Ultrastructural morphology of the mammary gland with observations on the size distribution of fat droplets in milk of the Weddell seal *Leptonychotes weddelli* (Pinnipedia). *J. Zool., Lond.* **200**: 131–141.

Tedman, R. A. & Bryden, M. M. (1981). The mammary gland of the Weddell seal, *Leptonychotes weddelli* (Pinnipedia). 1. Gross and microscopic anatomy. *Anat. Rec.* **199**: 519–529.

Trillmich, F. & Lechner, E. (1986). Milk of the Galapagos fur seal and sea lion, with a comparison of the milk of eared seals (Otariidae). *J. Zool., Lond.* (A) **209**: 271–277.

Turner, J. J., Anderson, B. D. & Blintz, G. L. (1989). Whole-body lipids and fatty acid synthetase activity in Richardson's ground squirrels, *Spermophilus richardsonii*. *Physiol. Zool.* **62**: 1383–1397.

U.S.D.A. (1987). The composition of foods: finfish and shellfish products. *Agric. Handb. U.S. Dep. Agric.* No. 8–15: 19–190.

Williamson, D. H. (1980). Integration of metabolism in tissues of the lactating rat. *FEBS Lett.* **117**: K93–K105.

World Health Organization (1985). *The quantity and quality of breast milk.* World Health Organization, Geneva.

Young, R. A. (1976). Fat, energy and mammalian survival. *Am. Zool.* **16**: 699–710.

Zinder, O., Hamosh, M., Clary Fleck, T. R. & Scow, R. O. (1974). Effect of prolactin on lipoprotein lipase in mammary gland and adipose tissue in rats. *Am. J. Physiol.* **226**: 774–748.

Symp. zool. Soc. Lond. (1993) No. 66: 293–314

The relationship between reproductive and foraging energetics and the evolution of the Pinnipedia

Daniel P. COSTA

*Department of Biology
and the Institute of Marine Science
University of California
Santa Cruz, CA 95064, USA*

Synopsis

Marine feeding and terrestrial parturition are fundamental components of the life history of the Pinnipedia and present a series of trade-offs between the conflicting needs to feed the pup ashore and to maximize energy acquisition at sea. The Otariidae, eared seals, and the Phocidae, earless seals, have evolved different solutions to this problem. The phocids are capable of storing what is required for the entire lactation interval, whereas otariids must feed during the lactation interval. The evolution of these different patterns is related to co-variant traits such as metabolic rate, energy and oxygen storage capacity and body size. The interaction of these traits sets wider or narrower limits to such parameters as dive duration, duration of lactation, fasting ability, efficiency and amount of maternal investment and rate of prey energy acquisition. These factors may have been important in the evolution of the different life-history patterns of otariids and phocids. The phocid breeding pattern is quite economical, but one that limits the total nutrient and energy investment in the pup. In contrast, the otariid breeding pattern is quite expensive, but it allows the mother to provide more energy and nutrients to her pup. However, the separation of lactation from feeding may have enabled phocids to utilize a patchy, low-production, highly dispersed or distant prey resource as well as an unstable breeding substrate.

Introduction

Life-history patterns evolve in response to trade-offs between beneficial traits and the deleterious traits that are often linked to them (Brodie 1975; Stearns 1983, 1989; Read & Harvey 1989). Two fundamental components of the life history of the Pinnipedia, marine feeding and terrestrial parturition, offer an excellent example of such a series of trade-offs. The initial utilization of the marine environment by pinnipeds occurred at a time

ZOOLOGICAL SYMPOSIUM No. 66
ISBN 0-19-854069-8
Copyright © 1993 The Zoological Society of London
All rights of reproduction in any form reserved

when coastal upwelling was at a cyclic high and thus presented an abundant, diverse and essentially untapped food resource (Lipps & Mitchell 1976). However, the need to return to shore to feed the young required a spatial and/or temporal separation of feeding from lactation (Bartholomew 1970). The ideal solution required a balance between the conflicting demands of optimal provisioning of the pup on shore and the need to maximize energy acquisition at sea. Within the Pinnipedia the Otariidae, the eared seals (sea lions and fur seals), and the Phocidae, or earless seals ('true seals') exhibit strikingly different solutions to this apparent conflict (Bonner 1984; Kovacs & Lavigne 1986; Oftedal, Boness & Tedman 1987; Bowen 1991). All conceive young during the previous reproductive season and exhibit a period of delayed implantation that usually lasts two to three months. Actual foetal development occurs over a nine-month period, while the mother feeds at sea, continuously in the case of phocids, or intermittently in the case of many of the otariids, which have not yet weaned their pups. Phocid mothers remain on or near the rookery continuously from the birth of their pup until it is weaned; milk is produced from body reserves stored prior to parturition. Although some phocids, most notably harbour (*Phoca vitulina*), ringed (*Phoca hispida*) and Weddell (*Leptonychotes weddelli*) seals, feed during lactation, most of the maternal investment is derived from body stores (Testa, Hill & Siniff 1989). Weaning is abrupt and occurs after a minimum of four days of nursing (hooded seal, *Cystophora cristata*: Bowen, Oftedal & Boness 1985) to a maximum of six to seven weeks (Weddell seal: Kaufman, Siniff & Reichle 1975; Thomas & DeMaster 1983).

In contrast, otariid mothers only stay with their pups for the first week or so after parturition and then periodically go to sea to feed, intermittently returning to suckle their pup on the rookery (Bonner 1984). Feeding trips vary from one to seven days, depending on the species, and shore visits to the pup, which has been fasting, last one to three days (Gentry, Costa *et al.* 1986). The age at which the pups are weaned varies from a minimum of four months in the sub-polar fur seals (Antarctic, *Arctocephalus gazella*, and northern, *Callorhinus ursinus*) to up to three years in the equatorial Galapagos fur seal (*A. galapagoensis*) (Gentry, Costa *et al.* 1986). The remaining otariids occur in temperate latitudes. In these species, pups are usually weaned within a year of birth (Gentry, Costa *et al.* 1986), although weaning age can vary both within and between species as a function of seasonal and site-specific variations in environmental conditions (Trillmich 1990). In summary, most phocids are capable of storing what is required for the entire lactation period, whereas all otariids feed during lactation.

Adaptation to marine feeding and terrestrial parturition has resulted in different reproductive strategies in otariids and phocids that reflect trade-offs between traits that have different optima in different marine environments. Traits such as metabolic rate and oxygen storage capacity co-vary with body size in ways that determine dive duration and the rate and ability to acquire prey (Costa 1991a, b). The relationship between metabolic rate, body energy and nutrient stores determines how long and how efficiently the mother can feed her pup while she is fasting. This paper will examine the interaction between these traits in an attempt to illustrate how they have influenced or even driven the evolution of the Pinnipedia and their ability to inhabit different oceanic regimes.

Maternal body mass

Body mass has a profound influence on the evolution of life-history patterns (Ralls 1976; Peters 1983; Lindstedt & Boyce 1985; Elgar & Harvey 1987; Read & Harvey 1989; Millar & Hickling 1990); it is therefore appropriate to examine body mass trends in otariids and phocids. A summary of maternal body mass for all extant phocids and otariids indicates that otariid females (mean = 80 kg; median = 55 kg) are significantly smaller than phocid females (mean = 229 kg; median = 141 kg). In fact the majority of otariid females are smaller than 100 kg, whereas phocid females are predominantly larger than 100 kg (Appendix). The following sections will explore what factors favour large body size in phocid females and smaller body size in otariid females.

Acquisition of maternal resources

Relationship to foraging ecology

It has been proposed that the phocid breeding pattern, in which most, if not all, of the nutrients and energy necessary to rear the young successfully are acquired and stored in advance of parturition, is quite economical (Costa, Le Boeuf et al. 1986). In contrast, otariids have an expensive reproductive pattern that relies on food resources immediately offshore and requires many trips between the foraging grounds and the rookery (Costa 1991a, b). For example, lactating northern fur seal females consume 80% more food than non-lactating females (Perez & Mooney 1986). Northern elephant seal females need only to increase their daily food intake by 12% because the acquisition of the energy necessary to cover the entire cost of lactation is spread over many months at sea (Costa, Le Boeuf et al. 1986).

It may be that the increased energy requirements of lactating fur seals (and probably otariid mothers in general) can only be sustained in highly productive areas such as upwelling regions. Otariids have a reproductive pattern that is optimal for prey that is concentrated and predictable, whereas some phocids have a reproductive pattern better suited for dispersed or unpredictable prey or prey that is located at great distances from the rookery. The long-distance foraging ability of these phocids, which would allow them to utilize a more dispersed or distant food resource, is achieved by reducing the importance of feeding during lactation. However, fasting during lactation places a limit on the duration of investment and this limits the total amount of energy that a phocid mother can invest in her pup (Costa 1991a).

Foraging energetics

Do data on rates of energy intake and expenditure support the hypothesis that the costs of foraging are low in phocids and higher in otariids? Direct measurements on four species of otariids indicate that, while at sea, otariids exhibit metabolic rates that are 4.8–7.3 times the predicted basal metabolic rate (BMR) (Costa & Gentry 1986; Costa, Croxall & Duck 1989; Costa, Kretzmann & Thorson 1989; Costa, Antonelis & DeLong 1991). Comparable measurements are not available for phocids. However, estimates of the diving metabolic rate of elephant seals suggest that they expend 1.3 times the predicted BMR while diving (Le Boeuf, Costa et al. 1988). Similarly, oxygen consumption measurements of Weddell seals diving from a restricted ice-hole indicate that they expend 1.5 to 3 times BMR while diving (Kooyman, Kerem et al. 1973).

Diving behaviour

What are the implications of such different metabolic rates for the diving and therefore foraging capability of pinnipeds? Phocids are exceptional divers with respect to their long and deep dive patterns. Otariid dives are shallow and brief (Fig. 1). What allows phocids to make such long deep dives and what restricts otariids to such short and therefore shallow dives? The maximum depth of a dive is directly proportional to its duration, since the deeper an animal dives the farther it must travel. Although animals can increase the duration of individual dives by using anaerobic metabolism, it is more efficient to rely on aerobic metabolism (Kooyman, Wahrenbrock et al. 1980; Kooyman, Castellini et al. 1983; Kooyman 1989). This maximum dive limit has been defined as the aerobic dive limit (ADL) and can be calculated from the total oxygen store (l oxygen) divided by the rate of oxygen utilization (ml oxygen · min^{-1}: Kooyman 1985).

The greater diving ability of phocids is not unexpected since they can store 60 ml oxygen · kg^{-1} body mass compared to 40 ml oxygen · kg^{-1} body mass for otariids (Kooyman 1985). However, this only accounts for a 50% increase in dive duration, while phocids dive ten times longer than otariids (Fig. 2). This can be explained by the difference in metabolism at sea; the rate measured in otariids is greater than that estimated for phocids. A comparison of the importance of oxygen stores and metabolic rate in

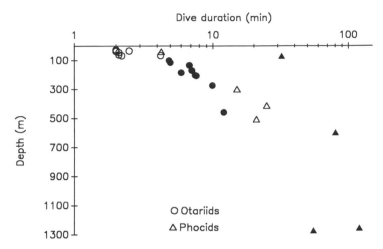

Fig. 1. Maximum (solid symbols) and mean (open symbols) dive duration is plotted as a function of dive depth for eight otariids and four phocids. Data are from Galapagos, Antarctic, Cape and northern fur seals (Gentry, Costa *et al.* 1986); California sea lion (Feldkamp *et al.* 1989); Australian sea lion (Costa, Kretzmann *et al.* 1989); Hooker's sea lion (Gentry, Roberts *et al.* 1987); Weddell seal (Kooyman 1981; M. A. Castellini pers. comm.); northern elephant seal (Le Boeuf, Costa *et al.* 1988 and unpubl. data); grey seal (Thompson *et al.* 1991); southern elephant seal (Hindell *et al.* 1991).

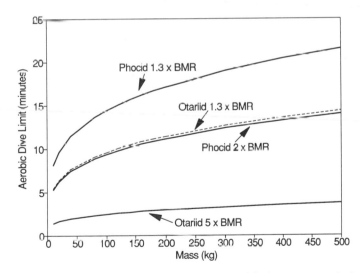

Fig. 2. The variation in aerobic dive limit as a function of body mass was calculated for a phocid seal operating at 1.3 and 2 × BMR and an otariid operating at 1.3 and 5 × BMR. Oxygen stores were assumed to be 60 ml oxygen·kg^{-1} for a phocid and 40 ml oxygen·kg^{-1} for an otariid (Kooyman 1985).

determining ADL as a function of body mass can be seen in Fig. 2. Variation in ADL due to body oxygen stores alone can be seen by comparing the lines predicted for both groups at 1.3 times BMR. Although differences between otariid and phocid oxygen storage capacity have a significant effect on ADL, differences in at-sea metabolism have a greater impact (Fig. 2). Also notice that ADL varies with body mass (Fig. 2). This is because metabolic rate scales to body mass$^{0.75}$ whereas oxygen storage capacity scales to mass$^{1.0}$. Thus larger mammals have a lower mass-specific metabolism for a relatively constant proportion of oxygen storage capacity (Kooyman, Castellini et al. 1983; Gentry, Costa et al. 1986; Kooyman 1989). Large mammals should be able to dive longer and therefore deeper than small ones. This suggests that, from differences in body size alone, phocids as a group should be better divers than otariids.

It is interesting that the influence of body size on ADL decreases as the metabolic rate increases. This implies that with respect to diving ability (ADL) the advantage of large size is less in otariids because their metabolism is higher (Fig. 2). There may even be an advantage to small body size, since although small animals have a higher mass-specific metabolism their absolute food energy requirements are lower (Peters 1983; Millar & Hickling 1990) (Fig. 3). These differences have implications for the evolution of optimal body size in the Pinnipedia. Large body size would be favoured in phocids because it confers the benefits of increased diving capability. In otariids, on the other hand, smaller size would be favoured, because they gain more from reductions in absolute energy requirements than from minimal increases in dive duration resulting from greater mass.

Observations of travelling behaviour are consistent with the economical life-style of phocids and the costly life-style of otariids. Otariids typically travel at the surface, porpoising, while phocids use stereotypic travelling dives that are more cost-effective (Gentry, Costa et al. 1986; Feldkamp, DeLong & Antonelis 1989; Le Boeuf, Naito et al. 1992). Although surface swimming is more costly than sub-surface swimming (Vogel 1981) it is likely to be faster and may therefore reduce the time otariid mothers spend in transit between the foraging grounds and the rookery. This is a more expensive strategy but it may be optimal when resources are abundant as in offshore upwelling regions.

Differences in the cost and efficiency of foraging

How do these different diving patterns and energy expenditures affect the ability of pinnipeds to obtain prey and the efficiency with which they acquire energy? Differences in dive performance for a sea lion, a fur seal and an elephant seal are presented in Table 1. Although elephant seals obtain more prey energy per dive than either northern fur seals or California sea

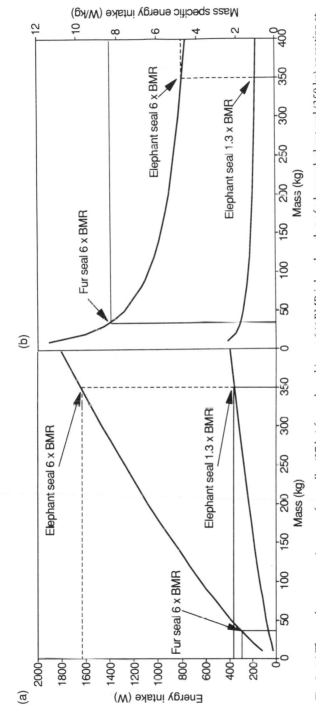

Fig. 3. (a) The total energy requirement of a smaller (37 kg) fur seal working at 6 × BMR is less than that of a larger elephant seal (350 kg) operating at 1.3 or 6 × BMR. The actual energy requirement of an elephant seal is likely to be that needed when operating at 1.3 × BMR; whereas (b) the mass-specific energy requirement of a smaller fur seal operating at 6 × BMR is higher than that of the larger elephant seal operating at either 6 or 1.3 × BMR.

Table 1. Dive rate and duration, rate of prey energy acquired, energy expenditure and metabolic rate of two otariids and a phocid are summarized. Data on prey intake were estimated from water influx and metabolic rate of the sea lion and fur seals from 0–18 doubly-labelled water measurements (Costa & Gentry 1986; Costa 1988; Costa, Antonelis et al. 1991). Metabolic rate of northern elephant seal was estimated from dive behaviour (Le Boeuf, Costa et al. 1988)

	Mass kg	Dive rate n/d	Dive duration min	Energy acquisition		Energy expended		Ratio of acquired/ expended	Metabolic rate FMR/BMR
				kJ/dive	kJ/min	kJ/dive	kJ/min		
Northern elephant seal (prey: squid)	350	65	19.2	1770	92	403	21	4	1.3
California sea lion (prey: fish)	85	202	2.0	224	112	52	26	4	4.8
Northern fur seal (prey: squid)	37	38	2.1	853	406	38	18	23	6.0

lions, they take more time to do it, and acquire less energy per unit time (Table 1). However, the ratio of energy acquired to energy expended is the same for the sea lion and the elephant seal, but is significantly greater for the fur seal. These data imply that a high metabolic rate may be advantageous in achieving a high rate of energy acquisition. There is a cost to this strategy, however, in terms of the animal's absolute energy requirements. For example, even though the elephant seal is ten times larger than the fur seal and four times larger than the sea lion, its absolute energy expenditure is almost identical and accordingly its absolute energy intake is equivalent to that of the sea lion and only one quarter that of the fur seal. If these species are representative, this analysis suggests that phocids expend less energy to obtain a similar amount of prey, which may enable them to subsist on a poorer or more dispersed prey resource than otariids require. One may consider otariids to be energy maximizers, that is, they maximize the net rate of prey energy intake irrespective of the cost (i.e. a high metabolic rate). In the right environment, like upwelling zones, this strategy is successful because it allows a greater net rate of prey energy intake (as shown by the fur seal data in Table 1). In contrast, phocids maximize the efficiency of prey energy intake. Phocids can maximize the efficiency of prey acquisition because they can take considerably longer foraging trips than otariids. One result of maximizing the efficiency of energy gain is that it either requires or enables a lower metabolic rate while diving (as shown in the elephant seal data in Table 1). A lower diving metabolic rate enables longer and thus deeper dives, which potentially opens up a new or different prey resource to phocids that would not be available to otariids. Alternatively, utilization of a deep prey resource may promote the development of a lower diving metabolic rate, which limits prey acquisition, thus in turn affecting the phocid breeding strategy. However, the dramatic performance of northern fur seals also suggests that in the right circumstances otariids may be able to make better use of resources when prey is plentiful. Such an interpretation of the data in Table 1 is consistent with the hypothesis that the phocid breeding pattern is economical.

Comparisons with terrestrial vertebrates

There are a number of examples where higher rates of energy expenditure allow a greater acquisition of energy. For example, a series of investigations of the foraging and reproductive energetics of predatory lizards show that those which employ a more costly, highly active, widely foraging behaviour expend more energy but in so doing acquire proportionately more energy than lizards that utilize a more economical, sit-and-wait foraging behaviour (Anderson & Karasov 1981; Karasov & Anderson 1984; Nagy, Huey & Bennett 1984; Anderson & Karasov 1988). These investigations also show that widely foraging predators are able to devote more energy to reproduction than the sit-and-wait predators. Similarly, McNab (1980, 1984, 1986) argues that mammals reproduce as fast as their metabolic rates allow. That is, mammals with high metabolic rates are able to invest more energy in reproduction than animals with low metabolic rates. Such a pattern holds true when marine mammals are included in this analysis (Schmitz & Lavigne 1984). These arguments are consistent with the hypothesis that otariids expend more energy foraging but get more for their effort than phocids. There must be sufficient resources available to support such an expensive life-style, but this is likely to be the case in the upwelling environments that otariids typically inhabit (Repenning & Tedford 1977; King 1983). The ultimate advantage and/or cost of these divergent life-history patterns lies in differences in reproductive success and pup production. Unfortunately, such data are not available.

Allocation of maternal resources

The role of metabolic overhead

If the phocid reproductive pattern is a derived trait that facilitates exploitation of resources that would not otherwise be available, what factors shape this pattern? Analogous to dive duration, the ability of the female to remain ashore fasting, while providing milk to her pup, is related to the size of her energy and nutrient reserves and the rate at which she utilizes them. In this case the optimal solution is to maximize the amount of energy and nutrients provided to the pup and to minimize the amount expended on herself. Fedak & Anderson (1982) coined the term 'metabolic overhead' to define the amount of energy the female expends on herself while ashore suckling her pup. Metabolic overhead can be reduced in two ways. The first is to increase the rate of milk energy transfer and thereby reduce the time spent ashore with the pup. The second is to reduce maternal maintenance metabolism and thus increase the relative amount of maternal reserves that are available for milk production. Apparently phocid mothers apply both strategies. For example, relative to otariids, phocids have a very short lactation interval that is facilitated by a rapid transfer of milk energy made possible by producing a lipid-rich milk (Bonner 1984; Oftedal, Boness & Tedman 1987; Costa 1991a). The extreme is the hooded seal, which lactates for only four days, producing milk that is 64% lipid (Oftedal, Boness & Bowen 1988).

The relationship between metabolic overhead, milk production and lactation duration was modelled with data from northern elephant seals (Fig. 4). In this model the total amount of energy available for maternal investment was assumed to be constant, and metabolic overhead was calculated for each lactation duration. Milk production was calculated as

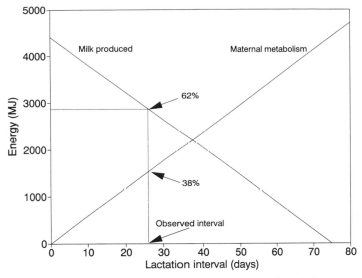

Fig. 4. Maintenance metabolism (as measured using the 0–18 doubly-labelled water method) and milk production data were taken from nine northern elephant seal mother-pup pairs (see Costa, Le Boeuf *et al.* 1986; D. P. Costa & B. J. Le Boeuf unpubl.) and were used to model changes in the energy budget as lactation duration varied. Actual data are shown by the vertical line starting at the abscissa marked as the observed lactation interval (26.4 days) where 38% of the energy went to maternal maintenance metabolism and 62% to pup milk intake. Total energy available for lactation (4400 MJ) was assumed to be the sum of the energy expended on maintenance (1670 MJ) and the milk energy provided to the pup (2730 MJ). Milk production was calculated for each time point as the difference between total energy available and maternal maintenance (4400 MJ − (lactation duration ·63.3 MJ/d)).

the proportion of maternal resources remaining after the cost of the metabolic overhead was met for a given lactation duration. Not surprisingly, the net amount of energy utilized for maternal maintenance increases with increasing lactation duration as the energy available for milk production steadily decreases (Fig. 4).

The effect of size on fasting duration

Metabolic overhead can also be reduced by attaining larger body mass and it also varies non-linearly with mass. Over an equivalent lactation period small pinnipeds experience a higher metabolic overhead than large ones (Costa 1991a). This is because female maintenance metabolism scales as mass$^{0.75}$, whereas energy stores scale as mass$^{1.0}$ or as mass$^{1.19}$ (Calder 1984; Lindstedt & Boyce 1985; Millar & Hickling 1990). Again using the data presented in Fig. 4, we can determine the proportion of total energy that would be available for mothers of different body mass. This analysis

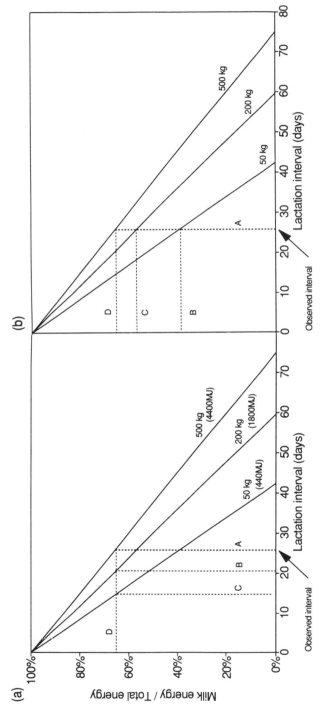

Fig. 5. The data presented in Fig. 4 were used to model how the relative energy budgets of 50-, 200- and 500-kg female northern elephant seals would vary as a function of maternal mass. Numbers in parenthesis are the total energy available for lactation and were calculated as a constant proportion of maternal body mass (8.8 MJ/kg) whereas maintenance metabolism was calculated as lactation duration in days times the daily maternal maintenance metabolism given as 0.560 kg$^{0.75}$. (a) The lactation duration necessary to be able to devote 62% of the available energy to milk production regardless of body size. A 62% level of maternal investment in milk energy is shown by dotted line D and the lactation interval necessary to achieve this is shown as dotted lines C for a 50-kg female, B for a 200-kg female and A for a 500-kg female. (b) The reduction in milk production if all females, regardless of body mass, had a lactation duration of 26 days. A 26-day lactation interval is shown as dotted line A and the energy available for milk production for a 50-kg female is given by dotted line B, for a 200-kg female by dotted line C and for a 500-kg female by dotted line D.

shows that over the same lactation period a small female has less energy available for milk production than a large female, owing to the increased metabolic overhead (Fig. 5b). However, a small female can reduce her metabolic overhead by reducing the duration of lactation (Fig. 5a). The relationship between lactation duration, body mass and the proportion of total energy available for milk production, for a given lactation interval, is summarized in Fig. 6. Here we find that the greatest increase in the proportion of energy made available for milk production is achieved by reducing lactation duration, whereas body size becomes increasingly important when the duration of lactation is long. The validity of this analysis can be seen by a comparison of lactation duration and body size for 12 phocid females (Fig. 7). We see that lactation is shortest for small females that fast during lactation and that the longest lactations are observed in females that are able to supplement their energy budget by feeding during lactation.

Are otariids placed at a significant disadvantage by their small body mass? Utilizing equivalent data for a representative otariid, the northern fur seal, we can make a similar model (Fig. 8). Here again we see that short shore visits make the greatest proportion of maternal energy available for milk production. However, the advantage of large body size is minimal, especially over the two-day or shorter shore visit typical of otariids.

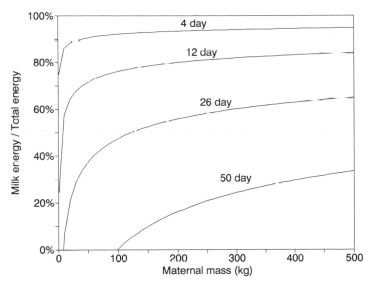

Fig. 6. The variation in the relative amount of energy available for milk production as a function of maternal body mass is estimated for hypothetical elephant seal females with lactation durations of 4, 12, 26 and 50 days. The model uses the data and methods described in Figs 4 and 5.

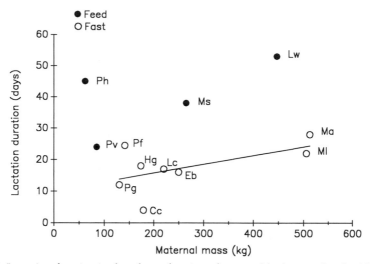

Fig. 7. Lactation duration is plotted as a function of maternal body mass for phocids. Solid symbols denote females that are known to do some feeding during lactation; open symbols denote females that fast. Cc = *Cystophora cristata*, Eb = *Erignathus barbatus*, Hg = *Halichoerus grypus*, Lc = *Lobodon carcinophagus*, Lw = *Leptonychotes weddelli*, Ma = *Mirounga angustirostris*, Ml = *Mirounga leonina*, Ms = *Monachus schauinslandi*, Pf = *Phoca fasciata*, Pg = *Phoca groenlandica*, Ph = *Phoca hispida*, Pv = *Phoca vitulina*. Data from Appendix.

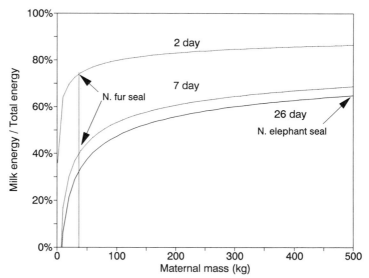

Fig. 8. The effect of body mass on the theoretical proportion of milk energy to total energy expended over lactation for northern fur seal and northern elephant seal females. Northern fur seal females remain ashore for 7 days immediately after parturition and thereafter stay ashore for 2 days after feeding, whereas the entire lactation period lasts 26 days in northern elephant seals. Data derived from Costa, Le Boeuf *et al.* (1986) and D. P. Costa & B. J. Le Boeuf (unpubl.) for elephant seals and from Costa & Gentry (1986) for fur seals.

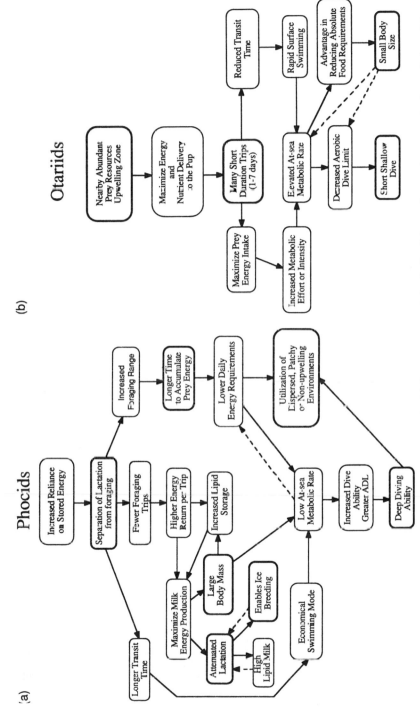

Fig. 9. Flow charts summarizing the relationships between the life-history traits and the feedback loops that are important in the evolution of the phocid breeding pattern (a) and the otariid breeding pattern (b). Boxes with thick outlines show those life-history traits that define the breeding pattern. Dotted lines represent secondary positive feedback.

(a)

Phocids

- Increased Reliance on Stored Energy
- Separation of Lactation from foraging
- Increased Foraging Range
- Longer Time to Accumulate Prey Energy
- Lower Daily Energy Requirements
- Utilization of Dispersed, Patchy or Non-upwelling Environments
- Fewer Foraging Trips
- Higher Energy Return per Trip
- Increased Lipid Storage
- Maximize Milk Energy Production
- Large Body Mass
- Enables Ice Breeding
- Attenuated Lactation
- High Lipid Milk
- Economical Swimming Mode
- Low At-sea Metabolic Rate
- Increased Dive Ability Greater ADL
- Deep Diving Ability
- Longer Transit Time

(b)

Otariids

- Nearby Abundant Prey Resources Upwelling Zone
- Maximize Energy and Nutrient Delivery to the Pup
- Many Short Duration Trips (1-7 days)
- Reduced Transit Time
- Rapid Surface Swimming
- Advantage in Reducing Absolute Food Requirements
- Small Body Size
- Maximize Prey Energy Intake
- Increased Metabolic Effort or Intensity
- Elevated At-sea Metabolic Rate
- Decreased Aerobic Dive Limit
- Short Shallow Dive

Therefore there is no advantage to large body mass in otariids, whereas large body size is favoured in phocids because of their longer shore visits. Figure 9 summarizes the major traits and factors important in the derivation of phocid (Fig. 9a) and otariid (Fig. 9b) life-history patterns.

Implications for the evolution of pinniped reproductive patterns

Previous hypotheses

It is interesting to ponder how these different life-history patterns have evolved. Stirling (1975, 1983) speculated that this breeding pattern was the result of all phocids descending from an ice-breeding ancestor. He argued that a common pagophilic existence would require an attenuated lactation interval since it would be critical in enabling pagophilic seals to rear their young successfully on unstable pack ice. However, island-breeding monk seals are one of the oldest extant phocid groups and apparently never had an ice-breeding ancestor, yet they exhibit the attenuated lactation duration typical of phocids (Repenning & Ray 1977; Repenning, Ray & Grigorescu 1979).

Evolution of pinniped breeding patterns

An alternative hypothesis for the evolution of the Pinnipedia from a monophyletic origin (see Wyss 1989), which incorporates energetic arguments, is as follows. First, an ancestral pinniped similar to *Enaliarctos* sp. (Berta, Ray & Wyss 1989) takes advantage of the increased upwelling in the North Pacific (Lipps & Mitchell 1976) and utilizes a primitive form of the otariid breeding pattern (Fig. 9b). This implies that the otariid pattern is the one most closely linked to its terrestrial ancestry. As these ancestral otariids refine their foraging and diving skills, they begin to forage further offshore, increase the duration of their foraging trips and decrease the number and frequency of foraging trips (Fig. 9a). Increases in trip duration coupled with decreases in trip frequency or number require a greater energy return per trip and, to achieve it, an increased reliance on stored maternal reserves. This relationship can be optimized by both increasing the lipid content of the milk and decreasing the metabolic overhead. Metabolic overhead can be reduced by decreasing the time spent on shore suckling the pup and/or by increasing body size (Fig. 9). Such a pattern might have been employed by *Desmatophoca* spp. and/or *Allodesmus* spp., fossil pinnipeds thought to be phylogenetically close to the transitional group between otariids and phocids (Wyss 1987, 1988, 1989). Especially relevant to this hypothesis is that desmatophocines were the earliest otariid lineage to evolve large body size (Barnes 1987). This trend is consistent with the importance of large body size in enabling separation of feeding from lactation, a fundamental component of the phocid breeding pattern (Fig. 9a). Once separation of foraging and lactation was achieved these ancestral phocids could inhabit and reproduce in non-upwelling environments like the Hawaiian Islands or the Caribbean. This would represent a transition between the desmatophocines and the monachine groups. This is consistent with the Hawaiian monk seal being a Pacific Ocean relic of the early phocid lineage (Wyss 1989). Ancestral monk seals could then pass through the Central American Seaway, and establish breeding populations in the Caribbean, because their life-history pattern allowed utilization of highly dispersed prey or environments of lower productivity. Otariids, however, either could not venture through the seaway or could not reproduce in the Caribbean owing to the lack of pronounced upwelling regimes. Once in the Atlantic, phocids could easily move north and south as they would have no competition from otariids. Upon reaching the high latitudes, the attenuated lactation interval would have pre-adapted them to breeding on unstable substrates such as ice. Once ice breeding was initiated, the lactation interval would be further shortened as they expanded their range to pack ice (following the arguments of Stirling 1975, 1983).

This paper has highlighted generalized features in the evolution of life-history patterns in the Pinnipedia. Deviations from the broad patterns delineated here are to be expected since individual species adapt to specific pressures or specialized habitats. It is hoped that such a broad examination of pinniped life-history patterns will help to further stimulate research into this area. Research into species that do not fit the model is especially warranted as it will test the ideas presented here as well as provide insight into reasons for divergence from the 'typical' phocid or otariid pattern.

Acknowledgements

This work was supported by grants from the Office of Naval Research (N 0014–91–MD24G40) and the Biological Oceanography Program of the National Science Foundation (OCE 9018626). Many of the ideas presented here are the distillation of numerous conversations with colleagues who have freely shared their ideas and in so doing have stimulated my thinking in this area. I would especially like to thank D. Boness, J. Croxall, M. Fedak, S. Feldkamp, R. Gentry, B. J. Le Boeuf, G. Kooyman, K. Nagy and F. Trillmich. Special thanks go to Dave Lavigne who opened my eyes to pinniped monophyly and the possibility that desmatophocids were ancestral phocids.

References

Anderson, R. A. & Karasov, W. H. (1981). Contrasts in energy intake and expenditure in sit-and-wait and widely foraging lizards. *Oecologia* **49**: 67–72.

Anderson, R. A. & Karasov, W. H. (1988). Energetics of the lizard *Cnemidophorus tigris* and life history consequences of food-acquisition mode. *Ecol. Monogr.* **58**: 79–110.

Barnes, L. G. (1987). An early Miocene pinniped of the genus *Desmatophoca* (Mammalia: Otariidae) from Washington. *Contr. Sci. Los Angeles* No. 382: 1–20.

Bartholomew, G. A. (1970). A model for the evolution of pinniped polygyny. *Evolution, Lancaster, Pa* **24**: 546–559.

Berta, A., Ray, C. E. & Wyss, A. R. (1989). Skeleton of the oldest known pinniped, *Enaliarctos mealsi. Science* **244**: 60–62.

Bonner, W. N. (1984). Lactation strategies in pinnipeds: problems for a marine mammalian group. *Symp. zool. Soc. Lond.* No. 51: 253–272.

Bowen, W. D. (1991). Behavioural ecology of pinniped neonates. In *The behaviour of pinnipeds*: 66–127. (Ed. Renouf, D.). Chapman and Hall, London etc.

Bowen, W. D., Oftedal, O. T. & Boness, D. J. (1985). Birth to weaning in 4 days: remarkable growth in the hooded seal, *Cystophora cristata. Can. J. Zool.* **63**: 2841–2846.

Brodie, P. F. (1975). Cetacean energetics, an overview of intraspecific size variation. *Ecology* **56**: 152–161.

Calder, W. A. (1984). *Size, function and life history.* Harvard University Press, Cambridge, Mass. & London.

Costa, D. P. (1988). Assessment of the impact of the California sea lion and northern elephant seal on commercial fisheries. In *California Sea Grant: biennial report of completed projects 1984–86*: 36–43. California Sea Grant College Program, University of California, La Jolla. Publication # R–CSCP–024.

Costa, D. P. (1991a). Reproductive and foraging energetics of pinnipeds: implications for life history patterns. In *The behaviour of pinnipeds*: 300–344. (Ed. Renouf, D.). Chapman and Hall, London etc.

Costa, D. P. (1991b). Reproductive and foraging energetics of high latitude penguins, albatrosses and pinnipeds: implications for life history patterns. *Am. Zool.* **31**: 111–130.

Costa, D. P., Antonelis, G. P. & DeLong, R. (1991). Effects of El Niño on the foraging energetics of the California sea lion. In *Pinnipeds and El Niño: responses to environmental stress*: 156–165. (Eds Trillmich, F. & Ono, K.). Springer-Verlag, Berlin. (*Ecol. Stud.* **88**.)

Costa, D. P., Croxall, J. P. & Duck, C. D. (1989). Foraging energetics of Antarctic fur seals in relation to changes in prey availability. *Ecology* **70**: 596–606.

Costa, D. P. & Gentry, R. L. (1986). Free-ranging energetics of northern fur seals. In *Fur seals: maternal strategies on land and at sea*: 79–101. (Eds Gentry, R. L. & Kooyman, G. L.). Princeton University Press, Princeton, N. J.

Costa, D. P., Kretzmann, M. & Thorson, P. H. (1989). Diving pattern and energetics of the Australian sea lion, *Neophoca cinerea*. *Am. Zool.* **29**: 71A.

Costa, D, P., Le Boeuf, B. J., Huntley, A. C. & Ortiz, C. L. (1986). The energetics of lactation in the Northern elephant seal, *Mirounga angustirostris*. *J. Zool., Lond.* **209**: 21–33.

Elgar, M. A. & Harvey, P. H. (1987). Basal metabolic rates in mammals: allometry, phylogeny and ecology. *Funct. Ecol.* **1**: 25–36.

Fedak, M. A. & Anderson, S. S. (1982). The energetics of lactation: accurate measurements from a large wild mammal, the grey seal (*Halichoerus grypus*). *J. Zool., Lond.* **198**: 473–479.

Feldkamp, S. D., DeLong, R. L. & Antonelis, G. A. (1989). Diving patterns of California sea lions, *Zalophus californianus*. *Can. J. Zool.* **67**: 872–883.

Gentry, R. L., Costa, D. P., Croxall, J. P., David, J. H. M., Davis, R. W., Kooyman, G. L., Majluf, P., McCann, T. S. & Trillmich, F. (1986). Synthesis and conclusions. In *Fur seals: maternal strategies on land and at sea*: 229–264. (Eds Gentry, R. L. & Kooyman, G. L.). Princeton University Press, Princeton.

Gentry, R. L. & Kooyman, G. L. (Eds) (1986). *Fur seals: maternal strategies on land and at sea*. Princeton University Press, Princeton.

Gentry, R. L., Roberts, W. E. & Cawthorn, M. W. (1987). Diving behavior of the Hooker's sea lion. *Abstr. bienn. Conf. Biol. mar. Mammals* 7. Society for Marine Mammalogy. Unpublished.

Hindell, M. A., Slip, D. J. & Burton, H. R. (1991). The diving behaviour of adult male and female southern elephant seals, *Mirounga leonina* (Pinnipedia: Phocidae). *Aust. J. Zool.* **39**: 595–619.

Karasov, W. H. & Anderson, R. A. (1984). Interhabitat differences in energy acquisition and expenditure in a lizard. *Ecology* **65**: 235–247.

Kaufman, G. W., Siniff, D. B. & Reichle, R. (1975). Colony behavior of Weddell seals, *Leptonychotes weddelli*, at Hutton Cliffs, Antarctica. *Rapp. P.-v. Réun. Cons. perm. int. Explor. Mer* **169**: 228–246.

King, J. E. (1983). *Seals of the world.* (2nd edn). British Museum (Natural History), London & Oxford University Press, Oxford. (*Publs Br. Mus. nat. Hist.* No. 868: 1–240.)

Kooyman, G. L. (1981). *Weddell seal: consummate diver.* Cambridge University Press, Cambridge, London & New York.

Kooyman, G. L. (1985). Physiology without restraint in diving mammals. *Mar. Mamm. Sci.* **1**: 166–178.

Kooyman, G. L. (1989). *Diverse divers: physiology and behavior.* Springer-Verlag, Berlin & London.

Kooyman, G. L., Castellini, M. A., Davis, R. W. & Maue, R. A. (1983). Aerobic diving limits of immature Weddell seals. *J. comp. Physiol (B)* **151**: 171–174.

Kooyman, G. L., Kerem, D. H., Campbell, W. B. & Wright, J. J. (1973). Pulmonary gas exchange in freely diving Weddell seals. *Respir. Physiol.* **17**: 283–290.

Kooyman, G. L., Wahrenbrock, E. A., Castellini, M. A., Davis, R. W. & Sinnett, E. E. (1980). Aerobic and anaerobic metabolism during voluntary diving in Weddell seals: evidence of preferred pathways from blood chemistry and behavior. *J. comp. Physiol. (B)* **138**: 335–346.

Kovacs, K. M. & Lavigne, D. M. (1986). Maternal investment and neonatal growth in phocid seals. *J. Anim. Ecol.* **55**: 1035–1051.

Le Boeuf, B. J., Costa, D. P., Huntley, A. C. & Feldkamp, S. D. (1988). Continuous, deep diving in female northern elephant seals, *Mirounga angustirostris*. *Can. J. Zool.* **66**: 446–458.

Le Boeuf, B. J., Naito, Y., Asaga, T., Crocker, D. & Costa, D. P. (1992). Swim velocity and dive patterns in a northern elephant seal, *Mirounga angustirostris*. *Can. J. Zool.* **70**: 786–795.

Lindstedt, S. L. & Boyce, M. S. (1985). Seasonality, fasting endurance, and body size in mammals. *Am. Nat.* **125**: 873–878.

Lipps, J. H. & Mitchell, E. (1976). Trophic model for the adaptive radiations and extinctions of pelagic marine mammals. *Paleobiology* **2**: 147–155.

McNab, B. K. (1980). Food habits, energetics, and the population biology of mammals. *Am. Nat.* **116**: 106–124.

McNab, B. K. (1984). Basal metabolic rate and the intrinsic rate of increase: an empirical and theoretical re-examination. Commentary. *Oecologia* **64**: 423–424.

McNab, B. K. (1986). The influence of food habitats on the energetics of eutherian mammals. *Ecol. Monogr.* **56**: 1–19.

Millar, J. S. & Hickling, G. J. (1990). Fasting endurance and the evolution of mammalian body size. *Funct. Ecol.* **4**: 5–12.

Nagy, K. A., Huey, R. B. & Bennett, A. F. (1984). Field energetics and foraging mode of Kalahari lacertid lizards. *Ecology* **65**: 588–596.

Oftedal, O. T., Boness, D. J. & Bowen, W. D. (1988). The composition of hooded seal (*Cystophora cristata*) milk: an adaptation for postnatal fattening. *Can. J. Zool.* **66**: 318–322.

Oftedal, O. T., Boness, D. J. & Tedman, R. A. (1987). The behavior, physiology, and anatomy of lactation in the Pinnipedia. *Curr. Mammal.* **1**: 175–245.

Perez, M. A. & Mooney, E. E. (1986). Increased food and energy consumption of lactating northern fur seals, *Callorhinus ursinus*. *Fish. Bull. U.S. Fish Wildl. Serv.* **84**: 371–381.

Peters, R. H. (1983). *The ecological implications of body size*. Cambridge University Press, Cambridge, London etc.

Ralls, C. M. (1976). Mammals in which females are larger than males. *Q. Rev. Biol.* **51**: 245–276.

Read, A. F. & Harvey, P. H. (1989). Life history differences among the eutherian radiations. *J. Zool., Lond.* **219**: 329–353.

Repenning, C. A. & Ray, C. E. (1977). The origin of the Hawaiian monk seal. *Proc. biol. Soc. Wash.* **89**: 667–688.

Repenning, C. A., Ray, C. E. & Grigorescu, D. (1979). Pinniped biogeography. In *Historical biogeography, plate tectonics, and the changing environment*: 357–369. (Eds Gray, J. & Boucot, J.). Oregon State Univ. Press, Corvallis.

Repenning, C. A. & Tedford, R. H. (1977). Otarioid seals of the Neogene. *Prof. Pap. U.S. geol. Surv.* No. 992: 1–93.

Schmitz, O. J. & Lavigne, D. M. (1984). Intrinsic rate of increase, body size, and specific metabolic rate in marine mammals. *Oecologia* **62**: 305–309.

Stearns, S. C. (1983). The influence of size and phylogeny on patterns of covariation among life-history traits in the mammals. *Oikos* **41**: 173–187.

Stearns, S. C. (1989). Trade-offs in life history evolution. *Funct. Ecol.* **3**: 259–268.

Stirling, I. (1975). Factors affecting the evolution of social behaviour in the Pinnipedia. *Rapp. P.-v. Réun. Cons. perm. int. Explor. Mer* **169**: 205–212.

Stirling, I. (1983). The evolution of mating systems in pinnipeds. *Spec. Publs. Am. Soc. Mammal.* No. 7: 489–527.

Testa, J. W., Hill, S. E. B. & Siniff, D. B. (1989). Diving behavior and maternal investment in Weddell seals (*Leptonychotes weddelli*). *Mar. Mamm. Sci.* **5**: 399–405.

Thomas, J. A. & DeMaster, D. P. (1983). Diel haul-out patterns of Weddell seal (*Leptonychotes weddelli*) females and their pups. *Can. J. Zool.* **61**: 2084–2086.

Thompson, D., Hammond, P. S., Nicholas, K. S. & Fedak, M. A. (1991). Movements, diving and foraging behaviour of grey seals (*Halichoerus grypus*). *J. Zool., Lond.* **224**: 223–232.

Trillmich, F. (1990). The behavioral ecology of maternal effort in fur seals and sea lions. *Behaviour* **114**: 3–20.

Vogel, S. (1981). *Life in moving fluids.* Willard Grant Press, Boston.

Wyss, A. R. (1987). The walrus auditory region and the monophyly of pinnipeds. *Am. Mus. Novit.* No. 2871: 1–31.

Wyss, A. R. (1988). Evidence from flipper structure for a single origin of pinnipeds. *Nature, Lond.* **334**: 427–428.

Wyss, A. R. (1989). Flippers and pinniped phylogeny: has the problem of convergence been overrated? *Mar. Mammal Sci.* **5**: 343–360.

Appendix

Maternal mass and lactation interval are summarized for all extant phocids and otariids. Data are from Bonner (1984); Kovacs & Lavigne (1986); Gentry & Kooyman (1986); Oftedal, Boness & Tedman (1987); Bowen (1991); Costa (1991a).

Phocid species	Maternal mass (kg)	Lactation duration (days)	Feeds while lactating	Breeding substrate
Caspian seal, *Phoca caspica*	55	20–25	?	Fast ice
Baikal seal, *P. siberica*	94	60–75	Yes	Fast ice lairs
Ring seal, *P. hispida*	62	41–48	Yes	Fast ice lairs
Harbour seal, *P. vitulina*	85	24	Yes	Land
Spotted seal, *P. largha*	86	14–21	?	Pack ice
Harp seal, *P. groenlandica*	130	12	No	Pack ice
Ribbon seal, *P. fasciata*	141	21–28	?	Pack ice
Grey seal, *Halichoerus grypus*	174	18	No	Land and ice
Hooded seal, *Cystophora cristata*	179	4	No	Pack ice
Ross seal, *Ommatophoca rossi*	186	30	?	Pack ice
Crabeater seal, *Lobodon carcinophagus*	220	17	Probably not	Pack ice

Appendix (*cont.*)

Phocid species	Maternal mass (kg)	Lactation duration (days)	Feeds while lactating	Breeding substrate
Bearded seal, *Erignathus barbatus*	250	12–18	Probably not	Pack ice
Hawaiian monk seal, *Monachus schauinslandi*	265	39–41	No	Land
Mediterranean monk seal, *M. monachus*	275	42–49	Yes	Land, caves
Weddell seal, *Leptonychotes weddelli*	447	53	Yes	Fast ice
Leopard seal, *Hydrurga leptonyx*	450	30	?	Pack ice
Southern elephant seal, *Mirounga leonina*	506	22	No	Land
Northern elephant seal, *M. angustirostris*	513	26	No	Land
Mean	229			
Median	141			

Otaviid species	Maternal mass (kg)	Lactation duration (days)
Galapagos fur seal, *Arctocephalus galapagoensis*	27	720
New Zealand fur seal, *A. forsteri*	35	300–365
Northern fur seal, *Callorhinus ursinus*	37	118
Antarctic fur seal, *A. gazella*	39	117
Juan Fernandez fur seal, *A. philippii*	40	300–365?
South American fur seal, *A. australis*	55	300–365
Guadalupe fur seal, *A. townsendi*	45	300–365?
Sub-Antarctic fur seal, *A. tropicalis*	36	300–330
South African fur seal, *A. pusillus*	57	300–330
Galapagos sea lion, *Z. c. wollebaeki*	80	180–365
Australian sea lion, *Neophoca cinerea*	82	532
California sea lion, *Zalophus californianus*	85	300–365
South American sea lion, *Otaria flavescens*	121	360
Hooker's sea lion, *Phocarctos hookeri*	183	365
Steller's sea lion, *Eumetopias jubatus*	273	300–365
Mean	80	
Median	55	

Physiology and bioenergetics

Symp. zool. Soc. Lond. (1993) No. 66: 317–332

To what extent can heart rate be used as an indicator of metabolic rate in free-living marine mammals?

P. J. BUTLER

School of Biological Sciences
The University of Birmingham
Edgbaston
Birmingham B15 2TT, UK

Synopsis

Physiologists and ecologists alike are in great need of a method by which they can obtain accurate information on the metabolic rate of free-ranging animals and relate metabolic rate to specific types of behaviour in the field. Under steady-state conditions, there is a good linear relationship between heart rate (f_H) and the rate of oxygen consumption ($\dot{V}_{O_2}$) in most terrestrial mammals that have been studied and data storage systems exist that could acquire f_H from free-ranging animals over long periods. Marine mammals spend large proportions of their time (up to 90% in elephant seals) under water, and f_H varies between the elevated rates at the surface and the much lower rates during submersion. Both physiological and behavioural evidence from a number of marine mammals indicates that the vast majority of the dives they perform are aerobic in nature. On occasions, excessively long dives are performed and in some species, such as the Weddell seal, accumulation of high levels of lactate after such dives may be related to subsequently long periods at the surface, when the animals appear to be exhausted. For elephant seals and grey seals, however, recovering from long dives does not require a longer period at the surface. It is suggested that during extended dives in these species, there is a reduction in overall metabolic rate, rather than a switch to anaerobic metabolism. Certainly in the grey seal f_H can reach very low levels (4 beats min ^{1}) during extended dives. Thus, the conditions may be right in marine mammals for f_H to be a good indicator of $\dot{V}_{O_2}$. They are probably in an overall steady state at the tissue level.

An important observation in the present context was that, if f_H is averaged over a complete dive cycle, there is a linear relationship between f_H and $\dot{V}_{O_2}$ for marine mammals. This has been demonstrated for a number of species and an attempt is now being made with fur seals to validate the use of f_H as an indicator of $\dot{V}_{O_2}$ in the field.

ZOOLOGICAL SYMPOSIUM No. 66
ISBN 0–19–854069–8

Copyright © 1993 The Zoological Society of London
All rights of reproduction in any form reserved

Introduction

Physiologists and ecologists alike have long desired to obtain reasonably accurate information on the metabolic rate of animals in the field and, what is more, to relate metabolic rate to specific types of behaviour in the field. The advent of the doubly labelled water (DLW) technique (Lifson, Gordon, Visscher & Nier 1949; Lifson, Gordon & McClintock 1955) has enabled data on average energy expenditure over a given period to be obtained from a wide range of animals (see Nagy 1989 for review), including marine mammals (Costa & Gentry 1986; Costa, Croxall & Duck 1989). There are, however, two major limitations of this technique: it will only give an *average* value for energy expenditure over the experimental period and the experimental period is limited to a few days, depending on the level of the initial enrichment in the animal and its metabolic rate, i.e. the rate at which the enrichment is reduced. Thus, it is necessary to capture the animal before the enrichment is too low. It is, no doubt, for this reason that the studies on marine mammals quoted above (Costa & Gentry 1986; Costa *et al.* 1989) were performed on lactating female fur seals which return periodically to their pups.

   A number of workers have investigated the relationship between heart rate (f_H) and the rate of oxygen consumption ($\dot{V}_{O_2}$) with a view to using the former to estimate the latter in free-range animals (see Butler 1989 for a review). Under steady-state conditions there is a good linear relationship between the two variables in most terrestrial mammals that have been studied, including man (Lundgren 1946; Malhotra, Sen Gupta & Rai 1963). Under natural conditions, however, animals are probably not always in a steady state. An extreme example of this is seen in aquatic birds and mammals during diving when there are large excursions in f_H associated with diving and surfacing and, maybe, a switch to anaerobic metabolism with the production of lactate (see Butler & Jones 1982 for a review). However, if f_H could be used as an indicator of $\dot{V}_{O_2}$ in free-ranging marine mammals, modern data storage systems (Hill 1986; Woakes 1992) would enable the aerobic metabolism of an animal to be determined over long periods (several months). Also, if there is a close temporal relationship between $\dot{V}_{O_2}$ and overall metabolic rate in marine mammals, metabolic rate (as indicated by f_H) could also be related to specific types of behaviour. Data could be gathered for a certain period and held until the animal is recaptured (provided that this is before the battery runs down). Such a technique would provide enormous opportunities to physiologists and ecologists to understand more fully the ways in which these most fascinating of animals have adapted so well to an aquatic life style.

   This article will discuss the factors that affect the relationship between f_H

and $\dot{V}_{O_2}$ and review how these factors vary in terrestrial and marine mammals during exercise. It will then consider our current understanding of the physiological and metabolic adjustments that occur in marine mammals during diving and what influence these adjustments may have on the relationship between f_H and $\dot{V}_{O_2}$. Finally, it will present recent attempts to determine the $f_H/\dot{V}_{O_2}$ relationship in marine mammals and to validate f_H as a useful indicator of $\dot{V}_{O_2}$ in the field.

Factors affecting the relationship between f_H and $\dot{V}_{O_2}$

The relationship between f_H and $\dot{V}_{O_2}$ is given by the Fick equation:

$$\dot{V}_{O_2} = (f_H \cdot V_s \, (CaO_2 - C\bar{v}O_2)),$$

where V_s = cardiac stroke volume, CaO_2 = oxygen content in arterial blood and $C\bar{v}O_2$ = oxygen content in mixed venous blood. Thus, the amount of oxygen used by the tissues per heart beat is $V_s \, (CaO_2 - C\bar{v}O_2)$ which is known as the oxygen pulse (Henderson & Prince 1914). If the oxygen pulse remains constant over a range of $\dot{V}_{O_2}$, f_H will clearly be a very accurate indicator of $\dot{V}_{O_2}$ (Fig. 1), and the line relating f_H to $\dot{V}_{O_2}$ passes through the origin. If oxygen pulse varies over a range of $\dot{V}_{O_2}$, there may be a complex

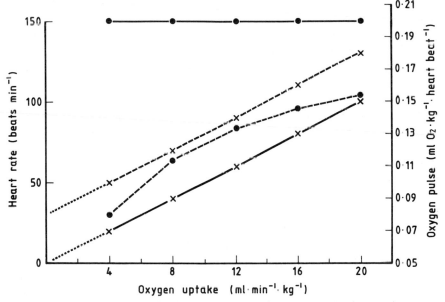

Fig. 1. The change in oxygen pulse (●) when there is a linear relationship between heart rate (X) and oxygen uptake when the regression line passes through zero (——) and when it does not (- - -).

relationship between the two variables and f_H may be a poor indicator of $\dot{V}_{O_2}$. However, if oxygen pulse varies in a systematic (curvilinear) fashion, f_H has a linear relationship with $\dot{V}_{O_2}$ (see Fig. 1), but the line relating the two variables does not pass through the origin. This is the usual situation, with the intercept being positive (see Fig. 8).

The response to exercise of V_s among terrestrial mammals tends to vary between species, and also between studies for the same species. It remains constant in horses, steers, ponies, goats and dogs (Horstman *et al.* 1974; Taylor *et al.* 1987; Jones *et al.* 1989) whereas it increases in dogs (Taylor *et al.* 1987) and in rats (Gleeson & Baldwin 1981). For most species, CaO_2 increases during progressively more intense exercise as a result of a rise in haemoglobin concentration [Hb] (Taylor *et al.* 1987; Jones *et al.* 1989), but in the rat it remains constant (Gleeson & Baldwin 1981). There is invariably a decrease in $C\bar{v}O_2$ which gives rise to an overall increase in $CaO_2 - C\bar{v}O_2$.

The relatively few data obtained from marine mammals indicate that the situation is to some extent even more complicated. Ponganis, Kooyman, Zornow *et al.* (1990) found that in common (harbour) seals, *Phoca vitulina*, V_s while the animals are at the surface is approximately twice that when they are submerged, over a wide range of $\dot{V}_{O_2}$ (Fig. 2), although the surface and submerged values do not themselves vary substantially. For California sea lions, *Zalophus californianus*, it was not possible to obtain separate data for the periods under water and at the surface as the latter were too short (1–2s), and in this case, V_s does not change substantially during exercise from the resting value of 2 ml kg^{-1} (Ponganis, Kooyman & Zornow 1991). In freely diving Weddell seals, *Leptonychotes weddelli*, a 60–70% increase in [Hb] does not completely counteract the large reduction in the partial

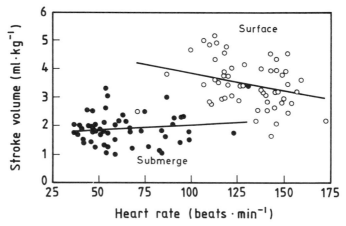

Fig. 2. The variation in cardiac stroke volume with increasing heart rate during surface (○) and submerged (●) swimming in common (harbour) seals (redrawn from Ponganis, Kooyman, Zornow *et al.* 1990).

pressure of oxygen in the arterial blood (Pao_2) that occurs during diving, so there is still a 30–40% reduction in Cao_2 (Qvist *et al.* 1986).

Physiological and metabolic adjustments during diving in marine mammals

For many years it was thought that the limited oxygen supply, substantial though it is in many marine mammals (Butler 1991), had to be conserved during diving for the oxygen-sensitive tissues, the central nervous system and heart, by reducing blood supply to the rest of the body, which would metabolize anaerobically and accumulate lactic acid (Scholander 1940; Butler & Jones 1982). Associated with these circulatory and metabolic changes is a dramatic reduction in heart rate, the classical diving bradycardia. The first indication that this may be an extreme response, rather than the norm during natural dives, was presented by Kooyman, Wahrenbrock *et al.* (1980), who showed that, in free-range adult Weddell seals, there is no accumulation of lactate in the vast majority (95+%) of dives (Fig. 3a), which are thought to be feeding dives. For those few dives

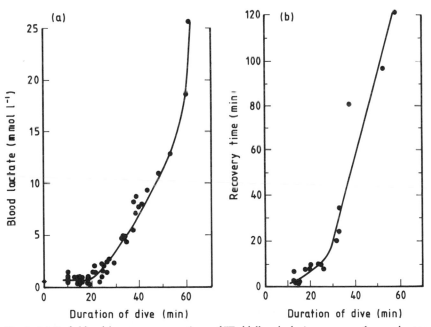

Fig. 3. (a) Peak blood lactate concentrations of Weddell seals during recovery from voluntary dives of various durations. ◆ (on the y axis) indicates average resting values. (b) Time required for blood lactate of freely diving Weddell seals to return to a concentration of 0.55m mol l⁻¹ (or when the recovery curve was extrapolated as a straight line to the resting level) following dives of various durations (redrawn from Kooyman, Wahrenbrock *et al.* 1980).

that do exceed the so-called 'aerobic dive limit', ADL, (thought to be exploratory dives) lactate is accumulated and appears in the blood upon surfacing (Fig. 3a). As might be expected, the length of the ADL is related to the size of the animal (Kooyman, Castellini *et al.* 1983), but the more the dive exceeds the ADL, the larger the accumulation of lactate and the longer it takes for the level of lactate to return to the resting levels (Fig. 3b). After very long dives (1 h or more) the seals are exhausted and sleep for several hours. This pioneering study indicated, therefore, that the vast majority of the dives, in the Weddell seal at least, are completely aerobic in nature, with little or no accumulation of lactate. This means that f_H could possibly be used as an indicator of the metabolic requirements associated with specific bouts of diving behaviour as the complications of post-exercise lactate metabolism (see Gaesser & Brooks 1984) would not be a problem.

Another indication that physiological functions are not disrupted during most natural dives, which is contrary to the conclusions drawn from earlier studies on restrained, forcibly submerged animals (Scholander 1940), is the fact that glomerular filtration rate and hepatic blood flow during relatively short (< ADL) voluntary dives of Weddell seals do not change from the resting levels (Davis, Castellini, Kooyman & Maue 1983).

Despite the above findings, there is a clear reduction in f_H, below the resting level (although not to the level seen during enforced dives), during

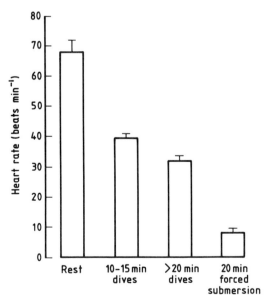

Fig. 4. Mean values of heart rate (± S.E. of mean) of Weddell seals at rest, during relatively short (10–15 min) and long (>20 min) natural dives (data from Hill *et al.* 1987), and at the end of enforced dives of 20 min duration (data from Liggins *et al.* 1980).

natural dives in Weddell seals (Kooyman & Campbell 1972; Hill *et al.* 1987). The former authors claimed that the initial diving heart rate was lower for (eventually) longer dives. The latter authors noted that a relatively high f_H at the start of a dive was always associated with a (very) short (< 5 min) dive, although there was no relationship between initial f_H and dive duration for dives longer than 5 min. There is no doubt, however, that f_H is lower during the longer (> 20 min) dives (Fig. 4). This corresponds with the metabolic data obtained by Castellini, Kooyman & Ponganis (1992) which indicate that $\dot{V}_{O_2}$ for relatively short (< 14 min for 355 kg seals) dives is only 25% above the resting value, whereas that for longer dives is not significantly different from the resting value (Fig. 5). These data indicate that the metabolic cost of diving is very low in Weddell seals.

As well as the reduction in f_H during aerobic dives of Weddell seals, there are also indications from studies on metabolite turnover patterns in freely diving animals (Guppy *et al.* 1986) that the pattern of circulation during diving (even during short, presumably aerobic dives) is different from that in resting animals (Fig. 6), and similar to that seen in animals prevented from surfacing during exercise (Castellini, Murphy *et al.* 1985).

What do all these apparently conflicting data mean? It seems as if dives below a certain duration, and these are the vast majority of natural dives, are completely aerobic in nature with no net accumulation of lactate, and yet there are bradycardia and circulatory adjustments during such dives, compared with the situation at the surface. What is most probably happening during these dives (Kooyman 1985; Fedak, Pullen & Kanwisher 1988) is that when the animal is at the surface between dives, high

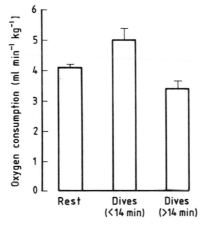

Fig. 5. Mean values (± S.E. of mean) of oxygen consumption for five Weddell seals at rest, after dives of <14 min duration and after dives of >14 min duration (data from Castellini, Kooyman *et al.* 1992).

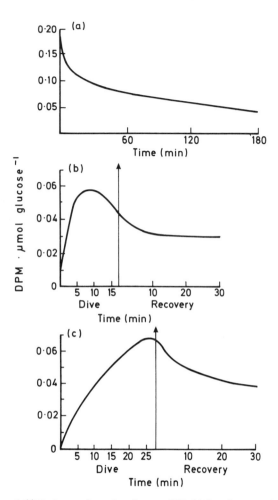

Fig. 6. Clearance of [^{14}C] glucose from the plasma of Weddell seals at rest (a), during a short (17 min) feeding dive (b) and during a longer (28 min) exploratory dive (c). The vertical lines in (b) and (c) indicate when the animal surfaced. The fact that it took a long time to reach peak activity during the dives is taken to indicate a slow circulation time resulting from peripheral vasoconstriction. DPM, disintegrations per minute (1 DPM = 0.0167 Bq) (modified from Guppy *et al.* 1986).

ventilation rates, cardiac output and perfusion of the tissues ensure quick and complete loading of the oxygen stores, which include the highly concentrated myoglobin (Mb) in the skeletal muscles (see Butler 1991 for a review). Upon submersion, cardiac output, and therefore f_H, fall and perfusion of parts of the body declines; these parts may include the skeletal muscles, as they have a substantial oxygen store associated with the Mb. The precise relationship between perfusion of the active muscles and depletion of the oxygen stores within the muscle remains to be determined but heart rate data indicate that, at least during the late stages of a dive, perfusion could be pulsatile. The low aerobic requirements of underwater activity mean that the rate at which the oxygen is used is not excessively high. The animal surfaces while there is still sufficient oxygen for aerobic metabolism and quickly replenishes the stores. Thus, at the tissue level the animal *is* in a steady state. The discontinuous nature of lung ventilation and uptake of oxygen from the atmosphere, compared with terrestrial mammals, causes associated fluctuations in circulatory performance, but does not necessarily reflect similar oscillations in the supply of oxygen to the tissues.

It will not have escaped the reader's notice that the Weddell seal is the only example of marine mammal used in this section. This is simply because it is the animal from which most behavioural and physiological data have been obtained. However, data, mostly behavioural, from a range of other species of marine mammals are consistent with this model (Kooyman 1989). Worthy of special mention are the northern and southern elephant seals, *Mirounga angustirostris* and *M. leonina* respectively. These animals spend approximately 90% of their time at sea under water (Le Boeuf, Costa *et al.* 1988; Le Boeuf, Naito *et al.* 1989; Hindell, Slip & Burton 1991) which would indicate that all of their bodily functions must be performing close to the average level while they are submerged. In addition, there is no relationship between the duration of the previous dive and the subsequent time at the surface, which could be taken to indicate that all dives are completely aerobic in nature. It could also mean that any lactate accumulated during an exceptionally long dive is cleared during subsequent dives. Certainly Castellini, Davis & Kooyman (1988) described such an occurrence in a Weddell seal, although in this case the dives following the long, lactate-accumulating dive were of relatively short duration, whereas in the northern elephant seal they may still be long. Also, this is a very rare occurrence in Weddell seals. Similar data have been obtained from free-range grey seals, *Halichoerus grypus* (Thompson & Fedak 1993). Here again, the behavioural data indicate that recovery from unusually long dives (6% were longer than 10 min) does not require a longer period at the surface. In this species, it is known that during such long dives, the animals are largely motionless and have exceptionally low values of f_H (< 4 beats min^{-1}). The authors suggest that there is a reduction in overall metabolic rate, rather than a switch to anaerobic metabolism, with the seals adopting a 'wait and ambush' strategy for prey capture rather than active pursuit. This deduction is based on the behavioural data, but it does imply that heart rate is, in some way, related to overall metabolism.

Recent data from Antarctic fur seals, *Arctocephalus gazella*, indicate that, although less than 1% of individual dives exceed the estimated ADL (Boyd & Croxall in press), some dive bouts (groups of dives separated by relatively long intervals) are of significantly longer durations than others and they are succeeded by longer surface intervals (I. L. Boyd, J. P. Y. Arnould, T. Barton & J. P. Croxall in prep.). These authors suggest that fur seals exceed their ADL when averaged over all dives in some bouts. Even so, this does not have any dramatic effect on their subsequent diving behaviour. However, data from two northern fur seals, *Callorhinus ursinus*, indicate that individual deep (> 75 m) dives, which may be fairly regular, involve anaerobic metabolism (Ponganis, Gentry *et al.* 1992).

The relationship between f_H and $\dot{V}_{O_2}$ in marine mammals and attempts to validate the use of f_H as an indicator of $\dot{V}_{O_2}$ in the field

Fedak (1986) made the important observation that, if f_H and $\dot{V}_{O_2}$ are averaged over a complete dive cycle, there is a linear relationship (Fig. 7)

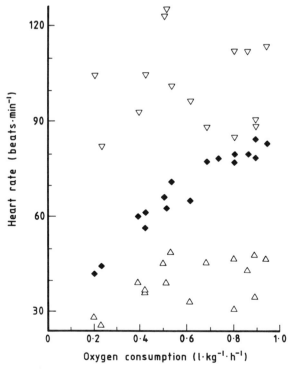

Fig. 7. The relationship between heart rate ($\triangledown$, at the surface; $\triangle$, while submerged; ◆, averaged over complete dive-breathing cycles) and oxygen consumption in a grey seal swimming in a water channel (redrawn from Fedak 1986).

between f_H and $\dot{V}_{O_2}$ for a marine mammal, the grey seal, at different levels of exercise, just as there is for terrestrial mammals. Such a relationship has also been described for two common seals, a California sea lion (Williams, Kooyman & Croll 1991) and two bottlenose dolphins, *Tursiops truncatus* (T. M. Williams, W. A. Friedl & J. E. Haun in prep.).

As part of a study, in collaboration with the British Antarctic Survey (BAS), to explore the utility of f_H as an indicator of metabolic rate, and to develop and deploy an implantable data storage device to record f_H, over many days, from animals in the field, the relationship between f_H and $\dot{V}_{O_2}$ has been determined in six subadult California sea lions (Butler *et al.* 1992). The activity levels ranged from resting to what would seem to be routinely reached in the wild, and there was a good linear relationship between the two variables in all six animals (see Fig. 8). The mean formula for estimating $\dot{V}_{O_2}$ from f_H in sea lions is: $\dot{V}_{O_2} = 0.26 \pm 0.02\,f_H - 13.48 \pm 1.44$, $r^2 = 0.89$. The data loggers, together with DLW, are currently being deployed with Antarctic fur seals at the BAS base at Bird Island. Clearly it will be necessary to establish the details of the relationship between $\dot{V}_{O_2}$ and f_H in this species of otariid but if, on the basis of the sea lion data, linearity is assumed, this could be achieved with two sets of data points (rest and relatively high level of activity). A data logger has been recovered from a female fur seal which spent a total of 50 days at sea. The logger contained 36.4 days worth of heart rate data on it. The data from one 24-h period are shown in Fig. 9.

Thus, the technology exists to enable f_H to be collected continuously over extended periods and to be stored, for a year if necessary, until the animal is recaptured. Variations in f_H with time of day and in association with different types of behaviour can be obtained. All the indications are that f_H is a good indicator of $\dot{V}_{O_2}$ in marine mammals provided that it is averaged over complete dive cycles. From the behaviour of some species, such as northern and southern elephant seals and grey seals, all dives, including exceptionally long ones, seem to be aerobic in nautre. In other species, such as the Weddell seal, anaerobiosis is probably a component of the small proportion (5%) of exceptionally long dives and in these situations, the linear relationship between f_H and overall metabolic rate may break down (see Elsner 1986). Evidence from the common seal indicates that during moderate exercise marine mammals oxidize 20–30% of the produced lactate (Davis, Castellini, Williams & Kooyman 1991), whereas terrestrial mammals, such as the dog, may oxidize as much as 75% (Depocas, Minaire & Chatonnet 1969). Davis, Castellini, Williams & Kooyman (1991) emphasize the minor role that glucose plays in the aerobic and anaerobic metabolism of seals during voluntary activity under water. So, except in extreme situations, maybe only in some species, anaerobic carbohydrate metabolism with the net production of lactate may be of minor importance in marine mammals. Antarctic fur seals may exceed their ADL over several dives

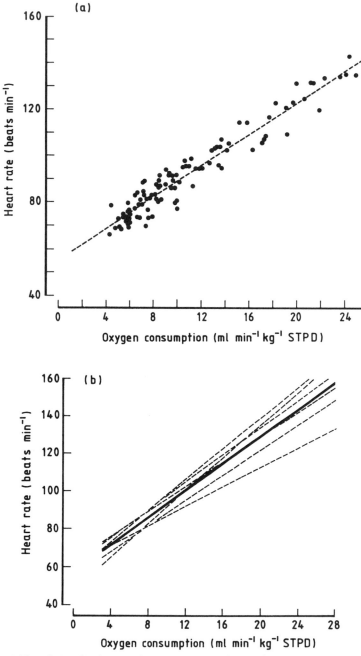

Fig. 8. (a) The relationship between heart rate (HR) and rate of oxygen consumption ($\dot{V}_{O_2}$) in a subadult, female California sea lion (29.5 kg). HR = 55.4 + 3.34 $\dot{V}_{O_2}$, $r^2 = 0.93$, $n = 96$. (b) Regression lines of the relationship between HR and $\dot{V}_{O_2}$ for six individual subadult California sea lions (- - - -). The mean regression line for all six sea lions is also given (——). For the mean data ($\pm$ S.E.) HR = (57.4 $\pm$ 1.98) + (3.56 $\pm$ 0.23) $\dot{V}_{O_2}$, $r^2 = 0.89$ (Butler *et al.* 1992).

(a)

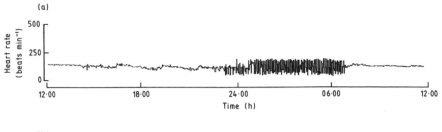

(b)

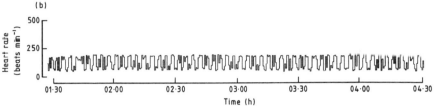

Fig. 9. (a) A 24-h record of heart rate from a female Antarctic fur seal (mass 39 kg) while foraging at sea on 14 and 15 Jan. 1991. (b) Expanded section of above record giving details of heart rate from 01.30 to 04.30 on 15 Jan. 1991 when the animal is thought to be feeding. Heart rate is counted over 30-s periods. (A. J. Woakes, R. M. Bevan, I. L. Boyd & P. J. Butler in prep.).

(a bout) but this does not appear to have a dramatic effect on subsequent dive behaviour, whereas northern fur seals may be more routinely anaerobic.

We await data from laboratory and field validation studies before it is possible categorically to state that f_H is a good indicator of metabolism in free-range marine mammals, but the signs are encouraging. Experiments on aquatic and semi-aquatic birds have indicated that f_H is a good predictor of $\dot{V}_{O_2}$ if grouped data from a number of individuals are used (Bevan, Keijer & Butler 1992; Nolet *et al.* in press). Bevan *et al.* (1992) have also demonstrated that the technique is accurate when the relationship between f_H and $\dot{V}_{O_2}$ is derived from one group of animals and f_H is used to predict $\dot{V}_{O_2}$ in another group of the same species. This is also the case even if the conditions under which the data were obtained, e.g. environmental temperature, are different (Bevan & Butler 1992). It is likely that the same constraints and advantages of the f_H technique exist with marine mammals.

Acknowledgements

The author's work on this topic is supported by SERC and NERC.

References

Bevan, R. M. & Butler, P. J. (1992). The effects of temperature on the oxygen consumption, heart rate and deep body temperature during diving in the tufted duck *Aythya fuligula*. *J. exp. Biol.* **163**: 239–151.

Bevan, R. M., Keijer, E. & Butler, P. J. (1992). A method for controlling the feeding behaviour of aquatic birds: heart rate and oxygen consumption during dives of different duration. *J. exp. Biol.* **162**: 91–106.

Boyd, I. L. & Croxall, J. P. (In press). Diving behaviour of lactating Antarctic fur seals. *Can. J. Zool.*

Butler, P. J. (1989). Telemetric recording of physiological data from free-living animals. In *Toward a more exact ecology*: 63–84. (Eds Grubb, P. J. & Whittaker, J. B.). Blackwell Scientific Publications, London.

Butler, P. J. (1991). Respiratory adaptations to limited oxygen supply during diving in birds and mammals. In *Physiological strategies for gas exchange and metabolism*: 235–257. (Eds Woakes, A. J., Grieshaber, M. K. & Bridges, C. R.). Cambridge University Press, Cambridge.

Butler, P. J. & Jones, D. R. (1982). The comparative physiology of diving in vertebrates. *Adv. comp. Physiol. Biochem.* **8**: 179–364.

Butler, P. J., Woakes, A. J., Boyd, I. L. & Kanatous, S. (1992). Relationship between heart rate and oxygen consumption during steady-state swimming in California sea lions. *J. exp. Biol.* **170**: 35–42.

Castellini, M. A., Davis, R. W. & Kooyman, G. L. (1988). Blood chemistry regulation during repetitive diving in Weddell seals. *Physiol. Zool.* **61**: 379–386.

Castellini, M. A., Kooyman, G. L. & Ponganis, P. J. (1992). Metabolic rates of freely diving Weddell seals: correlations with oxygen stores, swim velocity and diving duration. *J. exp. Biol.* **165**: 181–194.

Castellini, M. A., Murphy, B. J., Fedak, M., Ronald, K., Gofton, N. & Hochachka, P. W. (1985). Potentially conflicting metabolic demands of diving and exercise in seals. *J. appl. Physiol.* **58**: 392–399.

Costa, D. P., Croxall, J. P. & Duck, C. D. (1989). Foraging energetics of Antarctic fur seals in relation to changes in prey availability. *Ecology* **70**: 596–606.

Costa, D. P. & Gentry, R. L. (1986). Free-ranging energetics of northern fur seals. In *Fur seals: maternal strategies on land and at sea*: 79–101. (Eds Gentry, R. L. & Kooyman, G. L.). Princeton University Press, Princeton.

Davis, R. W., Castellini, M. A., Kooyman, G. L. & Maue, R. (1983). Renal glomerular filtration rate and hepatic blood flow during voluntary diving in Weddell seals. *Am. J. Physiol.* **245**: R743–R748.

Davis, R. W., Castellini, M. A., Williams, T. M. & Kooyman, G. L. (1991). Fuel homeostasis in the harbor seal during submerged swimming. *J. comp. Physiol.* **160**: 627–635.

Depocas, F., Minaire, Y. & Chatonnet, J. (1969). Rates of formation and oxidation of lactic acid in dogs at rest and during moderate exercise. *Can. J. Physiol. Pharmacol.* **47**: 603–610.

Elsner, R. (1986). Limits to exercise performance: some ideas from comparative studies. *Acta physiol. scand.* **128**: 45–51.

Fedak, M. A. (1986). Diving and exercise in seals: a benthic perspective. In *Diving in animals and man*: 11–32. (Eds Brubakk, A. D., Kanwisher, J. W. & Sundnes, G.). Tapir Publishers, Trondheim.

Fedak, M. A., Pullen, M. R. & Kanwisher, J. (1988). Circulatory responses of seals to periodic breathing: heart rate and breathing during exercise and diving in the laboratory and open sea. *Can. J. Zool.* **66**: 53–60.

Gaesser, G. A. & Brooks, G. A. (1984). Metabolic bases of excess post-exercise oxygen consumption: a review. *Med. Sci. Sports Exerc.* **16**: 29–43.

Gleeson, T. T. & Baldwin, K. M. (1981). Cardiovascular response to treadmill exercise in untrained rats. *J. appl. Physiol.* **50**: 1206–1211.

Guppy, M., Hill, R. D., Schneider, R. C., Qvist, J., Liggins, G. C., Zapol, W. M. & Hochachka, P. W. (1986). Microcomputer-assisted metabolic studies of voluntary diving of Weddell seals. *Am. J. Physiol.* **250**: R175–R187.

Henderson, Y. & Prince, A. L. (1914). The oxygen pulse and the systolic discharge. *Am. J. Physiol.* **35**: 106–115.

Hill, R. D. (1986). Microcomputer monitor and blood sampler for free-diving Weddell seals. *J. appl. Physiol.* **61**: 1570–1576.

Hill, R. D., Schneider, R. C., Liggins, G. C., Schuette, A. H., Elliott, R. L., Guppy, M., Hochachka, P. W., Qvist, J., Falke, K. J. & Zapol, W. M. (1987). Heart rate and body temperature during free diving of Weddell seals. *Am. J. Physiol.* **253**: R344–R351.

Hindell, M. A., Slip, D. J. & Burton, H. R. (1991). The diving behaviour of adult male and female southern elephant seals, *Mirounga leonina*. *Aust. J. Zool.* **39**: 595–619.

Horstman, D. H., Gleser, M., Wolfe, D., Tryon, T. & Delehunt, J. (1974). Effects of hemoglobin reduction on $\dot{V}_{O_2max}$ and related hemodynamics in exercising dogs. *J. appl. Physiol.* **37**: 97–102.

Jones, J. H., Longworth, K. E., Lindholm, A., Conley, K. E., Karas, R. H., Kayar, S. R. & Taylor, C. R. (1989). Oxygen transport during exercise in large mammals. I. Adaptive variation in oxygen demand. *J. appl. Physiol.* **67**: 862–870.

Kooyman, G. L. (1985). Physiology without restraint in diving mammals. *Mar. Mamm. Sci.* **1**: 166–178.

Kooyman, G. L. (1989). *Diverse divers: physiology and behavior.* Springer-Verlag, Berlin etc. (*Zoophysiology* 23).

Kooyman, G. L. & Campbell, W. B. (1972). Heart rates in freely diving Weddell seals, *Leptonychotes weddelli. Comp. Biochem. Physiol.* **43A**: 31–36.

Kooyman, G. L., Castellini, M. A., Davis, R. W. & Maue, R. A. (1983). Aerobic diving limits of immature Weddell seals. *J. comp. Physiol.* **151**: 171–174.

Kooyman, G. L., Wahrenbrock, E. A., Castellini, M. A., Davis, R. W. & Sinnett, E. E. (1980). Aerobic and anaerobic metabolism during voluntary diving in Weddell seals: evidence of preferred pathways from blood chemistry and behavior. *J. comp. Physiol.* **138**: 335–346.

Le Boeuf, B. J., Costa, D. P., Huntley, A. C. & Feldkamp, S. D. (1988). Continuous, deep diving in female northern elephant seals, *Mirounga angustirostris. Can. J. Zool.* **66**: 446–458.

Le Boeuf, B. J., Naito, Y., Huntley, A. C. & Asaga, T. (1989). Prolonged, continuous, deep diving by northern elephant seals. *Can. J. Zool.* **67**: 2514–2519.

Lifson, N., Gordon, G. B. & McClintock, R. (1955). Measurement of total carbon dioxide production by means of $D_2{}^{18}O$. *J. appl. Physiol.* **7**: 704–710.

Lifson, N., Gordon, G. B., Visscher, M. B. & Nier, A. O. (1949). The fate of utilized molecular oxygen and the source of the oxygen of respiratory carbon dioxide, studied with the aid of heavy oxygen. *J. biol. Chem.* **180**: 803–811.

Liggins, G. C., Qvist, J., Hochachka, P. W., Murphy, B. J., Creasy, R. K., Schneider,

R. C., Snider, M. T. & Zapol, W. M. (1980). Fetal cardiovascular and metabolic responses to simulated diving in the Weddell seal. *J. appl. Physiol.* **49**: 424–430.

Lundgren, N. P. V. (1946). The physiological effects of time schedule work on lumber-workers. *Acta physiol. scand. Suppl.* **41**: 1–137.

Malhotra, M. S., Sen Gupta, J. & Rai, R. M. (1963). Pulse count as a measure of energy expenditure. *J. appl. Physiol.* **18**: 994–996.

Nagy, K. A. (1989). Doubly-labeled water studies of vertebrate physiological ecology. In *Stable isotopes in ecological research*: 268–287. (Eds Rundel, P. W., Ehleringer, J. R. & Nagy, K. A.). Springer-Verlag, New York. (*Ecol. Stud. Anal. Synth.* **68**.)

Nolet, B. A., Butler, P. J., Masman, D. & Woakes, A. J. (In press). Estimation of daily energy expenditure from heart rate and doubly-labelled water in exercising geese. *Physiol. Zool.*

Ponganis, P. J., Gentry, R. L., Ponganis, E. P. & Ponganis, K. (1992). Analysis of swim velocities during deep and shallow dives of two northern fur seals, *Callorhinus ursinus. Mar. Mamm. Sci.* **8**: 69–75.

Ponganis, P. J., Kooyman, G. L. & Zornow, M. H. (1991). Cardiac output in swimming California sea lions, *Zalophus californianus. Physiol. Zool.* **64**: 1296–1306.

Ponganis, P. J., Kooyman, G. L., Zornow, M. H., Castellini, M. A. & Croll, D. A. (1990). Cardiac output and stroke volume in swimming harbor seals. *J. comp. Physiol. B* **160**: 473–482.

Qvist, J., Hill, R. D., Schneider, R. C., Falke, K. J., Liggins, G. C., Guppy, M., Elliot, R. L., Hochachka, P. W. & Zapol, W. M. (1986). Hemoglobin concentrations and blood gas tensions of free-diving Weddell seals. *J. appl. Physiol.* **61**: 1560–1569.

Scholander, P. F. (1940). Experimental investigations on the respiratory function in diving mammals and birds. *Hvalråd. Skr.* No. 22: 1–131.

Taylor, C. R,. Karas, R. H., Weibel, E. R. & Hoppeler, H. (1987). Adaptive variation in the mammalian respiratory system in relation to energetic demand: II. Reaching the limits to oxygen flow. *Respir. Physiol.* **69**: 7–26.

Thompson, D. & Fedak, M. A. (1993). Cardiac responses of grey seals during diving at sea. *J. exp. Biol.* **174**: 139–164.

Williams, T. M., Kooyman, G. L. & Croll, D. A. (1991). The effect of submergence on heart rate and oxygen consumption of swimming seals and sea lions. *J. comp. Physiol.* **160**: 637–644.

Woakes, A. J. (1992). An implantable data logging system for heart rate and body temperature. In *Wildlife telemetry*: 120–127. (Eds Priede, I. G. & Swift, S. M.). Ellis Horwood Ltd, London.

Symp. zool. Soc. Lond. (1993) No. 66: 333–348

Behavioural and physiological options in diving seals

M. A. FEDAK[1]
and D. THOMPSON[1,2]

[1]Sea Mammal Research Unit
Natural Environment Research
 Council
High Cross, Madingley Road
Cambridge CB3 0ET, UK

[2]Department of Environmental and
 Evolutionary Biology
University of Liverpool
PO Box 147
Liverpool L69 3BX, UK

Synopsis

Phocid seals are animals which live at depth and only periodically return to the surface to breathe. They are not surface-dwellers which occasionally leave the surface to forage. This view involves more than a semantic distinction: its adoption is essential to the creation of more realistic conceptual models of the behaviour of diving animals and the understanding of the physiological constraints under which they operate.

In this paper, we develop such a conceptual model in terms of the behavioural decisions seals must make and how the physiological options available may constrain behavioural choices. We then consider a model of one of these choices in more detail (the choice of resorting to anaerobic metabolism to extend dives) to demonstrate the consequences such a decision would have on diving behaviour. And finally, we examine some of the field data available from grey, Weddell and elephant seals in terms of which behavioural and physiological options are chosen in particular circumstances.

It seems that many species of phocid seals perform extended dives but that they choose to use this capability only in certain situations. We conclude from the comparison of field data and the model that reduction of metabolic costs by behavioural and/or biochemical means is frequently used to accomplish long dives, but that in some cases there is clear evidence of anaerobic metabolism. We argue that the circumstance in which dives are performed will determine the combination of physiological and behavioural options chosen and that, because of the complexity of the possible interrelationships of behaviour and physiology, combined modelling of both is necessary to explain field and laboratory observations and to avoid conflicting interpretations of results.

ZOOLOGICAL SYMPOSIUM No. 66
ISBN 0–19–854069–8 Copyright © 1993 The Zoological Society of London
All rights of reproduction in any form reserved

Introduction

Recording and telemetry devices are now providing detailed and comprehensive information at an ever-increasing rate on the behaviour and physiology of seals while at sea (Kooyman 1966; Kooyman, Billups & Farwell 1983; Hill *et al.* 1987; Fedak, Pullen & Kanwisher 1988; Le Boeuf, Naito, Huntley *et al.* 1989; Thompson *et al.* 1991; McConnell, Chambers & Fedak 1992). It is timely, therefore, to construct conceptual models to structure the existing data and to build formal mathematical models which can be tested with this information. Both should help to focus future efforts to collect data.

We now know that phocids often dive in continuous bouts. They can spend 90% of their time in the water submerged (Fedak 1986; Le Boeuf, Naito, Huntley *et al.* 1989; Hindell *et al.* 1992; Thompson *et al.* 1991) and, while at the surface, are usually inactive except for vigorous breathing (Thompson *et al.* 1991; le Boeuf, Naito, Asaga *et al.* 1992), This information clearly demonstrates that phocid seals are animals which live at depth, periodically returning to the surface to breathe, and not surface-dwellers which occasionally leave the surface to forage. This view involves more than a semantic distinction: the adoption of this view is essential to the creation of more realistic conceptual models of the behaviour of these animals and the physiological constraints under which they operate.

While foraging and travelling (or even resting) underwater, seals must make a number of behavioural and physiological 'decisions' driven by their short- and medium-term needs. They must decide when to surface for air, and how long to remain there before returning below; how deep to dive and at what angles to ascend and descend and how fast to swim, while searching for prey, travelling to or from the surface or moving from place to place. If they encounter a wide range of potential prey species, they must choose among these prey and decide on appropriate capture strategies. While making these and other behavioural decisions, they also have to 'choose' among a number of physiological options (e.g. the proportional reliance on aerobic and anaerobic pathways and selective shut-down of body systems and possibly other unsuspected strategies). They could thereby defer maintenance and support processes and modify the metabolic requirements of dives.

Considered in this way, it becomes clear that the seal's situation and its immediate 'goals' may have a profound influence on both the behavioural and the physiological tactics chosen. For example, an animal travelling from place to place might employ a very different set of dive durations and surface intervals and might even utilize metabolic pathways to a different extent from one resting or sitting and waiting to ambush prey. It is crucial therefore to consider any data on physiological responses or mechanisms in the context of the animal's behaviour before making comparisons between species or animals in different situations.

In this paper, we consider such a conceptual model in terms of the behavioural decisions seals must make and the way in which the physiological options available may constrain behavioural choices. We then consider one of these choices in more detail (the choice of resorting to anaerobic metabolism to extend dives) to demonstrate the consequences such a decision would have for diving behaviour. And finally, we examine some of the field data available from three species of seals in terms of which behavioural and physiological options are actually chosen in particular circumstances. In the following paper, Thompson, Hiby & Fedak consider the choice of swimming speed and its consequences.

A conceptual model

Seal behaviour modelled as a set of choices

The life of a seal can be considered as two interconnected cycles, one of moving between land and sea and the second, while at sea, of surfacing and diving. Considering only the diving cycle, the seal's choices can be broken into three broad categories: (1) what to do while below; (2) when to go to the surface; and (3) when to return below.

1. How to behave while below? These choices reduce to deciding how much and how fast to swim and in which direction and at what angle to do so. Such decisions obviously depend on the goals of the activity (hunting, travelling, resting, exploration), the environmental situation and the characteristics of the seal in question such as its metabolic power/speed curve. If the seal is hunting, the characteristics of the sensory capabilities of both prey and seal, their speeds, the capture skills and abilities of the seal and evasion tactics of the prey will all be important. We can imagine a range of hunting tactics, from 'sit and wait' to active searching with high-speed chases, which will cover a range of exercise and activity levels that would make very different demands on the seal.

2. When to surface? This decision involves a cost/risk/benefit analysis and is perhaps one of the most complex and important decisions a seal has to make. It may be influenced by physiological state (oxygen stores, end-product accumulation etc.), the distance to the breathing location, prey availability and various drive states (hunger, fatigue etc.). Superimposed on these factors will be the need for the maintenance of a 'safety margin' with regard to metabolic requirements.

All these considerations will be interrelated. For example, it would be unrealistic to suppose that a single simple set-point of a single parameter (e.g. blood oxygen content) would act as a switch to cause the animal to decide to surface (see Thompson *et al.* this volume). This does not imply that a measure of arterial oxygen content would not be a good predictor of surfacing in a single repetitive foraging situation but rather that different animals in different situations would be expected to choose other levels of oxygen or even other parameters as being critical.

3. When to leave the surface? How long should the seal spend loading oxygen and dumping and recycling waste products before returning to feed or travel, keeping in mind that seals do little at the surface except breathe and recover from diving (Le Boeuf, Naito, Asaga *et al.* 1992; Thompson & Fedak 1993). Some of the same considerations as influenced surfacing such as drive state, prey availability, etc., will be involved. In addition, the time-course and shape of the oxygen-loading curves as well as those for carbon dioxide unloading and waste-product removal will presumably play an important role in the decision.

A set of physiological choices must be co-ordinated with all of the behavioural decisions the seal makes. The full range of these options is presumably under both hormonal and nervous control. The choice between them would be driven both by the immediate situation and by the animal's predictions of the near future. The choices involve a set of decisions about scheduling the activity levels of the various anatomical systems and the choice of metabolic pathways to be employed in each of them. We might expect that while at rest the seal will be relatively free from constraints but that its options will become more restricted as the demands of travel or foraging are imposed on the relevant systems or the requirements of recovery become acute.

Anaerobic metabolism or metabolic suppression

The capabilities which seals have to eliminate costs or defer energy and oxygen demand during dives can have a very profound effect on their behaviour. If parts of the seal's body are isolated from the circulation, as seems likely during diving, metabolic reductions (with their effect on oxygen demand) might be realized in some tissues in the short term but deferred costs may limit subsequent options. The extent and urgency of these deferred 'payments' could determine the timing and duration of subsequent dives.

Recent field data have shown that some phocid seals perform extended bouts of dives longer than can be explained by their estimated oxygen stores and metabolic rates. That is, seals seem to be able to dive for durations beyond their *estimated* aerobic dive limit or ADL. (The ADL (Kooyman 1989) is defined as the duration of dive which could be performed without net whole-body lactate production by anaerobic metabolism.) Parts of the seal's body may be producing energy by anaerobic pathways before this point but all lactate produced is oxidized or resynthesized elsewhere with no net increase. Dives exceeding the ADL are signalled operationally by the appearance of an increase in blood lactate in the central circulation after a dive. In practice, the ADL has only rarely been measured (Kooyman, Wahrenbrock et al. 1980; Kooyman, Castellini et al. 1983) but it has often been estimated (Kooyman 1981) from calculations of available oxygen stores and an expected metabolic rate.

If an animal uses anaerobic metabolism to extend the duration of a dive, this could have a significant effect on subsequent behaviour and the 'success' of subsequent dives. If, on the other hand, the animal can extend dives by means of metabolic reductions, whether behavioural or biochemical, the effect on subsequent dives could be very different. We have therefore modelled in detail the expected effect of an anaerobic increment during one dive on future dive schedules to see if observed dive sequences are consistent with this option or if the use of some physiological capability for metabolic suppression is required.

The results of anaerobic metabolism

The model predicts the time course of recovery from an anaerobic period of diving from a balance sheet of energy production and substrate and product flux. The model considers the following four potential metabolic strategies: (1) lactate is oxidized at the surface; (2) lactate is recycled to glucose at the surface; (3) lactate is oxidized during subsequent dives; (4) lactate is recycled to glucose during subsequent dives. Though oversimplified, the model's general conclusions should hold for more realistic situations.

Consider a seal with an ADL of 15 min which performs one dive of 20 min. It achieves this with 5 min of anaerobic metabolism.

We make the following assumptions:

1. The animal derives all energy from glucose, a 6-carbon carbohydrate (6C). During aerobic metabolism, 1 'unit of glucose' (U_{6C}) will yield 1 unit of high-energy phosphates (1 $U_{36\ ATP}$). During anaerobic metabolism, 1 U_{6C} will yield only 1/18 $U_{36\ ATP}$ (equivalent to 2 ATP) and will produce 2 units of lactate (2 U_{3C}). It will certainly not be the case in a real seal that all energy comes from glucose—fatty acids or glycogen may be involved as substrates—but this assumption allows us to simplify the book-keeping.

2. The animal uses ATP at a *constant* rate (call it 1 $U_{36\ ATP}$/min) and this rate is equivalent to an oxygen consumption rate of 1 U_{O2}/min.

3. The animal can store (and always does) 1 'load' of oxygen equivalent to 15 U_{O2}.

4. The animal can fully replenish its oxygen stores in 2 min. Thus, the maximum loading rate is 17 U_{O2} in 2 min (15 U_{O2} stored + 2 U_{O2} required to support metabolism at the surface), or 8.5 U_{O2}/min.

(a) Initial dive with 5 min anaerobic increment

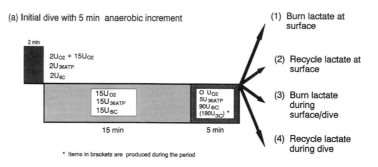

(1) Burn lactate at surface

(2) Recycle lactate at surface

(3) Burn lactate during surface/dive

(4) Recycle lactate during dive

* Items in brackets are produced during the period

(b) Burn lactate at surface

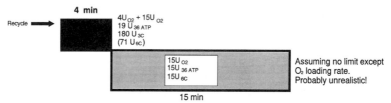

(c) Recycle lactate at surface

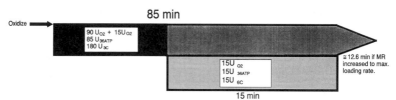

Assuming no limit except O_2 loading rate. Probably unrealistic!

(d) Burn lactate during surface/dive

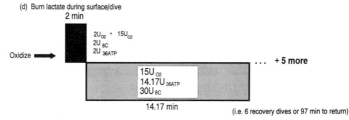

(i.e. 6 recovery dives or 97 min to return)

(e) Recycle lactate during dive

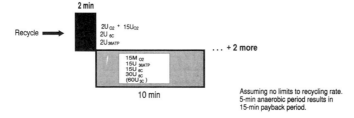

Assuming no limits to recycling rate. 5-min anaerobic period results in 15-min payback period.

We consider the dive cycle as beginning with a 2-min breathing period where the seal takes on board 15 U_{O2} while using 2 U_{O2}. In the process, it uses 2 $U_{36\ ATP}$ and 2 U_{6C}. The seal then begins a dive which will last a total of 20 min (33% longer than its ADL). Fifteen minutes of the dive can be supported aerobically and an additional 5 min must be supported by anaerobic pathways (Fig. 1a).

During the 15-min aerobic period, the seal uses 15 $U_{36\ ATP}$ which it produces by oxidizing 15 U_{6C} using 15 U_{O2}. During the 5-min anaerobic period it uses 5 $U_{36\ ATP}$ but to produce this it must use 90 U_{6C}. It will of course use no oxygen but will produce 180 U_{3C} which have to be either oxidized or reconverted to glucose (recycled) if the animal is to return to its starting state. The key point to consider here is that, because only two molecules of ATP are generated for each molecule of glucose split and pair of lactate molecules produced, a short period of anaerobiosis will result in large amounts of lactate in tissues which will take *both* oxygen and time to oxidize or recycle (see below).

If a lactate molecule is oxidized to carbon dioxide and water, 17 molecules of ATP are created; but only one molecule of ATP is created for each lactate molecule produced by glycolysis. Therefore, if ATP demand remains constant, the seal would require a period 17 times as long as the anaerobic period in order to clear the lactate by oxidation. If lactate is reconverted to glucose, extra energy will be needed to side-step the physiologically irreversible reaction steps of glycolysis. Therefore, the seal will use three times the energy released during the anaerobic period to clear the lactate by resynthesis (because each molecule of glucose resynthesized requires six molecules of ATP compared with only two molecules of ATP produced by its breakdown via glycolysis). Either biochemical route will have significant consequences for future dives.

The seal can return to its pre-anaerobic state in four ways (Fig. 1b–e):

1. Oxidize lactate at the surface. If the seal remained at the surface to oxidize all the lactate it produced and did not increase its metabolic rate, it would take approximately 85 min to recover from the dive (Fig. 1b). Even if it could increase its metabolic rate to the limit imposed by its oxygen loading rate of 8.5 U_{O2}/min, this would take about 13 min, a period far longer than a typical inter-dive interval.

2. Recycle lactate at the surface. Recycling the lactate back to glucose

---

Fig. 1. A diagrammatic representation of the consequences of a brief anaerobic period in a dive of a seal. The lengths of boxes are drawn proportional to the durations of dive-cycle events. (a) The initial dive, which includes a 5-min anaerobic component during which lactate is produced to support ATP production. (b)–(e) Options and results of recovery: (b), (c) at the surface; (d), (e) during subsequent dives. Assumptions and explanation are given in the text.

could take as little as 4 min if the only limitation was the oxygen loading rate (Fig. 1c). (This follows from the maximum loading rate and the fact that two molecules of ATP are produced for each molecule of glucose converted to lactate while six molecules of ATP are required for the reaction creating glucose from lactate). This is highly unlikely. Other factors associated with the recycling pathways may be limiting, as might the transport of glucose to the liver or other site where recycling takes place.

High rates of recycling could well be advantageous for divers, but there is no information on the maximum rates at which recycling can be maintained after long dives. Data on lactate disappearance after long dives in Weddell seals (*Leptonychotes weddelli*) diving from an isolated ice hole suggest that the process of lactate clearance takes 40–80 min (Kooyman, Wahrenbrock *et al.* 1980; Castellini, Davis & Kooyman 1988). Data from grey seals (*Halichoerus grypus*) taken after 12–20 min forced dives in the laboratory also suggest low rates of disappearance (Castellini, Murphy *et al.* 1985). Davis *et al.* (1991) examined glucose recycling in swimming harbour seals (*Phoca vitulina*) during very short dives. They also found rates which would suggest an extended recovery period, although these measurements were made under very different conditions to those faced by wild seals.

3. Oxidize lactate during subsequent dives. If the animal used the accumulated lactate as the sole substrate for oxidative metabolism until it was all consumed, 15 U_{O2} would support a dive of about 13.5 min, about 10% shorter than one in which glucose was used, and it would take about six dives of this duration before concentrations of lactate were reduced to those at the start (Fig. 1d). However, dives based on the oxidation of fatty acids by 15 U_{O2} would also be shorter, and this is likely to be the typical substrate for aerobic dives (Davis *et al.* 1991). Nonetheless, approximately 97 min of aerobic activity using lactate as a substrate would elapse before the animal could repeat the anaerobic interval under the initial conditions.

4. Recycle lactate during subsequent dives. Three times the ATP yielded by the production of lactate is required to regenerate the glucose from which it came. Therefore, 5 min of anaerobic time during a dive will result in 15 min of diving lost, because the oxygen needed for the recycling must come from the oxygen stores taken down on subsequent dives (Fig. 1e). An animal could exercise this option over a wide range of time-scales until the 15 min was used (e.g. if three dives were shortened to 10 min, pre-dive levels would be regained in 36 min). Again, this assumes that there are no limits to the recycling rate, which is probably unrealistic.

A seal could choose to exercise more than one of these options, but the model clearly shows that some combination of increased surface time, reduced dive time or a series of short dives would have to occur after dives which exceeded the ADL.

Discussion

Observations of seals in their natural environment

Most of the data available to test the ideas presented in this model come from studies of grey, Weddell and elephant (*Mirounga* sp.) seals. Therefore, we now discuss a set of behavioural and physiological observations of grey seals and consider the implications of their choice of behavioural and physiological options. In particular, we consider the implications of the choice between anaerobic metabolism and metabolic reductions as a means of extending dives and consider the influence of these choices on diving schedules and behaviour. We then compare these data with data from Weddell seals, for which additional information on blood chemistry is available, and with the data on the extraordinary dive bouts displayed by elephant seals.

Behavioural and physiological data were collected from grey seals as they rested, travelled and fed in the seas around western Scotland. We were able to monitor each seal's position and collect information on its swimming speed, depth and heart rate continuously (Thompson & Fedak 1993). These data therefore provide an opportunity to examine variation in behaviour and physiological response in terms of several aspects of this model.

Each activity—rest, travel and foraging—had characteristic dive schedules and swimming behaviour.

Resting

Seals spent the majority of their time resting in the water adjacent to areas where they or other seals hauled out. Interestingly, seals did not appear to rest at feeding locations but returned to sites adjacent to haul-outs. Dives made near haul-outs were variable in duration, rarely involved swimming but were usually short (1–7 min) and shallow (< 20 m).

Travelling

Seals travelled between resting sites, sometimes stopping to feed on route, or travelled away from land to feeding sites 3–40 km away from resting areas. While they were travelling, diving became much more regular. Dives were typically 3–9 min in duration with a characteristic U- or V-shaped profile and were separated by short breathing intervals (Thompson *et al.* 1991). Animals remained stationary while at the surface and swam continuously at 1–2 m/s while submerged.

Foraging

The seals sometimes paused in their travels, apparently to feed. They also would repeatedly travel to specific locations, some well out to sea, to feed.

While feeding, animals moved little between subsequent surfacings and invariably dived to the bottom or very near it. During ascent and descent, seals swam at 1–2 m/s but swam little or not at all while on the bottom.

Circulatory responses to diving in relation to behaviour

We now review some of the heart-rate data from this study which emphasize the importance of the context in which dives are made to the behavioural and physiological choices the seals apparently make.

While levels of activity, behaviour schedules and circulatory responses varied from activity to activity, consistent patterns were observed within particular activities. However, while resting, seals displayed very variable dive schedules and behaviour. Nonetheless, the heart-rate responses during rest dives followed the general pattern observed in travelling or foraging dives described below. Figure 2 presents heart rate, swim speed and depth data from a series of dives which took place during both travelling and

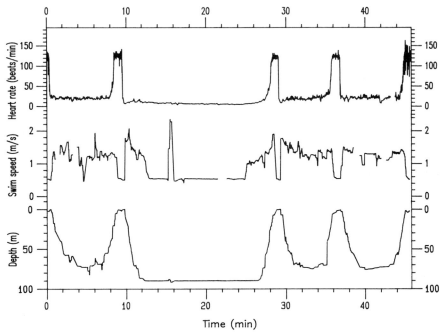

Fig. 2. Typical traces of heart rate (top), swim speed (centre) and dive depth (bottom) obtained from grey seals. Data from a 150-kg ♀ moving from the Isle of Skye to Heiskier swimming to depths near the bottom at the midpoint of each dive. Four complete dive cycles are shown. The first dive shows a typical 'travelling' dive; in the second dive, the animal paused in her travels and then resumed travelling in the last two dives. Note that the dive profile is U-shaped in the travelling dives and that speed is maintained throughout the dive. Also, heart rate does not fall as low during the travelling dives. See Thompson & Fedak (1993) for more detail.

feeding periods. We can use the data from Thompson & Fedak (1993) to provide some clues to the 'physiological decisions' the seals make to facilitate their diving habit. Periods at the surface were short (mean, 0.77 min) with no swimming activity; heart rates were uniformly high (119 ± 9 beats/min) and did not vary with duration of surface period. For very short dives, those lasting less than 7 min, the period that the seal spent on the surface after each dive increased as the length of the dive increased. But it did not increase any further for dives lasting more than 7 min. As soon as the dive began, and as the animal started to swim, heart rate fell dramatically, implying a rapid change in blood distribution. Heart rate remained low, but was often highly arhythmic, for the duration of the dive. Before surfacing, heart rate increased, as if in anticipation of surfacing. That is, the heart response to swimming activity at the end of the dive is different from what it was at the beginning of the dive. Clearly, different constraints may apply at the beginning and end of the dive.

The lowest heart rates were observed in the longest dives. In these dives, seals went to the bottom and remained motionless. While animals were travelling, the HR never fell to the low levels seen when animals remained motionless on the bottom, where it fell as low as 2 beats/min (Fig. 3). During bradycardia, stroke volume decreased by up to 50% (Ponganis, Kooyman *et al.* 1990). Given that blood pressure is maintained, these low heart rates imply that very little of the seal's body is being circulated with

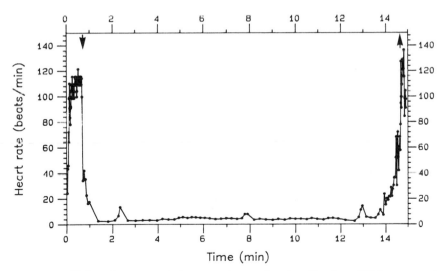

Fig. 3. Beat-by-beat heart rate during a stationary foraging dive in a 210-kg grey seal. Swimming was confined to ascent and descent. Each dot represents a single beat. ↓ marks the start of the dive and ↑ represents surfacing. The figure shows the most extreme bradycardia recorded for any seal. For the entire voluntary dive, HR averaged 6.5 beats/min and was below 4 beats/min for 90% of the time (Thompson & Fedak 1993).

blood during long dives. Shorter dives, including those where the seal is travelling between surfacings, show a much less dramatic response. These radical changes in heart rate according to the nature of a dive suggest, therefore, that the seals are changing their physiological tactics in response to the conditions encountered or expected in a dive.

We interpret the function of these long dives to be to wait for and ambush prey. We further suggest that this strategy plays a role in improving the seal's chance of encountering certain types of prey (Thompson *et al.* this volume).

If we assume that all oxygen stores are fully available and metabolic rate is equivalent to standard mammalian basal metabolic rate (BMR) (Kleiber 1961), the ADL for grey seals of the sizes studied here is only 10–15 min. Many (6% overall) of the dives of these seals were longer than this (Thompson & Fedak 1993). Dives exceeding the estimated ADL occurred in sequence or with only one shorter intervening dive (Fig. 4), and yet these dives involved swimming, at least to the bottom and back, and occurred while the seal was 'foraging', rather than resting or sleeping. Therefore they

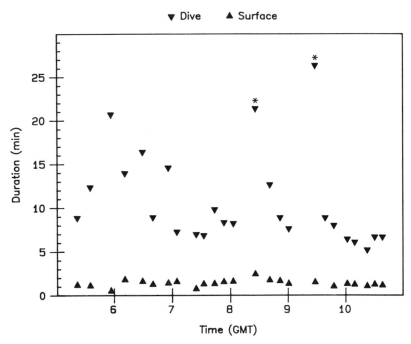

Fig. 4. A 6-h sequence of surface and dive times from a grey seal foraging trip with the animal repeatedly diving to near the bottom in 70–80 m of water. Note that long dives (marked *) are not preceded by or followed by extended times at the surface. From Thompson & Fedak (1993).

occurred at a time when the animal would be expected to be metabolizing at a higher rate than BMR.

Dive circumstances compared for grey and Weddell seals

In some situations, 6% of the dives of Weddell seals are also longer than ADL but these dives are almost always followed by extended periods at the surface (Kooyman, Wahrenbrock *et al.* 1980; Castellini, Davis *et al.* 1988) and involve the appearance in the blood of lactate, which slowly returns to pre-dive levels while the animal rests at the surface. In the few cases where diving continued after dives longer than ADL (only four in more than 23 000 dives) (Castellini, Davis *et al.* 1988), subsequent dives were short. Indeed, these Weddell seal data suggest that surface intervals increase exponentially with dive duration for long dives. In the grey seals, surface intervals remained relatively constant.

However, the circumstances in which these particular grey and Weddell seals performed long dives were markedly different and their dives may well have had different objectives. The grey seals seemed to be foraging in a limited area. Swimming was confined to that needed to reach the bottom and return to the surface and they were inactive while near the bottom. In contrast, the Weddell seals in the studies summarized by Kooyman (1981) usually were diving from an ice hole far removed from any other breathing site. While they may have been foraging at some times, the longest dives were considered to be 'exploratory' (the animals may have been looking for alternative breathing holes (Kooyman 1981)). These long dives may have involved considerable swimming activity (Castellini, Kooyman & Ponganis 1992; Ponganis, Castellini & Kooyman 1991). Thus the exercise component of the long dives by Weddell seals was very different from that of the long dives by the grey seals and the requirement (and indeed the rewards) for diving again soon after surfacing were almost certainly different. When the grey seals swam continuously during dives, they were usually travelling from place to place. These dives were short, as were the breathing intervals.

There is no information about blood lactate from any free-swimming seals other than Weddell seals and therefore the question of lactate production during long dives in other species must remain open, as must the question of which metabolic options were used to accomplish the dive schedules observed in grey seals. It is not unreasonable to suggest that, although the species might have the same physiological capabilities, different circumstances will make different demands on muscles and other organs and different metabolic options and physiological tactics might be chosen.

It is certainly the case that grey seals take behavioural steps which minimize the rate of utilization of oxygen stores while they are under water, and that this could have implications for foraging success (see Thompson *et al.* this volume). Whether metabolic reductions of a cellular or biochemical nature were used to augment the effects of behaviour also remains an open question.

The extended bouts of long dives seen in both elephant seal species (Hindell *et al.* 1992; Le Boeuf *et al.* this volume; Stewart & De Long this volume) also suggest that cellular or metabolic steps are taken to reduce metabolic demands during diving. A southern elephant seal (*M. leonina*) with an estimated ADL of 30 min made a 2-h dive to > 600 m followed by a 3-min surface interval and then a series of 40-min dives also with short surface intervals (3 min) between them (Hindell *et al.* 1992). Such sequences of long dives were not exceptional. One animal carried out 29 foraging dives during 23 h, all of which were longer than the estimated ADL. Another seal displayed 63 dives (all longer than estimated ADL) in a sequence covering more than 48 h. Another record indicates that one seal went 40 days with no surface interval extending over more than 6 min (Hindell *et al.* 1992).

These records are inconsistent with the use of anaerobic metabolism to extend dives, given the results of the model presented above. They seem to indicate that the actual ADL of southern elephant seals is significantly longer than the estimates. Estimates of oxygen stores are not likely to be sufficiently in error to account for this discrepancy. Therefore, metabolic depression to levels below predicted resting values is likely to be involved in producing these dramatic dive records. However, it is important to point out that, even in southern elephant seals, as in grey (Thompson & Fedak 1993) and Weddell seals (Castellini, Davis *et al.* 1988), dives longer than the theoretical ADL make up only a small fraction of all dives monitored (only 10% of the dives were longer than 20 min—McConnell *et al.* 1992). Indeed, one seal in this study spent 30 days diving, during which time 90% of the dives were shorter than 15 min and no dives were longer than 25 min.

It seems, therefore, that many species of phocid seals perform extended dives but that they may choose to use this capability only in certain situations. We conclude, from the comparison of field data and the model, that reduction of metabolic costs by behavioural and/or biochemical means is used to accomplish long dives. Grey seals, Weddell seals and elephant seals all perform dives in a wide variety of situations. The highly variable dive characteristics seen in dives by 'resting' grey seals may be indicative of the plasticity of the responses possible when interlocking constraints on behaviour and physiology are relaxed. We argue that, when this is not the case, the circumstances in which the dive is performed will determine which combination of physiological and behavioural options is chosen. It is only under particular conditions that such bouts of long dives are advantageous (Thompson *et al.* this volume; Ydenberg & Clark 1989) and even in these particular conditions different behavioural, physiological or metabolic options may best be chosen to accomplish them. Clearly, consideration of the circumstances and behavioural goals of diving must be taken into account when attempting to understand the physiology involved. Because of the complexity of the possible interrelationships, combined modelling of behaviour and physiological constraints is necessary to explain observations and avoid conflicting interpretations of results.

References

Castellini, M. A., Davis, R. W. & Kooyman, G. L. (1988). Blood chemistry regulation during repetitive diving in Weddell seals. *Physiol. Zool.* **61**: 379–386.

Castellini, M. A., Kooyman, G. L. & Ponganis, P. J. (1992). Metabolic rates of freely diving Weddell seals: correlations with oxygen stores, swim velocity and diving duration. *J. exp. Biol.* **165**: 181–194.

Castellini, M. A., Murphy, B. J., Fedak, M. A., Ronald, K., Gofton, N. & Hochachka, P. W. (1985). Potentially conflicting metabolic demands of diving and exercise in seals. *J. appl. Physiol.* **58**: 392–399.

Davis, R. W., Castellini, M. A., Williams, T. M. & Kooyman, G. L. (1991). Fuel homeostasis in the harbor seal during submerged swimming. *J. comp. Physiol. B* **160**: 627–635.

Fedak, M. A. (1986). Diving and exercise in seals: a benthic perspective. In *Diving in animals and man*: 11–32. (Eds Brubakk, A.O., Kanwisher, J. W. & Sundnes, G.). Tapir, Trondheim.

Fedak, M. A., Pullen, M. R. & Kanwisher, J. (1988). Circulatory responses of seals to periodic breathing: heart rate and breathing during exercise and diving in the laboratory and open sea. *Can. J. Zool.* **66**: 53–60.

Hill, R. D., Schneider, R. C., Liggins, G. C., Shuette, A. H., Elliott, R. L., Guppy, M., Hochachka, P. W., Qvist, J., Falke, K. J. & Zapol, W. M. (1987). Heart rate and body temperature during free diving of Weddell seals. *Am. J. Physiol.* **253**: 344–351.

Hindell, M. A., Slip, D. J., Burton, H. R. & Bryden, M. M. (1992). Physiological implications of continuous, prolonged, and deep dives of the southern elephant seal (*Mirounga leonina*). *Can. J. Zool.* **70**: 370–379.

Kleiber, M. (1961). *The fire of life: an introduction to animal energetics.* Wiley & Sons, New York & London.

Kooyman, G. L. (1966). Maximum diving capacities of the Weddell seal (*Leptonychotes weddelli*). *Science* **151**: 1553–1554.

Kooyman, G. L. (1981). *Weddell seal: consummate diver.* Cambridge University Press, Cambridge etc.

Kooyman, G. L. (1989). *Diverse divers: physiology and behaviour.* Springer Verlag, Berlin & London. (*Zoophysiology* 23).

Kooyman, G. L., Billups, J. O. & Farwell, W. D. (1983). Two recently developed recorders for monitoring diving activity of marine birds and mammals. In *Experimental biology at sea*: 197–214. (Eds Macdonald, A. G. & Priede, I.). Academic Press, London.

Kooyman, G. L., Castellini, M. A., Davis, R. W. & Maue, R. A. (1983). Aerobic diving limits of immature Weddell seals. *J. comp. Physiol.* **151**: 171–174.

Kooyman, G. L., Wahrenbrock, E. A., Castellini, M. A., Davis, R. W. & Sinnett, E. E. (1980). Aerobic and anaerobic metabolism during voluntary diving in Weddell seals; evidence of preferred pathways from blood chemistry and behavior. *J. comp. Physiol.* **138**: 335–346.

Le Boeuf, B. J., Naito, Y., Asaga, T., Crocker, D. & Costa, D. P. (1992). Swim speed in a female northern elephant seal: metabolic and foraging implications. *Can. J. Zool.* **70**: 786–795.

Le Boeuf, B. J., Naito, Y., Huntley, A. C. & Asaga, T. (1989). Prolonged, continuous, deep diving by northern elephant seals. *Can. J. Zool.* **67**: 2514–2519.

McConnell, B. J., Chambers, C. & Fedak, M. A. (1992). Foraging ecology of southern elephant seals in relation to the bathymetry and productivity of the Southern Ocean. *Antarct. Sci.* **4**: 393–398.

Ponganis, P. J., Kooyman, G. L., Zornow, M. H., Castellini, M. A. & Croll, D. A. (1990). Cardiac output and stroke volume in swimming harbor seals. *J. comp. Physiol. B* **160**: 473–482.

Ponganis, P. J., Castellini, M. A. & Kooyman, G. L. (1991). Metabolic rate, swim velocity and dive durations in Weddell seals. *Abstr. bienn. Conf. Biol. mar. Mammals* **9**: 54. Society for Marine Mammalogy. Unpublished.

Thompson, D. & Fedak, M. A. (1993). Cardiac responses of grey seals during diving at sea. *J. exp. Biol.* **174**: 139–164.

Thompson, D., Hammond, P. S., Nicholas, K. S. & Fedak, M. A. (1991). Movements, diving and foraging behaviour of grey seals (*Halichoerus grypus*). *J. Zool., Lond.* **224**: 223–232.

Ydenberg, R. C. & Clark, C. W. (1989). Aerobiosis and anaerobiosis during diving by western grebes: an optimal foraging approach. *J. theor. Biol.* **139**: 437–447.

Symp. zool. Soc. Lond. (1993) No. 66: 349–368

How fast should I swim? Behavioural implications of diving physiology

D. THOMPSON[1,2],
A. R. HIBY[1]
and M. A. FEDAK[1]

[1]*Sea Mammal Research Unit,
Natural Environment Research
Council,
High Cross, Madingley Road,
Cambridge CB3 0ET UK*

[2]*Department of Environmental and
Evolutionary Biology
University of Liverpool
PO Box 147
Liverpool L69 3BX, UK*

Synopsis

The extraordinary abilities of seals to survive long, apparently life-threatening, periods of anoxia have interested comparative physiologists for much of this century, with the emphasis on physiological responses as defensive measures to conserve oxygen. The advent of miniature recording devices and transmitters in the 1960s and 1970s dramatically changed the way we perceived diving. It is the steady-state behaviour of seals at sea. Seals can dive continually, spending up to 90% of their time submerged over extended periods.

In the previous paper (Fedak & Thompson, this volume) we described how combinations of behavioural and physiological adaptations enable seals to dive continually, often beyond estimated aerobic capacities, and suggested how these adaptations might restrict their choice of hunting tactics. Here we switch from regarding diving as a physiological event and look at it as primarily a behavioural event. Diving patterns are constrained by rigid physiological limits, but, within these limits, seals can vary the patterns to most efficiently exploit their environment. Seals must choose the duration and depth of dives, the swim speeds and angles of ascent/descent and their hunting tactics while at depth.

Any attempt to examine the likely strategies of diving requires information on the metabolic costs in various behavioural states. In this paper we examine some aspects of swimming activity during dives and postulate how it would be expected to alter under different conditions. We look at two particular aspects of swimming activity: (1) ascent/descent rates, which alter the cost of travel to and from foraging patches: (2) swimming speed during foraging, which directly affects prey encounter rates. We present simple models for these two behaviours and compare predictions with

ZOOLOGICAL SYMPOSIUM No. 66
ISBN 0–19–854069–8

Copyright © 1993 The Zoological Society of London
All rights of reproduction in any form reserved

swimming activity data from wild seals. We suggest that hunting strategies are sensitive to prey swim speeds but in a counter-intuitive way, requiring slower swim speeds for faster-moving prey with the maximum search speeds for sedentary targets.

Introduction

The extraordinary abilities of many species of diving birds and mammals to survive long, apparently life-threatening, periods of anoxia have interested comparative physiologists for much of this century. Pioneering work on seals during the 1940s (Irving 1934; Scholander 1940) identified a set of physiological responses to diving which included intense bradycardia, dramatic changes in blood circulation and a rapid switch to anaerobic metabolism in much of the body. All of these could be seen as defensive measures to conserve oxygen for the obligatorily aerobic tissues such as brain and heart.

The advent of miniature recording devices and transmitters in the 1960s and 1970s (Kooyman 1981, 1989; Le Boeuf, Costa *et al.* 1988) dramatically changed the way diving was perceived. Some seal species are now known to dive continuously, spending up to 90% of their time submerged over extended periods. The extreme physiological defence measures against anoxia are only rarely utilized; less intense bradycardia combined with behavioural adaptations appears to be important in keeping the animals primarily aerobic (Kooyman & Campbell 1973; Hill *et al.* 1987; Castellini, Davis & Kooyman 1988; Fedak, Pullen & Kanwisher 1988). Fedak & Thompson (this volume) have described how grey seals use a combination of behavioural and physiological adaptations to enable them to dive continuously while at sea, and suggested how these adaptations may restrict their choice of foraging tactics.

In this paper we consider diving as a behavioural event rather than as a physiological event. Diving is constrained by a set of physiological limits but, within these limits, seals have the option to alter such variables as duration and depth of dives, swim speeds and angles of ascent/descent and hunting tactics in order to most effectively exploit their environment.

In its simplest form diving can be seen as a form of foraging in patches where the relationship between dive depth, dive duration, surface recovery and foraging success can be modelled in the context of the marginal value theorem (Kramer 1988). However, any attempt to examine diving in this way requires information on the metabolic costs in various behavioural states. In previous modelling exercises these had, for simplicity, been assumed to be constant throughout any particular activity irrespective of duration (Kramer 1988). In this paper we examine how the swimming activity of seals during foraging dives under different conditions may be varied in order to maximize some aspect of encounters with prey. The discussion will be limited to foraging dives in which we can compare the likely costs and benefits of particular tactics. In other behaviours, such as transit swimming, it is not as easy to identify the aims of particular actions. We will concentrate on two aspects of swimming activity: (1) the swimming speed during ascent and descent, which will alter the energetic cost of travel and hence the available oxygen resources at depth: (2) the swimming speed during foraging, which will directly affect prey encounter rates.

Studies of dive behaviour show that many species of phocid seal dive either to some hundreds of metres in the ocean (Le Boeuf, Costa *et al*. 1988; Le Boeuf, Naito *et al*. 1989; McConnell, Chambers & Fedak 1992) or to the sea bed in shallower waters (Elsner *et al*. 1989; D. Thompson *et al*. 1991; D. Thompson & Fedak 1993). A proportion of these deep dives have been interpreted as foraging (Le Boeuf, Naito *et al*. 1989; D. Thompson *et al*. 1991). We assume that during such foraging dives seals will attempt to maximize some aspect of their energy intake at depth. We consider three objectives for these dives: (1) to maximize the rate of gross energy intake; (2) to maximize the efficiency of energy intake; (3) to maximize the rate of net energy intake.

We use a simple graphical model to describe the short-term energetic implications of choosing particular swim speeds during ascent and descent, and of choosing particular levels of energy consumption while foraging. We describe the implications of these choices for the three optimization strategies. Finally we consider how swimming speed during foraging may affect the rate of encounters with prey, taking into account both the metabolic costs of swimming and the effect of prey swim speed.

These simple models are not expected to provide accurate descriptions of the behaviour of seals in any particular set of circumstances. Rather they are designed to provide a context in which the observed behaviour patterns of free-ranging animals can be evaluated.

General swimming speed diving model

Assumptions

We assume that during a sequence of foraging dives a seal feeds at a depth D (for a glossary of terms used in the following equations, see p. 367) and that its time can be divided into time spent in transit to and from depth (t_t), time spent foraging (t_f) or time spent at the surface replenishing oxygen stores and dealing with anaerobic products (t_r). Thus the duration of a complete dive cycle is $1/2t_t + t_f + 1/2t_t + t_r$. We assume that on any one dive a seal searches at a constant speed v_f and travels to and from depth at a constant speed v_t at some fixed angle Θ, so $v_t = 2(D/\sin\Theta)t_t^{-1}$. We assume that the rate of energy expenditure during a dive is directly proportional to the rate of oxygen consumption and that this is a function $c(v)$ of the swimming speed. For simplicity this is scaled so that $c(0) = 1$. Then the amount of energy expended during foraging at depth is $E_f = c(v_f)t_f$.

We assume that the energy acquired during a dive is related to the number of prey items encountered while foraging at depth. This is related to the swimming speed of the seal, while foraging, by the function $p(v_f)$. The effect of the precise form of $p(v_f)$ and of other variables such as prey density and, particularly, prey swimming speed will be discussed later. In this section we consider the implications of variations of t_t, t_r, v_f and E_f for the three optimization objectives defined above.

1. Maximizing gross energy intake

During a dive cycle the gross energy intake (G) is given by:

$$G = p \ (V_f) \ t_f \tag{1}$$

The complete dive cycle has duration $t_f + t_t + t_r$ therefore rate of gross energy intake is given by:

$$G' = \frac{p \ (v_f) \ t_f}{t_t + t_f + t_r} \tag{2}$$

Since $t_f = E_f/c(v_f)$ we obtain

$$G' = \frac{p \ (v_f)}{\dfrac{1 + c \ (v_f) \ (t_t + t_r)}{E_f}} \tag{3}$$

This function must be maximized with respect to the choice of t_t, t_f and v_f. For a given v_f, rate of gross energy intake will be maximized by minimizing $(t_t + t_r)/E_f$, i.e. by maximizing $E_f/(t_t + t_r)$. The relationship between the amount of energy or oxygen available for foraging at depth (E_f), and the combined travel and recovery times $(t_t + t_r)$ can be examined by means of a simple graphical model.

The function $c(v)$ relating oxygen consumption to swimming speed is assumed to have positive first and second derivatives. The amount of oxygen used in descent/ascent is $h(t_t) = c(v_t) \ t_t$; this relationship is convex downwards with respect to t_t. Thus the value t_t determines both the swim speed (because $v_t = 2(D/\sin\Theta) \ t_t^{-1}$) and the amount of oxygen used in transit. The recovery time (t_r) is determined by the choices of t_t and the amount of oxygen used at depth (E_f). We assume that t_r is an increasing function of the amount of oxygen consumed in the preceding dive with a positive second derivative. If this is the case, then the function $r(t_r)$ relating amount of oxygen loaded to the recovery time (t_r) will be convex upwards.

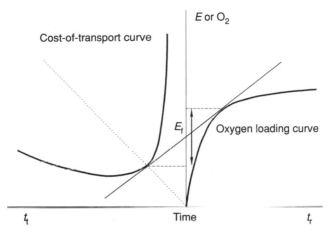

Fig. 1. Graphical model of the relationship between travel time t_t, energy used in foraging E_f and the recovery time t_r. The $h(t_t)$ curve represents the cost of travelling to a particular depth D, the $r(t_r)$ curve represents the amount of oxygen loaded as a function of time at the surface t_r and the vertical distance between the intersections of the common tangent with $h(t_t)$ and $r(t_r)$ represents the amount of energy or oxygen available for foraging at depth.

In Fig. 1 the oxygen loading against recovery time $(r(t_r))$ and cost of transport to a particular depth against travel time $(h(t_t))$ are plotted on the same axes so that we can examine the energetic consequences of different dive schedules. The x axis in Fig. 1 represents time, with recovery time (t_r) increasing to the right and travel time (t_t) increasing to the left. The slope of any line joining the two curves indicates the value of $E_f/(t_t + t_r)$. We can see from Fig. 1 that the curves have a common tangent. The slope of the common tangent represents the maximum value of $E_f/(t_t + t_r)$ and the contact points of this tangent represent the values of t_t and t_r which maximize the amount of oxygen available for foraging at depth with respect to the total travel and recovery time.

These choices of travel and recovery times will therefore maximize both the rate of oxygen delivery to the foraging area and the proportion of time spent foraging, irrespective of the choice of foraging tactics which the seal pursues while there.

If the seal dives to a greater depth, the $h(t_t)$ curve will move up to the left (Fig. 2), i.e. the amount of oxygen used to reach the foraging area at any particular speed will increase in proportion to the depth. The dotted line in Fig. 2 shows the level of oxygen depletion resulting from swimming to any depth at a particular speed. Since this line passes through the intersection of the common tangent with the lower $h(t_t)$ curve it represents the t_t and hence v_t which maximizes rate of oxygen delivery to the foraging area at depth D_1. The curve representing cost of transport to depth D_2 will again have a

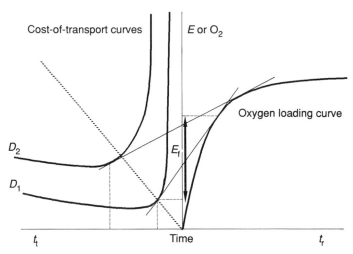

Fig. 2. Same graphical model as Fig. 1 with a second $h(t_r)$ curve representing cost of travel to greater depth.

common tangent with the oxygen loading curve. It is apparent from Fig. 2 that, in deeper dives, the tangent will touch the cost-of-transport curve to the left of its intersection with the dotted line. Thus the swimming speeds required to maximize rate of oxygen delivery to the foraging area decrease as depth increases. This result is independent of the choice of foraging tactics and holds for any system where the rate of oxygen consumption increases curvilinearly with speed and the rate of oxygen uptake declines as a function of time at the surface.

So, if seals are trying to maximize their gross energy intake, the above graphical model suggests that they should alter swim speed in response to the depth of a particular dive, swimming slower to reach greater depths. Also, since $v_t = 2(D/\sin\Theta)\, t_t^{-1}$ the rate of gross energy intake will be maximized with respect to Θ at $\Theta = \pi/2$, i.e. vertical swimming.

The foraging speed, v_f, chosen to maximize the rate of gross energy intake will also depend on the value of $E_f/(t_t + t_r)$. From Equation 3 we can see that if $c(v_f)\,(t_t + t_f)/E_f \gg 1$, then v_f will be chosen to maximize $p(v_f)/c(v_f)$, i.e. the seal should choose foraging speeds which maximize the energy gained per unit energy expended. This will occur if $E_f/(t_t + t_r)$ becomes small, which will only be the case in very deep dives. In shallower dives $E_f/(t_t + t_r)$ will be large and maximizing rate of gross energy intake will be equivalent to simply maximizing $p(v_f)$, the rate of intake.

2. Maximizing efficiency of energy intake

The efficiency with which a seal acquires energy (F) is the energy gained divided by the energy expended, and is given by:

$$F = \frac{p\,(v_f)\,t_f}{c\,(v_f) + c\,(v_t)\,t_t + c\,(0)\,t_r} \tag{4}$$

If we scale $c(v)$ so that $c(0) = 1$ and substitute for t_f we obtain:

$$F = \frac{p\,(v_f)/c(v_f)}{1 + \left(c\,(v_t)\,\dfrac{t_t + t_r}{E_f}\right)} \tag{5}$$

Again we must maximize Equation 5 with respect to t_t, E_f and v_f. For a given v_f, F will be maximized by minimizing $c(v_t)\,(t_t + t_r)/E_f$. This is minimized with respect to travel time when $c(v_t)\,t_t$ is also minimized, i.e. at the minimum-cost-of-transport (MCT) speed. Thus, to maximize efficiency of energy intake, ascent and descent speeds should be held constant at MCT.

From Equation 5 we can see that foraging swim speed (v_f) should be chosen to maximize $p(v_f)/c(v_f)$, the rate of prey acquisition per unit of energy expended.

Thus in order to maximize efficiency seals should swim vertically to depth at the MCT speed and then forage at the speed which provides the most energy gain per unit expended.

3. Maximizing net energy intake rate

Net energy intake (N) is the gain from a foraging period t_f minus the cost of the whole dive cycle. The rate of net energy gain (N') is given by:

$$N' = \frac{[p\,(v_f)\,t_f] - [c\,(v_f)\,t_f + c\,(v_t)\,t_t + t_r]}{t_f + t_t + t_r} \tag{6}$$

Rearranging Equation 6 and substituting for t_f we obtain:

$$N' = \frac{p\,(v_f) - c\,(v_f) - \left[c(v_t)\,\dfrac{t_t + t_r}{E_f}\,c\,(v_f)\right]}{1 + \left[\dfrac{t_t + t_r}{E_f}\,c(v_f)\right]} \tag{7}$$

Again this function must be maximized with respect to t_t, E_f and v_f. For a fixed v_f, N will be maximized by simultaneously minimizing both $c(v_t)\,(t_t + t_r)/E_f$ and $(t_t + t_r)/E_f$. This can be demonstrated if Equation 7 is recast as:

$$S(x) = \frac{a - f_1\,(x)\,b}{1 + f_2\,(x)\,b} \tag{8}$$

Now if $f_1(x) = f_2(x) = f(x)$ we can see that $S(x)$ and $f(x)$ will be maximized at the same value of x because:

$$\frac{dS(x)}{dx} = \frac{-f'(x)b}{1 + f(x)b} - \frac{a - f(x)b}{(1 + f(x)b)^2} \; f'(x)b \; = 0$$

$$\Rightarrow \; f'(x)b \left(\frac{1 + a - f(x)b}{1 + f(x)b} \right) = 0$$

$$\Rightarrow \; f'(x) = 0 \qquad\qquad (9)$$

We can repeat this argument with $f_1(x)$ not equal to $f_2(x)$ and obtain:

$$f'_1(x)b + \frac{a - f_1(x)b}{1 + f_2(x)b} \; f'_2(x)b = 0 \qquad\qquad (10)$$

Given that the net energy intake is positive, the above sum can only be 0 if $f'_1(x$ and $f'_2(x)$ are of different sign, i.e. x is between values which make $f'_1(x)$ and $f'_2(x)$ equal to 0.

Thus, in order to maximize the net rate of energy intake seals should choose swim speeds for ascent and descent which are intermediate between the values which maximize gross energy intake and efficiency of energy intake. Thus swim speed should either decrease in deeper dives, or the seal should swim at the MCT speed. In either case the swim speed should not increase as depth of dive increases.

Swim speed predictions

The modelling exercise allows us to make three predictions about swim speed during transit between the foraging site and the surface which are independent of the foraging tactics. The first is obvious. In all foraging dives seals should swim as directly as possible to their feeding site. In most cases this would imply swimming as near to the vertical as possible. Second, in deep dives, in which the travel time represents a large proportion of the total dive duration, the seal should swim at the MCT speed. This result holds true for all three maximization strategies. Finally, in dives to shallower depths, seals can increase the proportion of time spent at the bottom by swimming faster, even though the oxygen consumption per metre increases and the proportion of time spent submerged decreases at higher swim speeds. This tactic should be employed if seals are attempting to maximize either gross energy intake or net energy intake rates.

Foraging tactics

So far we have examined how choices of ascent and descent speeds and recovery durations affect the amount of oxygen delivered to the foraging area. We have examined the implications of these choices for the three maximization strategies defined above. However, the preceding treatment has ignored the precise form of the function $p(v_f)$ relating energy gained during foraging to the swimming speed of the seal. Here we examine how variations in seal swimming speed and other factors such as prey movements and prey density affect the rate of energy acquisition and we look at the implications of these effects for the same three maximization strategies.

How then should seals hunt? Searching for prey can be considered as a specialized form of line transect survey. The number of prey items encountered is determined by the area searched, which is related to the distance covered. However, the likelihood of encounter is also determined by the behaviour of the target.

Sedentary prey will be encountered at a rate set by their density and the area searched. This is a linear function of the speed of the predator. Encounter rate (A_r) is given by:

$$A_r = B\, v(\sigma_1 + \sigma_2) \tag{11}$$

where B is the density of prey items, v is predator speed, σ_1 is the diameter of the prey and σ_2 is the diameter of the perceptual field of the predator (Laing 1938; Holling 1966; Curio 1976). A_r is equivalent to our gain function $p(v_f)$.

The encounter rate for active prey species will be affected by the movement of the target relative to the predator in several ways. If, for example, prey can detect and then avoid predators the effect of prey movement could be extremely complex. Here we assume that prey will not take avoiding action before they have come into an encounter region where they have a finite probability of detection by the predator. The shape of the encounter region will have an effect on the encounter rate when the targets are moving (Hiby 1982, 1985). Fortunately, if we can assume a circular encounter region the effect of target motion is easily accounted for.

The rate at which a searcher encounters randomly moving targets is a function of the velocities of both the searcher and targets (Yapp 1956). With the simplifying assumption that the motion of both predator and prey is rectilinear between encounters, the encounter rate A_r becomes a function of $(v^2 + w^2)^{1/2}$; where $w = $ prey swimming speed (Yapp 1956). Although the assumption of rectilinear motion is unrealistic, Yapp's formula is a reasonable approximation in many practical situations (Skellam 1958). Holling (1966) formulated a predator/prey encounter equation taking into account both the size of the encounter region (equivalent to the perceptual range of the predator and the effective size of the prey) and the motion of both predator and prey:

$$A_n = (2 \, (v^2 + w^2)^{1/2} \, t \, \sigma) + \pi \, \sigma^2)B \tag{12}$$

where: $A_n = p(v_f) \, t_f$ = number of encounters in a dive; $\sigma = \sigma_1 + \sigma_2$.

Thus we have an encounter rate per dive which is determined by the speeds of both predator and prey. But there is an energetic cost involved in swimming and therefore a trade-off between increased encounter rate and more rapid depletion of oxygen reserves, i.e. t_f is a function of the swim speed of the predator.

To determine the optimum swim speed when dealing with moving prey, time t in Equation 13 becomes a function of seal swim speed:

$$t_f = \frac{E_f}{c(v_f)} \tag{13}$$

where E_f is the amount of oxygen or energy available for foraging at the bottom and is assumed to be constant for a particular depth; $c(v_f)$ is the rate of oxygen utilization by seals as a function of swim speed. Substituting for t in Equation 13 we get:

$$A_n = 2 \, [\, (v^2 + w^2)^{1/2} \, \frac{E_f}{c(v_f)} \, \sigma + \pi\sigma^2] \, B \tag{14}$$

Metabolic rate (MR) as a function of swimming speed was determined for seals swimming in a flume tank (Fedak 1986). Data from two animals are given in Fig. 3. An exponential relationship was fitted by least squares regression, giving the following equations:

$$MR = 3.06 \, e^{0.807v} \tag{15}$$

and

$$MR = 4.14 \, e^{0.810v} \tag{16}$$

where MR is in ml $O_2.kg^{-1}.min^{-1}$.

From these data we can obtain the cost of transport (in ml $O_2.m^{-1}$), as a function of swim speed (Fig. 4). In an adult grey seal the MCT speed is

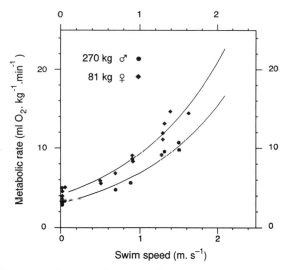

Fig. 3. Metabolic rate as a function of swim speed in two grey seals. Data were collected from seals swimming in a flume tank.

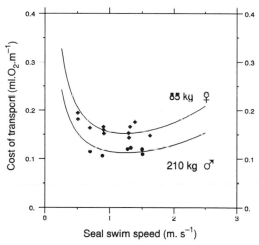

Fig. 4. Cost of transport as a function of swimming speed for two grey seals. Based on swimming flume study.

around 1.3 m.s^{-1}; costs increase rapidly as swim speed falls below MCT but increase only slowly as swim speed increases above MCT. The following discussion of swim speed and encounters uses the fitted $c(v_f)$ function for a 210 kg adult grey seal.

Again, because it is not clear which maximization strategy a foraging seal should be attempting to follow, we examine the implications of the relationship between swim speeds and encounter rates for the same three maximization strategies.

Maximizing energy efficiency

It is implicit in the above model that seals begin a dive with optimally loaded oxygen stores, and resurface when these are optimally depleted. The levels to which stores are loaded or depleted will be determined by the depth to which the seals have to swim. Since the energy expended while foraging is E_f, the efficiency of energy intake during foraging is:

$$A_e = \frac{p(v_f)\, t_f}{E_f} = \frac{p(v_f)}{c(v_f)} \tag{17}$$

However, since E_f is a constant for a particular depth (see above), A_e is directly proportional to A_n and we can examine hunting strategies in terms of efficiency by simply examining the numbers of encounters with prey per dive (A_n).

Figure 5 shows the expected number of prey encounters per dive as a function of seal swim speed for a range of prey swim speeds. The model predicts that when hunting stationary or very slow-moving prey, seals will maximize the number of encounters in a dive by swimming at around the MCT speed. This will enable them to search the maximum area possible. However, when hunting active prey, seals will increase the number of encounters per dive by reducing swimming activity. To maximize the

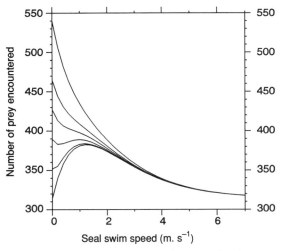

Fig. 5. Expected number of encounters per dive as a function of seal swim speed. A family of curves is shown representing a range of prey swim speeds from 0 to 1.5 m/s.

energetic efficiency when hunting active prey species, the model predicts that seals should sit and wait for prey to come into the encounter region. The parameter estimates used in the model suggest that seals should remain stationary when hunting prey which swim at more than 0.25 m.s^{-1}.

The form of the relationship is not sensitive to the amount of oxygen available for foraging (E_f). It is, however, sensitive to the perception range. If prey can be detected at greater ranges, the encounter rate is no longer as sensitive to either predator or prey swim speeds.

Maximizing gross energy intake rate

As described above, the choice of v_f which maximizes the gross energy intake rate will depend on the depth of dive. In deep dives v_f should be chosen to maximize $p(v_f)/c(v_f)$ while in shallow dives v_f should be chosen to maximize $p(v_f)$. Thus in deep dives maximizing gross energy intake tends towards maximizing energy efficiency. In shallow dives gross energy intake will continue to increase as a function of swim speed up to unrealistically high speeds.

Maximizing net energy intake rate

If seals are attempting to maximize the net rate of energy gain it is no longer sufficient to examine the number of encounters per dive. Active dives may give fewer encounters per unit oxygen expended but they are shorter and can be repeated more frequently. To determine the swim speed which optimizes net energy gain we must take into account the number of dives per unit time.

Equation 12 gives the number of encounters per dive, which is proportional to the energy intake per dive $p(v_f)/c(v_f)$. The rate of encounters with prey per unit time is obtained by dividing Equation 14 by the dive cycle duration:

$$A_i = \frac{[\,(2(v^2 + w^2)^{1/2}\dfrac{E_f}{c(v_f)}\,\sigma) + \pi\sigma^2]B}{t_f + t_t + t_r} \tag{18}$$

However, as pointed out above, there is an energetic cost involved in making more dives. The rate of net energy gain A_j is given by:

$$A_j = A_i - \frac{K}{t_t + t_f + t_r} \tag{19}$$

where K is the ratio of the energy expended per dive to the energy gained per encounter. Obviously the magnitude of K will have a profound effect on the relationship between A_j and predator swim speed.

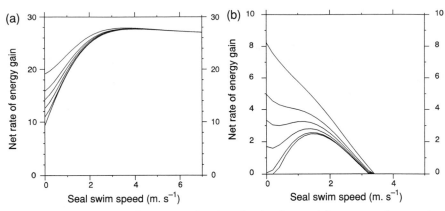

Fig. 6. The net energy intake rate as a function of seal swim speed for a range of prey swim speeds from 0–1.5 m/s. (a) with large K, representing a high prey density situation; (b) with small K, representing a low prey density situation.

If the cost of a dive is small compared with the gain per encounter, net energy gain is maximized by swimming more rapidly for all prey swim speeds (Fig. 6a). The marginal benefit of swimming faster levels off at predator speeds around twice the MCT speed. If the cost of a dive is high relative to the gain per encounter the pattern is different (Fig. 6b). For slow-swimming prey the net gain increases with seal swim speed, reaching a maximum at around the MCT, then declining at higher predator speeds. With fast-moving prey the net energy gain decreases with seal swim speed, a pattern similar to that shown by the energy efficiency model.

Discussion

Maximization strategy

To determine what may constitute the most effective diving tactics we need to identify the objectives of the diving animal. Obviously one aim of a foraging seal is to acquire food but it is not immediately apparent what aspect of energy intake should be maximized. We have presented three possible optimization strategies: which is the most appropriate?

Maximizing gross energy intake rate is probably the least likely strategy for a marine predator. A foraging pattern which maximized intake rate could leave the seal with a net energy deficit. Such strategies could develop where excluding competitors from food was advantageous. There is no evidence of this type of competition in seals of any species.

Most treatments of foraging theory have assumed that the variable to be maximized is the long-term average net rate of resource acquisition

(Charnov 1976; Stephens & Krebs 1986). However, several recent studies have shown that in practice foraging behaviour may more closely approximate to energy efficiency rather than net rate maximization (Schmid-Hempel 1987; Houston, Schmid-Hempel & Kacelnik 1988). Welham & Ydenberg (1988) suggested that where foraging animals can easily meet their energy requirements and there is some demonstrable fitness cost involved in energy expenditure they may be expected to follow a strategy of maximizing energy efficiency. They showed that ring-billed gulls (*Larus delawarensis*) feeding behind moving ploughs travelled between patches at speeds consistent with net-rate-maximizing models and altered patch residence times in line with energy-efficiency models. Although dealing with very different constraints, the moving plough experimental model, where extended patch residence time leads to increased travel costs, is functionally similar to the diving situation where extended time at depth leads to increased combined travel and recovery times.

The observed foraging behaviour of both grey seals (D. Thompson *et al.* 1991; Hammond *et al.* this volume) and common seals (P. M. Thompson, Fedak *et al.* 1989; P. M. Thompson, Pierce *et al.* 1991) suggests that they are able to obtain sufficient food during intermittent foraging bouts. These species may therefore fit Welham & Ydenberg's (1988) criteria for energy efficiency maximization.

Oxygen loading and cost-of-transport curves

The predicted form of the relationship between t_t, v_t, E_f, and t_r is based on data from laboratory studies of the cost of swimming in grey seals and field studies of the diving physiology of Weddell seals (Kooyman *et al.* 1980). However, the same relationship would result from any situation where the oxygen loading and utilization curves had similar general shapes. If the shapes of either the oxygen loading curve or the metabolic cost of swimming curves were significantly different from the examples we could obtain different predictions. Are the assumptions about oxygen reloading rates and the rate of oxygen utilization in relation to swim speed tenable?

Kooyman *et al.* (1980) showed that in Weddell seals the length of the post-dive recovery period at the surface is closely related to the post-dive blood lactate level. This pattern would result if ATP demand remained constant throughout a dive. Oxygen reserves would be depleted at a constant rate until exhausted. In longer dives there would be a net production of lactate which would have to be burnt or resynthesized to glucose or glycogen.

Data from grey and elephant seals show that this pattern of extended recovery after long dives does not appear in all phocids (Le Boeuf, Costa *et al.* 1988; Le Boeuf, Naito *et al.* 1989; Hindell, Slip & Burton 1991; Hindell,

Slip, Burton & Bryden 1992; D. Thompson & Fedak 1993). In grey seals the duration of post-dive surfacing increases with dive duration for dives of up to 7 min, but not for longer dives. A similar pattern is shown by both elephant seal species. Thus the simple relationships between lengths of dives and surface recovery periods used in previous models may not be appropriate (see, for example, Kramer 1988). The observed relationship between recovery times and dive durations in grey seals may be due to a reduction in metabolic rate in longer dives allowing seals to remain aerobic (D. Thompson & Fedak 1993). We have used the relationship between dive and surface durations in Weddell seals to provide an approximation to the oxygen loading curve. This will be reasonable if dive duration is a good index of total energy expenditure. This can be inferred from the fact that in Weddell seals the peak post-dive blood lactate levels are closely correlated with dive duration.

The swim-speed metabolism data used were taken from experiments performed using grey seals swimming in a flume. The effect of swim speed on metabolism has been monitored in several pinniped species (Davis, Williams & Kooyman 1985; Fedak 1986; Feldkamp 1987). The observed relationships are very similar in all species, suggesting that the model predictions may be applicable to a wide range of marine mammals.

Ascent/descent swim speeds

If, during foraging trips, seals are attempting to maximize the proportion of time spent at depth or the net rate of energy intake our model suggests that they should alter the speed at which they ascend and descend when diving to different depths. If they are maximizing the efficiency of energy intake, the models predict that they should always swim at the MCT speed. The models therefore predict that swimming speed may remain constant or even decline with increasing depth of dive, but should not increase with dive depth.

Two studies of diving in northern elephant seals (Le Boeuf, Costa *et al.* 1988) and northern fur seal (*Callorhinus ursinus*) (Ponganis *et al.* 1992) appear to show that swim speed increased significantly with dive depth. However, in the elephant seal study swimming speed was inferred from rate of change of depth. The lack of swimming speed data precluded differentiating between change in swim speed and change in angle of descent. In the fur seal study foraging and travelling dives were not distinguished so it is not possible to differentiate between an effect on swim speed induced by depth and one induced by a changed behavioural state.

Foraging swim speeds

We do not expect these simple models accurately to describe the behaviour of wild seals, but they may help to interpret apparently surprising observed behaviours. In grey seals feeding off the Hebrides, a striking feature of foraging dives was the relative lack of swimming activity (D. Thompson & Fedak 1993). Our models predict that both net rate of energy gain at low prey density and foraging efficiency will be maximized if seals remain stationary when hunting active prey.

Unfortunately, we do not have data from one seal species foraging on different prey types in different environments. But we can compare two sympatric species feeding on different prey. Foraging grey seals, diving to 60–100 m off the Hebrides, show sequences of dives where they swim slowly close to the bottom, interspersed with sequences of dives showing little or no active swimming at the bottom. Analysis of faecal samples indicates that grey seals in the study area prey mainly on ling (*Molva molva*), other gadoids and sand eels (*Ammodytes* spp.) (P. S. Hammond pers. comm.). All of these are mobile demersal or epibenthic animals during the time of the dive behaviour study. If grey seals were actively swimming along the bottom searching for suitable, static, foraging sites and then switching to a wait-and-ambush strategy when they found one, the observed foraging patterns would fit the model predictions.

In contrast, common seals foraging at similar depths in Froan, Norway (D. Thompson *et al.* in prep.), swam continuously, at around MCT speed, throughout their foraging dives. Seals were observed with their catch on several occasions; in each case the prey species were flat fish (pleuronectid species). These are cryptically coloured and fairly sedentary fish. Both the efficiency and net rate maximizing models show that the optimum hunting strategy for such prey is active, MCT swimming.

It is often difficult to define the constraints acting upon free-ranging foraging animals and often impossible to measure their effects. In the case of foraging seals some constraints are easily identified and we have in some cases been able to measure their effects. Thus we may be able to use the observed behaviour patterns to gain an insight into the strategies being followed.

Again it must be stated that these simple models are not expected to predict actual seal behaviour. However, when looked at in conjunction with models of other aspects of dive scheduling (e.g. Kramer 1988; Ydenberg & Clark 1989; Houston & Carbone 1992) they should both demonstrate the range of choices available to a diving predator and provide a context in which to interpret the rapidly accumulating dive behaviour data from wild seals.

References

Castellini, M. A., Davis, R. W. & Kooyman, G. L. (1988). Blood chemistry regulation during repetitive diving in Weddell seals. *Physiol. Zool.* **61**: 379–386.

Charnov, E. L. (1976). Optimal foraging: the marginal value theorem. *Theoret. Pop. Biol.* **9**: 129–136.

Curio, E. (1976). *The ethology of predation.* Springer-Verlag, Berlin. (*Zoophysiol. Ecol.* 7).

Davis, R. W., Williams, T. M. & Kooyman, G. L. (1985). Swimming metabolism of yearling and adult harbor seals *Phoca vitulina. Physiol. Zool.* **58**: 590–596.

Elsner, R., Wartzok, D., Sonafrank, N. B. & Kelly, B. P. (1989). Behavioral and physiological reactions of arctic seals during under-ice pilotage. *Can. J. Zool.* **67**: 2506–2513.

Fedak, M. A. (1986). Diving and exercise in seals: interactions of behaviour and physiology. *Prog. underwat. Sci.* **11**: 155–169.

Fedak, M. A., Pullen, M. R. & Kanwisher, J. (1988). Circulatory responses of seals to periodic breathing: heart rate and breathing during exercise and diving in the laboratory and open sea. *Can. J. Zool.* **66**: 53–60.

Feldkamp, S. D. (1987). Swimming in the California sea lion: morphometrics, drag and energetics. *J. exp. Biol.* **131**: 117–135.

Hiby, A. R. (1982). The effect of random whale movement on density estimates obtained from whale sighting surveys. *Rep. int Whal. Commn* **32**: 791–793.

Hiby, A. R. (1985). An approach to estimating population densities of Great Whales from sighting surveys. *IMA J. Math. appl. Med. Biol.* **2**: 201–220.

Hill, R. D., Schneider, R. C., Liggins, G. C., Shuette, A. H., Elliott, R. L., Guppy, M., Hochachka, P. W., Qvist, J., Falke, K. J. & Zapol, W. M. (1987). Heart rate and body temperature during free diving of Weddell seals. *Am. J. Physiol.* **253**: 344–351.

Hindell, M. A., Slip, D. J. & Burton, H. R. (1991). The diving behaviour of adult male and female southern elephant seals, *Mirounga leonina* (Pinnipedia: Phocidae). *Aust. J. Zool.* **39**: 595–619.

Hindell, M. A., Slip, D. J., Burton, H. R. & Bryden, M. M. (1992). Physiological implications of continuous, prolonged, and deep dives of the southern elephant seal (*Mirounga leonina*). *Can J. Zool.* **70**: 370–379.

Holling, C. S. (1966). The functional response of invertebrate predators to prey density. *Mem. ent. Soc. Can.* **48**: 5–86.

Houston, A. I. & Carbone, C. (1992). The optimal allocation of time during the diving cycle. *Behav. Ecol.* **3**: 255–265.

Houston, A., Schmid-Hempel, P. & Kacelnik, A. (1988). Foraging strategy, worker mortality, and the growth of the colony in social insects. *Am. Nat.* **131**: 107–114.

Irving, L. (1934). On the ability of warm blooded animals to survive without breathing. *Scient Mon., N.Y.* **38**: 422–428.

Kooyman, G. L. (1981). *Weddell seal: consummate diver.* Cambridge University Press, Cambridge etc.

Kooyman, G. L. (1989). *Diverse divers: physiology and behaviour.* Springer-Verlag, Berlin & London. (*Zoophysiology* 23.)

Kooyman, G. L. & Campbell, W. B. (1973). Heart rate in freely diving Weddell seals (*Leptonychotes weddelli*). *Comp. Biochem. Physiol.* **43**: 31–36.

Kooyman, G. L., Wahrenbrock, E. A., Castellini, M. A., Davis, R. W. & Sinnett, E. E. (1980). Aerobic and anaerobic metabolism during voluntary diving in

Weddell seals: evidence of preferred pathways from blood chemistry and behavior. *J. comp. Physiol.* **138**: 335–346.

Kramer, D. L. (1988). The behavioral ecology of air breathing by aquatic animals. *Can J. Zool.* **66**: 89–94.

Laing, J. (1938). Host-finding by insect parasites. II. The chance of *Trichogramma evanescens* finding its hosts. *J. exp. Biol.* **15**: 218–302.

Le Boeuf, B. J., Costa, D. P., Huntley, A. C. & Feldkamp, S. D. (1988). Continuous, deep diving in female northern elephant seals, *Mirounga angustirostris*. *Can. J. Zool.* **66**: 446–458.

Le Boeuf, B. J., Naito, Y., Huntley, A. C. & Asaga, T. (1989). Prolonged, continuous, deep diving by northern elephant seals. *Can J. Zool.* **67**: 2514–2519.

McConnell, B. J., Chambers, C. & Fedak, M. A. (1992). Foraging ecology of southern elephant seals in relation to the bathymetry and productivity of the Southern Ocean. *Antarct. Sci.* **4**: 393–398.

Ponganis, P. J., Gentry, R. L., Ponganis, E. P. & Ponganis, K. V. (1992). Analysis of swim velocities during deep and shallow dives of the two northern fur seals, *Callorhinus ursinus*. *Mar. Mamm. Sci.* **8**: 69–75.

Schmid-Hempel, P. (1987). Efficient nectar-collecting by honeybees. I. Economic models. *J. Anim. Ecol.* **56**: 209–218.

Scholander, P. F. (1940). Experimental investigations on the respiratory function in diving mammals and birds. *Hvalrad. Skr.* **22**: 1–131.

Skellam, J. G. (1958). The mathematical foundations underlying the use of line transects in animal ecology. *Biometrics* **14**: 385–400.

Stephens, D. W. & Krebs, J. R. (1986). *Foraging theory*. Princeton University Press, Princeton & Guildford.

Thompson, D. & Fedak, M. A. (1993). Cardiac responses of grey seals during diving at sea. *J. exp. Biol.* **174**: 139–164.

Thompson, D., Hammond, P. S., Nicholas, K. S. & Fedak, M. A. (1991). Movements, diving and foraging behaviour of grey seals (*Halichoerus grypus*). *J. Zool., Lond.* **224**: 223–232.

Thompson, P. M., Fedak, M. A., McConnell, B. J. & Nicholas, K. S. (1989). Seasonal and sex-related variation in the activity patterns of common seals (*Phoca vitulina*). *J. appl. Ecol.* **26**: 521–535.

Thompson, P. M., Pierce, G. J., Hislop, J. R. G., Miller, D. & Diack, J. S. W. (1991). Winter foraging by common seals (*Phoca vitulina*) in relation to food availability in the inner Moray Firth, N.E. Scotland. *J. Anim. Ecol.* **60**: 283–294.

Welham, C. V. J. & Ydenberg, R. C. (1988). Net energy versus efficiency maximizing by foraging ring-billed gulls. *Behav. Ecol. Sociobiol.* **23**: 75–82.

Yapp, W. B. (1956). The theory of line transects. *Bird Study* **3**: 93–104.

Ydenberg, R. C. & Clark, C. W. (1989). Aerobiosis and anaerobiosis during diving by western grebes: an optimal foraging approach. *J. theor. Biol.* **139**: 437–447.

Glossary

A_e efficiency of energy intake while foraging

A_i rate of encounters with prey per unit time

A_j net encounter rate, i.e. rate of encounters minus cost of foraging

A_n number of encounters per dive

A_r rate of encounters between seal and prey

B density of prey items

$c(v)$ function relating oxygen consumption to swim speed

D depth

E_f energy or oxygen used during foraging

F efficiency of energy intake, i.e. energy intake divided by energy expended

G gross energy intake

G' rate of gross energy intake

$h(t_t)$ function relating amount of oxygen used swimming to depth D as a function of t_t

K ratio of energy expended during a dive to the energy gained per encounter

N net energy intake, i.e. energy intake less energy expended

N' rate of net energy intake

$r(t_r)$ function relating amount of oxygen loaded as a function of t_r

σ radius of the encounter region

t_f time spent foraging at depth

t_r time spent recovering at the surface

t_t time spent travelling between surface and foraging site

v_f swim speed while foraging at depth

$v_t.$ swim speed while travelling

Symp. zool. Soc. Lond. (1993) No. 66: 369–382

Role of plasma and tissue lipids in the energy metabolism of the harbour seal

R. W. DAVIS,

*Department of Marine Biology
Texas A & M University
PO Box 1675
Galveston, TX 77553, USA*

W. F. BELTZ, F. PERALTA
and J. L. WITZTUM

*Department of Medicine
University of California
La Jolla, CA 92097, USA*

Synopsis

Lipid is an important source of energy for harbour seals and other carnivorous marine mammals. This results primarily from their diet, which is rich in fish oil, and the large lipid stores in the blubber. Measurements of respiratory quotient (RQ) indicate that whole-body lipid catabolism provides 87% of metabolic energy for resting and 95% for exercising harbour seals. In earlier studies, we showed that the catabolism of plasma free fatty acids (FFA) provides only about 20% of total energy production in harbour seals. Recent studies indicate that very low density lipoprotein triglycerides (VLDL-TG) may provide an additional 13% of energy production in postabsorptive harbour seals. However, the catabolism of fatty acids from triglyceride-rich chylomicrons may increase fourfold after a meal and represent the primary source of lipid for energy metabolism during foraging dives. The third source of lipids is endogenous triglyceride stores in the tissues, especially skeletal muscle. Although studies to quantify the catabolism of endogenous tissue triglycerides are just beginning, the prospect of assembling a complete fuel budget for resting and exercising harbour seals appears good. This information will contribute to our understanding of how marine mammals partition oxygen and fuel reserves to maintain an aerobic, fat-based metabolism during most voluntary dives.

Introduction

Harbour seals (*Phoca vitulina*) and other piscivorous marine mammals consume a diet that is high in fat (fish oil) and protein but has negligible carbohydrate (Bonner 1979; Riedman 1990: 139–175; Worthy 1990). Dietary lipid can be catabolized immediately or stored in adipose tissue and used later as a source of energy. Most marine mammals (with the exception

ZOOLOGICAL SYMPOSIUM No. 66
ISBN 0–19–854069–8

Copyright © 1993 The Zoological Society of London
All rights of reproduction in any form reserved

of sea otters) have a subcutaneous layer of blubber that provides thermal insulation and serves as the primary energy reserve during periods of fasting. Lipid stores in the blubber as well as intraperitoneal fat can constitute over 30% of a seal's total mass (Bryden 1969; Stirling & McEwan 1975; Worthy & Lavigne 1983).

Unlike fat, dietary amino acids cannot readily be stored as an energy source. Those that are not used for cellular maintenance and growth are catabolized immediately or synthesized into carbohydrate (gluconeogenesis) and fat. In carnivores, gluconeogenesis from amino acids and glycerol is the primary source of carbohydrate, which is conserved for obligate glucose-metabolizing tissues such as the brain and red blood cells (Roberts, Samuels & Reinecke 1943; Blazquez, Castro & Herrera 1971; Kettelhut, Foss & Migliorini 1980). Studies have shown that carbohydrate-deficient diets produce metabolic changes that conserve carbohydrate (i.e. through the recycling of three-carbon intermediates such as lactate and glycerol) and minimize its oxidation (Randle *et al.* 1963; Suzuki & Fuwa 1970; Eisenstein, Strack & Steiner 1974). Similar metabolic adjustments that conserve carbohydrate have been observed in harbour seals (Davis 1983; Davis, Castellini *et al.* 1991).

The purpose of this paper is to review available information on the fuel homeostasis of harbour seals. We include data on the metabolism of glucose, lactate and free fatty acids (FFA) that have been published previously as well as new data on the metabolism of lipoprotein triglycerides. In addition, we identify the need for new research on the role of endogenous muscle triglyceride stores as an immediate source of energy, especially during exercise. The ultimate goal of our research is to assemble a complete fuel and energy budget for resting and exercising harbour seals under postprandial and postabsorptive conditions. Although studies to quantify the catabolism of endogenous muscle triglycerides are just beginning, the prospect of assembling a complete fuel budget for harbour seals appears good. This information will contribute to our understanding of how marine mammals partition oxygen and fuel reserves to maintain an aerobic, fat-based metabolism during most voluntary dives.

Whole-body lipid metabolism in resting and exercising harbour seals

The respiratory quotient (RQ) is the ratio of metabolic carbon dioxide production ($\dot{M}CO_2$) and oxygen consumption ($\dot{M}O_2$). The RQ is of interest because it reveals information about the composition of materials catabolized. Catabolism of carbohydrate, lipids and protein leads to a characteristic RQ in each case. For the catabolism of carbohydrate, the RQ is 1.0; for fat it is 0.7; and for meat protein it is 0.83 on average.

The RQ of resting, postabsorptive seals is about 0.74 (Scholander 1940; Hart & Irving 1959; Kooyman, Kerem *et al.* 1973; Davis 1983), indicating that lipid is the primary metabolic fuel. The proportion of lipid and carbohydrate catabolized can be calculated from the nitrogen-free (N-free) RQ by measuring urinary nitrogen loss or by assuming that protein catabolism is negligible under postabsorptive conditions (Kleiber 1975: 60–93). The latter assumption is reasonable for seals that are more than 12 h postabsorptive. Under these circumstances, digestion is complete, the animal is catabolizing endogenous lipid and glucose, and protein oxidation contributes no more than 3% of total energy metabolism (Pernia, Hill & Ortiz 1980). The proportion of fat and carbohydrate catabolized can be estimated from the N-free RQ by using the following equation of Kleiber (1975):

$$\text{Fat percentage} = [3800\ (1 - \text{RQ})\ ]/[59.5\ \text{RQ} - 30.9]$$

$$\text{Carbohydrate percentage} = 100 - \text{fat percentage}$$

With a resting, postabsorptive N-free RQ of 0.74, the seal would catabolize 75% fat and 25% carbohydrate (i.e. for every 100 g of substance catabolized, 75 g are fat and 25 g are carbohydrate). However, because the energy content of fat (39.3 kJ/g) is more than twice that of carbohydrate (17.6 kJ/g), 87% of the seal's energy is derived from fat catabolism and 13% from carbohydrate. During steady-state swimming, the N-free RQ for harbour seals decreases to 0.72 (Davis, Castellini *et al.* 1991). Under these conditions, the seal derives 95% of its energy from fat and 5% from carbohydrate.

Plasma glucose, lactate and FFA as an energy source

How do these estimates of carbohydrate and fat catabolism based on the N-free RQ compare with measurements of substrate oxidation using isotopically labelled metabolites? For postabsorptive harbour seals that are resting or exercising at a submaximum level, the plasma concentrations and turnover rates of glucose, lactate and FFA (Table 1) are similar to those for terrestrial mammals when scaled by means of a weight-specific allometric coefficient of 0.75 for standard metabolism (Schmidt-Nielsen 1970; Davis 1983; Davis, Castellini *et al.* 1991). However, the oxidation rates of glucose and lactate are lower than those for terrestrial, omnivorous mammals. During swimming exercise, the plasma glucose concentration, turnover rate and oxidation rate do not change appreciably from resting levels (Table 1), which indicates the minor role that glucose plays in the aerobic energy metabolism of seal skeletal muscle. In harbour seals, much of the

Table 1. Rates of turnover and oxidation for glucose, lactate and FFA at rest, 35% $\dot{M}O_{2max}$ (1.3 m/s) and 50% $\dot{M}O_{2max}$ (1.9 m/s) for harbour seals (Davis, Castellini *et al.* 1991)

	Rest	35% $\dot{M}O_{2max}$ 1.3 m/s	50% $\dot{M}O_{2max}$ 1.9 m/s
Glucose metabolism			
[Glu] (mM)	8.6	7.7	7.8
Total turnover rate ($\mu mol \cdot min^{-1} \cdot kg^{-1}$)	23.2	21.4	23.0
Oxidation rate ($\mu mol \cdot min^{-1} \cdot kg^{-1}$)	1.5	1.2	0.8
Oxidation (%)	7	6	4
Lactate metabolism			
[Lac] (mM)	0.8	0.9	1.1
Total turnover rate ($\mu mol \cdot min^{-1} \cdot kg^{-1}$)	26.2	39.7	55.0
Oxidation rate ($\mu mol \cdot min^{-1} \cdot kg^{-1}$)	7.0	12.4	11.4
Oxidation (%)	27	31	21
FFA metabolism			
[FFA] (mM) $\times 10^{-1}$	7.32	7.83	9.10
Total turnover rate ($\mu mol \cdot min^{-1} \cdot kg^{-1}$)	7.5	8.8	10.3
Oxidation rate ($\mu mol \cdot min^{-1} \cdot kg^{-1}$)	2.5	5.2	8.7
Oxidation (%)	33	59	85

carbohydrate carbon is conserved through recycling rather than entering oxidative pathways for ATP production (Davis 1983; Davis, Castellini *et al.* 1991).

The direct oxidation of plasma glucose and lactate accounts for only 10% of total energy production (calculated as mmol ATP/min) in resting harbour seals (Table 2; Davis 1983; Davis, Castellini *et al.* 1991). In contrast, about 35% of energy production is derived from glucose oxidation in resting humans. During steady-state swimming at 50% $\dot{M}O_{2max}$, the oxidation of glucose and lactate in harbour seals accounts for only 5% of total energy production (Davis, Castellini *et al.* 1991). These measurements of the metabolic energy derived from glucose and lactate oxidation agree well with estimates of carbohydrate oxidation in harbour seals (13% at rest and 5% during submaximum swimming) based on the N-free RQ calculations described above.

For postabsorptive harbour seals resting in water, about one-third of the FFA turnover is oxidized (Table 1), which is similar to the proportion observed in humans and dogs; 67% of the resting FFA turnover rate in all three species does not undergo direct oxidation (Havel & Carlson 1963; Issekutz *et al.* 1964). The remainder is probably re-esterified into triglycerides in the liver, adipose tissue, or muscle. During swimming exercise at 50% $\dot{M}O_{2max}$, 85% of the FFA turnover in harbour seals is directly oxidized, which is similar to the oxidation rate observed for humans and dogs (Havel & Carlson 1963; Issekutz *et al.* 1964).

The oxidation of plasma FFA accounts for only 18% of total energy

Table 2. Rates of ATP production from aerobic and anaerobic metabolism and from the oxidation of glucose, lactate, FFA and VLDL-TGFA. The percentage contribution to total ATP production is shown in parentheses. Data for glucose, lactate and FFA are from Davis, Castellini *et al.* (1991)

	ATP production (mmol · min^{-1})		
	Rest	35% $\dot{M}O_{2max}$ 1.3 m/s	50% $\dot{M}O_{2max}$ 1.9 m/s
Aerobic metabolism[a]	72.0 (98)	138.0 (99)	206.0 (99)
Anaerobic metabolism[b]	1.1 (2)	1.6 (1)	2.2 (1)
Total	73.1 (100)	139.6 (100)	208.2 (100)
Glucose oxidation[c]	2.2 (3)	1.7 (1)	1.2 (1)
Lactate oxidation[c]	4.8 (7)	5.5 (4)	7.8 (4)
FFA oxidation[c]	13.1 (18)	27.2 (20)	45.6 (22)
VLDL-TGFA oxidation[c,d]	9 (13)	17 (12)	28 (14)
Total	29.1 (41)	51.4 (37)	82.6 (41)

[a] Assumes fat is the primary metabolic fuel (RQ $\approx$ 0.74) so that 5.65 mmol ATP are produced per mmol O_2 consumed.

[b] Assumes 1.5 mmol ATP are produced per mmol of lactate processed.

[c] The percentage of energy metabolism contributed by glucose, lactate, FFA or TGFA oxidation was calculated as follows:

$$\%P = \frac{(Ro \times A \times 100)}{\dot{M}O_2 \times 5.65 \text{ mmol ATP} \cdot \text{mmol } O_2^{-1}}$$

where $\%P$ is the percent energy contribution; Ro is the oxidation rate in mmol · min^{-1}; A is the ATP produced during oxidation of lactate (17 mmol ATP · mmol^{-1}), glucose (36 mmol ATP · mmol^{-1}) or FFA (117 mmol ATP · mmol^{-1}); $\dot{M}O_2$ is the oxygen consumption in mmol O_2 · min^{-1} (Lehninger 1970: 313–395).

[d] The oxidation rate of VLDL-TGFA was estimated from the turnover rate of VLDL-TGFA, assuming that the same percentage of the TG fatty acid turnover was oxidized as measured for plasma FFA shown in Table 1.

production in resting harbour seals (Table 2) (Davis 1983; Davis, Castellini *et al.* 1991). This percentage increases to only 22% during swimming at 50% $\dot{M}O_{2max}$, despite the fact that oxidation accounts for 85% of plasma FFA turnover (Davis, Castellini *et al.* 1991). Obviously, these values for FFA oxidation do not agree well with estimates of the proportion of total fat catabolized calculated from the N-free RQ. The oxidation of plasma FFA alone fails to account for 69–73% of the energy derived from fat catabolism. There must be another pool of lipid that is catabolized other than plasma FFA. Two possible sources are triglycerides (TG) in plasma lipoproteins and endogenous muscle TG.

Plasma triglycerides as an energy source

In pinnipeds, as in other mammals, plasma TG are transported in large, spherical particles in which nonpolar lipids (TG and cholesterol ester) are

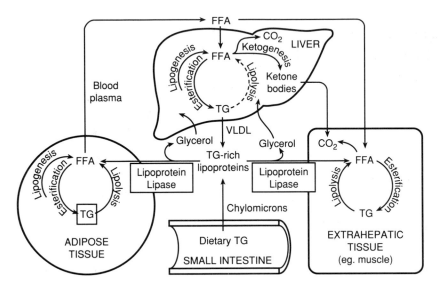

Fig. 1. Overview of lipid metabolism in the harbour seal (VLDL, very low density lipoproteins; FFA, free fatty acids; TG, triglycerides) (adapted from Martin *et al.* 1985).

surrounded by a monolayer of polar lipids (phospholipids and cholesterol) and specific apoproteins to form a hydrophilic lipoprotein complex (Puppione & Nichols 1970; Smith, Pownall & Gotto 1978). Four major groups of lipoproteins have been identified on the basis of their density or electrophoretic mobility. These are chylomicrons, very low density lipoproteins (VLDL), low density lipoproteins (LDL), and high density lipoproteins (HDL). Most of the TG derived from intestinal absorption of fat or from the liver are transported in the blood as chylomicrons or VLDL, respectively (Fig. 1). By mass, chylomicrons are about 88% TG and VLDL are about 56% TG (Martin *et al.* 1985: 194–256).

Chylomicrons are found in chyle formed only by the lymphatic system draining the intestine (Fig. 2). In addition to lipids, the chylomicrons contain apoproteins (i.e. Apo-A, Apo-B-48, Apo-C and Apo-E) that function as structural proteins and ligands for the receptor-mediated metabolism of triglycerides and the hepatic clearance of the chylomicron remnants (Bisgaier & Glickman 1983). The plasma concentration of chylomicron TG increases following a fatty meal, then decreases as the chylomicrons are cleared by extrahepatic and hepatic tissues. In a series of postprandial plasma samples from four harbour seals, TG reached a peak (mean = 115 mg/dl or 1.4 mM) 2–4 h after a meal of herring and mackerel and returned to postabsorptive levels (e.g. 16 mg/dl or 0.2 mM) after 14 h (Fig. 3).

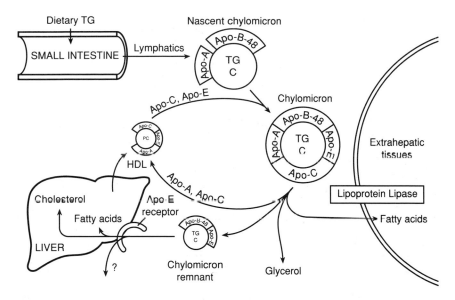

Fig. 2. Metabolism of chylomicrons. (Apo-A, apolipoprotein A; Apo-B, apolipoprotein B; Apo-C, apolipoprotein C; Apo-E, apolipoprotein E; HDL, high density lipoprotein; TG, triglyceride; C, cholesterol and cholesterol ester; P, phospholipid) (adapted from Martin *et al.* 1985).

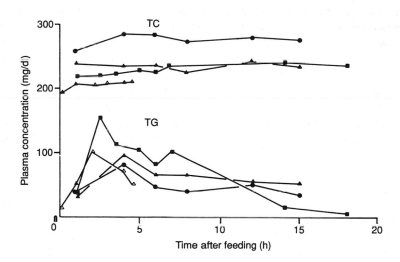

Fig. 3. Plasma concentration of total cholesterol (TC) and triglycerides (TG) in four harbour seals after eating 2 kg of herring and mackerel.

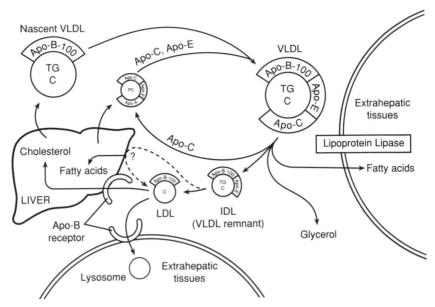

Fig. 4. Metabolic fate of very low density lipoproteins (VLDL). (Apo-A, apolipoprotein A; Apo-B, apolipoprotein B; Apo-C, apolipoprotein C; Apo-E, apolipoprotein E; HDL, high density lipoprotein; TG, triglyceride; IDL, intermediate density lipoprotein; LDL, low density lipoprotein; VLDL, very low density lipoprotein; C, cholesterol and cholesterol ester; P, phospholipid) (adapted from Martin *et al.* 1985).

Postabsorptive plasma TG are found primarily in the VLDL, the bulk of which is of hepatic origin (Puppione & Nichols 1970; Martin *et al.* 1985). VLDL is the primary vehicle of transport of TG from the liver to the extrahepatic tissues (Fig. 4). As with chylomicrons, hepatic VLDL contains apoproteins that mediate the receptor-dependent metabolism of TG by extrahepatic tissues. As the TG are removed, VLDL remnants are converted into LDL and finally cleared by the liver and extrahepatic tissues.

The fatty acids used in the synthesis of hepatic TG are derived from the uptake of FFA in the circulation and from synthesis within the liver from acetyl-CoA derived from amino acids and carbohydrate. The first source is probably predominant in well-fed harbour seals on a high-fat diet, where the level of circulating FFA is raised and more is abstracted into the liver. Under these conditions, FFA are most probably the main source of TG in the liver and plasma VLDL, because lipogenesis from acetyl-CoA is depressed (Martin *et al.* 1985).

Lipoprotein TG in chylomicrons or VLDL cannot be taken up intact by tissues, but must first undergo hydrolysis by lipoprotein lipase, an enzyme situated on the capillary endothelium of extrahepatic tissues (Cryer 1981). The TG are hydrolysed progressively through diglycerides and monoglycerides to FFA plus glycerol. Some of the FFA return to the circulation attached to albumin, but most are taken up into the tissues where they are re-esterified into TG or oxidized as fuel.

No critical qualitative differences between the catabolism of TG in chylomicrons and VLDL have been demonstrated (Havel & Kane 1975). When chylomicrons labelled in the TGFA are administered intravenously in laboratory animals, about 80% of the label is found in adipose tissue, heart and muscle and about 20% in the liver (Martin *et al.* 1985). As experiments with perfused organs have shown that the liver does not metabolize chylomicrons or VLDL significantly, the label in the liver must result secondarily from their metabolism in extrahepatic tissues.

To determine the role of TG in the energy metabolism of harbour seals, we measured the turnover rate of VLDL-TG. Isotopically labelled glycerol was used as a precursor for VLDL-TG in four postabsorptive harbour seals at rest (Davis, Beltz *et al.* in prep.). A multicompartmental model was used to analyse the kinetic data according to the method of Zech *et al.* (1979). The concentration of plasma TG (0.54 mM) and VLDL-TG (0.36 mM) and the percentage of plasma TG in VLDL (63%) (Table 3) were within the normal mammalian range (Martin *et al.* 1985). Although the postabsorptive VLDL-TG turnover for harbour seals (4.3 μmol $\cdot$ min^{-1} $\cdot$ kg$^{-0.75}$) is seven times faster than for humans (Steiner & Murase 1975; Zech *et al.* 1979), it is similar to that of rats (Wolfe & Durkot 1985) when scaled by a weight-specific allometric coefficient of 0.75 for standard metabolism (Schmidt-Nielsen 1970).

Although we did not measure the oxidation rate of VLDL-TGFA, we can make a reasonable estimate based on the turnover rate. Assuming that, as has been shown for rats (Wolfe & Durkot 1985), the same percentage of

Table 3. Plasma concentrations of TG, fractional turnover rates (FCR) for VLDL-TG and VLDL-TG turnover rates for four harbour seals under resting, postabsorptive conditions. The VLDL triglyceride fatty acid (VLDL-TGFA) turnover rate was calculated by multiplying the VLDL-TG turnover rate by three (each TG has three fatty acids)

Seal no. & body mass	Plasma TG (mM)	VLDL-TG (mM)	TG in VLDL (%)	FCR h^{-1}	VLDL-TG turnover (μmol/min/kg)	VLDL-TGFA turnover (μmol/min/kg)
1 (33 kg)	0.65	0.41	63	4.9	2.2	6.6
2 (53 kg)	0.28	0.15	54	3.6	0.6	1.8
3 (80 kg)	0.89	0.64	72	3.9	2.7	8.1
4 (49 kg)	0.35	0.22	62	2.8	0.7	2.1
Mean (54 kg)	0.54	0.36	63	3.8	1.6	4.8

VLDL-TGFA is oxidized as of plasma FFA (see Table 1 for percentage of FFA turnover oxidized), then the VLDL-TGFA oxidation rate for harbour seals would be 1.6 μmol · min^{-1} · kg^{-1} at rest and 4 μmol · min^{-1} · kg^{-1} at 50% $\dot{M}o_{2max}$. These calculations assume that the VLDL-TGFA turnover rate is 4.8 μmol · min^{-1} · kg^{-1} at rest and does not change during exercise. In actual fact, the FFA turnover rate increases 37% during exercise at 50% $\dot{M}o_{2max}$, so our calculation for the VLDL-TGFA oxidation rate may be under-estimated. The percentage contribution of VLDL-TGFA oxidation to total ATP production would be about 13%, which is comparable to the 20% for FFA oxidation (Table 2).

On the basis of current estimates of the oxidation of glucose, lactate, FFA and VLDL-TGFA, we can account for about 40% of total ATP production. Of this amount, three-quarters may come from the oxidation of VLDL-TGFA and plasma FFA. The indirect calorimetry data indicate that almost all of the energy derived from the oxidation of nonprotein sources is derived from the oxidation of fat. Therefore, endogenous lipids (such as the intramuscular pool) must be an important source of lipid for energy metabolism.

Endogenous tissue triglycerides as an energy source

The catabolism of endogenous fatty acids is probably an important source of energy, especially for muscle. Endogenous triglycerides in heart and skeletal muscle have been shown to be a major source of fuel for metabolism (Issekutz & Paul 1968; Neptune, Sudduth & Foreman 1969). Exercise decreases the concentration of intramuscular triglycerides, which may constitute 50–75% of the total fatty acids oxidized (Oscai, Caruso & Wergeles 1982). Histochemical studies on the muscles of seals have shown lipid droplets between the myofilaments (O. Mathieu-Costello pers. comm.) and a high lipase activity (George, Vallyathan & Ronald 1971). The TG stored in the muscle, which are not modelled by isotopically labelled plasma FFA, could be an important energy source for oxidative metabolism in the seal. Reduced blood flow to muscle during the dive response may require a greater reliance on endogenous muscle TG when plasma substrate availability is restricted (Issekutz & Paul 1968). Consequently, muscle TG may provide much of the ATP production not accounted for by the oxidation of plasma FFA and VLDL-TGFA. Verification of this hypothesis must await further studies.

The importance of lipid metabolism during diving

How are these results related to the fuel homeostasis of harbour seals making natural dives in the wild? Unfortunately, we have no information on the substrate turnover and oxidation rates in seals during voluntary dives (excluding short dives during submerged swimming in a water flume). Routine dive durations of 3–8 min have been measured for free-ranging harbour seals off the coasts of England and California (Fedak, Pullen & Kanwisher 1988; B. Stewart pers. comm.). If we assume that, as has been shown for Weddell seals (Kooyman, Wahrenbrock *et al.* 1980), most of the dives made by a harbour seal are aerobic, then most tissues, especially muscle, probably use lipid as a primary source of energy, as has been shown during submerged swimming (Davis *et al.* 1991). However, it is unclear whether plasma FFA and VLDL-TGFA continue to serve as a primary source of metabolic lipid, or whether the dive response forces a heavier reliance on endogenous muscle lipid stores.

All our studies of fuel homeostasis in harbour seals have been conducted under postabsorptive conditions in order to avoid non-steady state changes in blood metabolites associated with digestion and absorption. However, marine mammals must dive in order to forage, and they appear to digest their food during the foraging bout. In Weddell seals, a foraging bout may last 6–8 h and include 20–30 dives (Kooyman, Wahrenbrock *et al.* 1980). After the first hour of foraging, the concentration of chylomicrons increases and the plasma becomes very lipemic (R. W. Davis unpubl. obs.). During this time, chylomicron-TGFA may be the primary source of lipid for the energy metabolism of muscle. If we assume that the proportion of chylomicron-TGFA catabolized is proportional to the plasma concentration, then a fourfold increase during digestion and absorption could account for over 50% of ATP production. If the ATP production from the catabolism of glucose, lactate and FFA is similar to that shown in Table 2 (i.e. about 28%), then we can begin to account for most of the substrate used to support diving energetics. The remaining difference may be accounted for by the contribution from postprandial protein catabolism, which typically ranges from 11–21% of total energy production in non-ruminant mammals (Long 1961: 1119–1120).

Future research

Although our understanding of fuel homeostasis in harbour seals has increased greatly in the past ten years, much research is still needed before we can assemble a complete fuel and energy budget. Areas where immediate research should be focused are:

1. Measure the oxidation rate of VLDL-TGFA and determine the percentage contribution to total ATP production under resting and exercise conditions.
2. Measure the turnover and oxidation rates of endogenous muscle

triglyceride stores and determine the percentage contribution to total ATP production under resting and exercise conditions.

3. Measure the postprandial turnover and oxidation rates of chylomicron-TGFA and determine the percentage contribution to total ATP production.

4. Measure the postprandial turnover and oxidation of amino acids and determine the percentage contribution to total ATP production.

Most of these measurements are possible using existing methods of isotopically labelled metabolites (Wolfe 1984). Once we have acquired this information from laboratory studies, we can focus on the more challenging goal of measuring fuel homeostasis in seals during voluntary dives, including foraging dives. This will require infusing isotopically labelled metabolites and taking blood samples during the dive. Some of this technology has been developed (Guppy *et al.* 1986), but additional engineering will be required. Seals that are trained for open-ocean diving or Weddell seals making voluntary dives beneath the ice in Antarctica (Kooyman, Wahrenbrock *et al.* 1980) hold the greatest promise for obtaining these data. This information will contribute to our understanding of how marine mammals partition oxygen and fuel reserves to maintain an aerobic, fat-based metabolism during most voluntary dives.

Acknowledgements

We gratefully acknowledge the laboratory assistance of Lisa Mastro, Paul Jobsis and Vivian Casanas. We thank Drs Gerald Kooyman, Michael Castellini, Graham Worthy and Ian Boyd for their suggestions and comments on the manuscript.

References

Bisgaier, C. L. & Glickman, R. M. (1983). Intestinal synthesis, secretion, and transport of lipoproteins. *A. Rev. Physiol.* **45**: 625.

Blazquez, E., Castro, M. & Herrera, E. (1971). Effect of high-fat diet on pancreatic insulin release, glucose tolerance and hepatic gluconeogenesis in male rats. *Revta esp. Fisiol.* **27**: 297–304.

Bonner, W. N. (1979). Harbour (common) seal. In *Mammals in the seas* **2**: 58–62. Food and Agriculture Organization of the United Nations, Rome (*FAO Fish. Ser. No. 5*).

Bryden, M. M. (1969). Relative growth of the major body components of the southern elephant seal, *Mirounga leonina* (L). *Aust. J. Zool.* **17**: 153–177.

Cryer, A. (1981). Tissue lipoprotein lipase activity and its action in lipoprotein metabolism. *Int. J. Biochem.* **13**: 525.

Davis, R. W. (1983). Lactate and glucose metabolism in the resting and diving harbor seal (*Phoca vitulina*). *J. comp. Physiol.* **153**: 275–288.

Davis, R. W., Beltz, W. F., Peralta, F. & Witztum, J. L. (In preparation). *The turnover rate of plasma very low density lipoprotein fatty acids in the harbor seal.*

Davis, R. W., Castellini, M. A., Williams, T. M. & Kooyman, G. L. (1991). Fuel homeostasis in the harbor seal during submerged swimming. *J. comp. Physiol. B* **160**: 627–635.

Eisenstein, A. B. Strack, I. & Steiner, A. (1974). Increased hepatic gluconeogenesis without a rise of glucagon secretion in rats fed a high fat diet. *Diabetes* **23**: 869–875.

Fedak, M. A., Pullen, M. R. & Kanwisher, J. (1988). Circulatory responses of seals to periodic breathing: heart rate and breathing during exercise and diving in the laboratory and open sea. *Can. J. Zool.* **66**: 53–60.

George, J. C., Vallyathan, N. V. & Ronald, K. (1971). The harp seal, *Pagophilus groenlandicus* (Erxleben, 1777). VII. A histophysiological study of certain skeletal muscles. *Can. J. Zool.* **49**: 25–30.

Guppy, M., Hill, R. D., Schneider, R. C., Qvist, J., Liggins, G. C., Zapol, W. M. & Hochachka, P. W. (1986). Microcomputer-assisted metabolic studies of voluntary diving of Weddell seals. *Am. J. Physiol.* **250**: R175–R187.

Hart, J. S. & Irving, L. (1959). The energetics of harbor seals in air and in water with special consideration of seasonal changes. *Can. J. Zool.* **37**: 447–457.

Havel, R. J. & Carlson, L. A. (1963). Comparative turnover rates of free fatty acids and glycerol in blood of dogs under various conditions. *Life Sci.* No. 9: 651–658.

Havel, R. J. & Kane, J. P. (1975). Quantification of triglyceride transport in blood plasma: a critical analysis. *Fed. Proc.* **34**: 2250–2257.

Issekutz, B., Miller, H. I., Paul, P. & Rodahl, K. (1964). Source of fat oxidation in exercising dogs. *Am. J. Physiol.* **207**: 583–558.

Issekutz, B. & Paul, P. (1968). Intramuscular energy sources in exercising normal and pancreatectomized dogs. *Am. J. Physiol.* **215**: 197–204.

Kettelhut, I. C., Foss, M. C. & Migliorini, R. H. (1980). Glucose homeostasis in a carnivorous animal (cat) and in rats fed a high-protein diet. *Am. J. Physiol.* **239**: R437–R444.

Kleiber, M. (1975). *The fire of life: an introduction to animal energetics.* (2nd edn.) Robert E. Krieger Publishing Company, Huntington, New York.

Kooyman, G. L., Kerem, D. H., Campbell, W. B. & Wright, J. J. (1973). Pulmonary gas exchange in freely diving Weddell seals, *Leptonychotes weddelli. Respir. Physiol.* **17**: 283–290.

Kooyman, G. L., Wahrenbrock, E. A., Castellini, M. A., Davis, R. W. & Sinnett, E. E. (1980). Aerobic and anaerobic metabolism during voluntary diving in Weddell seals: evidence of preferred pathways from blood chemistry and behavior. *J. comp. Physiol.* **138**: 335–346.

Lehninger, A. L. (1970). *Biochemistry: the molecular basis of cell structure and function.* Worth Publishers Inc., New York.

Long, C. (1961). *Biochemist's handbook.* Van Nostrand, Philadelphia.

Martin, D. W., Mayes, P. A., Rodwell, V. W. & Granner, D. K. (1985). *Harper's review of biochemistry.* Lang Medical Publications, Los Altos, California.

Neptune, E. M., Sudduth, H. C. & Foreman, D. R. (1969). Labile fatty acids of rat diaphragm muscle and their possible role as the major endogenous substrate for maintenance of respiration. *J. biol. Chem.* **234**: 1659–1660.

Oscai, L. B., Caruso, R. A. & Wergeles, A. C. (1982). Lipoprotein lipase hydrolyzes endogenous triacylglycerols in muscle of exercised rats. *J. appl. Physiol. respir. envir. Exercise Physiol.* **52**: 1059–1063.

Pernia, S. D., Hill, A. & Ortiz, C. L. (1980). Urea turnover during prolonged fasting in the northern elephant seal. *Comp. Biochem. Physiol.* **65B**: 731–734.

Puppione, D. L. & Nichols, A. V. (1970). Characterization of the chemical and physical properties of the serum lipoproteins of certain marine mammals. *Physiol. Chem. Phys.* **2**: 49–58.

Randle, P. J., Garland, P. B., Hales, C. N. & Newsholme, E. A. (1963). The glucose fatty-acid cycle: its role in insulin sensitivity and the metabolic disturbances of diabetes mellitus. *Lancet* **1963** (1): 785–789.

Riedman, M. (1990). *The pinnipeds: seals, sea lions and walruses.* University of California Press, Berkeley, Los Angeles, Oxford.

Roberts, S., Samuels, L. T. & Reinecke, R. M. (1943). Previous diet and the apparent utilization of fat in the absence of the liver. *Am. J. Physiol.* **140**: 639–644.

Scholander, P. F. (1940). Experimental investigations on the respiratory function in diving mammals and birds. *Hvalråd. Skr.* No. 22: 1–131.

Schmidt-Nielsen, K. (1970). Energy metabolism, body size, and problems of scaling. *Fed. Proc.* **29**: 1524–1532.

Smith, L. C., Pownall, H. J. & Gotto, A. M. (1978). The plasma lipoproteins: structure and metabolism. *A. Rev. Biochem.* **47**: 751.

Steiner, G. & Murase, T. (1975). Triglyceride turnover: a comparison of simultaneous determinations using the radioglyceride and the lipolytic rate procedures. *Fed. Proc.* **34**: 2258–2263.

Stirling, I. & McEwan, E. H. (1975). The caloric value of whole ringed seals (*Phoca hispida*) in relation to polar bear (*Ursus maritimus*) ecology and hunting behavior. *Can. J. Zool.* **53**: 1021–1027.

Suzuki, H. & Fuwa, H. (1970). Influence of dietary composition on the capacity of glucose formation in the liver of rats. *Agric. biol. Chem.* **34**: 80–87.

Wolfe, R. R. (1984). *Tracers in metabolic research.* Alan R. Liss Inc., New York.

Wolfe, R. R. & Durkot, M. J. (1985). Role of very low density lipoproteins in the energy metabolism of the rat. *J. Lipid Res.* **26**: 210–217.

Worthy, G. A. J. (1990). Nutritional energetics of marine mammals. In *Handbook of marine mammal medicine*: 489–520. (Ed. Dierauf, L. A.). CRC Press, Boca Raton, Florida.

Worthy, G. A. J. & Lavigne, D. M. (1983). Energetics of fasting and subsequent growth in weaned harp seal pups, *Phoca groenlandica. Can. J. Zool.* **61**: 447–456.

Zech, L. A., Grundy, S. M., Steinberg, D. & Berman, M. (1979). Kinetic model for production and metabolism of very low density lipoprotein triglycerides. *J. clin. Invest.* **63**: 1262–1273.

Symp. zool. Soc. Lond. (1993) No. 66: 383–394

Balancing power and speed in bottlenose dolphins (*Tursiops truncatus*)

T. M. WILLIAMS,
W. A. FRIEDL, J. E. HAUN
and N. K. CHUN

Naval Ocean Systems Center
Kailua
HI 96734, USA

Synopsis

Hydrodynamic, energetic and physiological limitations dictate the swimming speed and duration of submergence of marine mammals. To determine the effects of these limitations on swimming and diving performance by cetaceans, we examined the relationships among aerobic transport costs, oxygen stores and locomotor speed of bottlenose dolphins. Metabolic rate and the cost of transport were assessed for two adult dolphins trained to swim next to a boat. v_{mr}, the maximum range speed, was determined from minimum transport costs and ranged from 1.67 to 2.27 m $\cdot$ s^{-1} for swimming animals. The duration of submergence in diving dolphins depended on the balance between oxygen stores and the energetic cost of swimming. Oxygen stores for a 145 kg dolphin were 33 mlo$_2$ $\cdot$ kg^{-1} body weight. These stores supported breath-hold durations of 4–4.5 min during rest and swimming speeds below v_{mr}. This duration agreed with changes observed for Po$_2$ and Pco$_2$ in blood samples taken from quiescent dolphins during trained breath-holds. The depletion rates of oxygen stores also indicated that aerobic dives by bottlenose dolphins are limited to depths shallower than 250 m. Dolphins trained to dive to depths of 100–535 m selectively swam at 1.60–2.23 m $\cdot$ s^{-1}. These speeds approximated v_{mr} and enabled the bottlenose dolphin to maximize submergence time through cost-efficient travel.

Introduction

The criteria for balancing power and speed differ between swimming and diving in the bottlenose dolphin. Drag forces have a profound effect on energetic cost and locomotor speed of swimming mammals (Williams 1989; Williams & Kooyman 1985). These forces increase exponentially with transit speed. Additional drag, and hence energetic cost, is incurred by swimming on or near the water surface (Hertel 1966), a necessity for respiration by marine mammals. As a result, the swimming speeds of

ZOOLOGICAL SYMPOSIUM No. 66
ISBN 0–19–854069–8

Copyright © 1993 The Zoological Society of London
All rights of reproduction in any form reserved

dolphins must account for costs associated with moving through a relatively viscous medium and with the act of surfacing to breathe.

Diving dolphins must balance the seemingly conflicting demands of energetic costs associated with exercise and energy conservation during submergence (Castellini *et al.* 1985; Hochachka 1986). Theoretically, fast transit times incur high energetic costs and shorten the duration of aerobic submergence. Slow speed during submergence reduces the immediate power requirements but prolongs the duration of the breath-hold. In both instances, oxygen stores may be depleted before the termination of the dive. When this occurs, the dolphin must rely on anaerobic processes or initiate unusual metabolic strategies (i.e. selective anoxia or hypometabolism of various organs).

This paper examines the relationships among energetic cost, oxygen stores and locomotor speed in the bottlenose dolphin. We investigated two related conditions; transit swimming, in which dolphins move near the water surface within access to air, and aerobic diving, in which locomotor costs must balance oxygen reserves. We predict locomotor speeds for these activities based on aerobic efficiency and the availability of oxygen. The predicted values are then compared to preferred speeds measured for swimming dolphins and diving dolphins. Lastly, we investigate the pattern of oxygen utilization during breath-hold and its influence on dive duration of bottlenose dolphins.

Aerobic costs during transit swimming

Many studies have speculated about the preferred swimming speed and energetic cost of locomotion in dolphins (reviewed by Fish & Hui 1991). Recently, we measured the physiological responses of swimming in these animals by training two bottlenose dolphins (body weight = 145 kg) to match their transit speed with that of a boat. Respiration rate, heart rate and blood lactate were measured and showed graded, though not linear, increases with speeds up to 2.9 m $\cdot$ s^{-1}. Metabolic rates and energetic costs determined for actively swimming dolphins (Fig. 1) were lower than those reported for other marine mammals. The minimum cost of transport, 1.29 $\pm$ 0.05 J $\cdot$ kg^{-1} $\cdot$ m^{-1}, was 2.1 times the predicted value for fish (Williams, Friedl, Fong *et al.* 1992; Williams, Friedl & Haun in press).

As found for other swimming mammals, the relationship between cost of transport and swimming speed for dolphins is a U-shaped curve (Fig. 1). The shape of this curve allows predictions about preferred swimming speeds. For example, the trough of the curve, as delineated by $\pm$ 10% of the minimum cost of transport, defines the theoretical maximum range speeds (v_{mr}) of the animal (Williams 1987). At these speeds the swimmer achieves the greatest distance per unit power input. Migration and daily movements

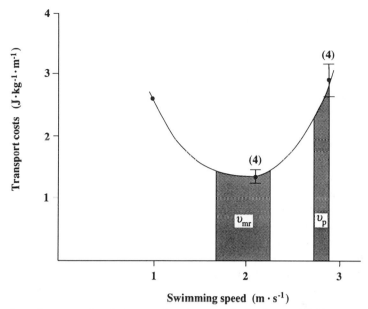

Fig. 1. Cost of transport in relation to swimming speed for bottlenose dolphins. Vertical lines are ± 1 S.D. for each point. Numbers in parentheses represent the number of swimming trials. Shaded areas show the predicted velocities for maximum range (v_{mr}) and porpoising (v_p) for the dolphins in this study.

between areas or food patches probably occur at these speeds in wild animals.

For the dolphins used in this study, v_{mr} ranged from 1.67 to 2.27 m · s^{-1} (Fig. 1). Although this range agreed with routine speeds observed by Würsig & Würsig (1979) for wild coastal dolphins, other studies have reported routine speeds that exceed the calculated range of v_{mr} (Lockyer & Morris 1987; Shane 1990). These studies suggest that the observed speeds of wild dolphins depend on a variety of factors including the duration of effort, wave-riding behaviour and water depth. Further research is needed before we are able to correlate these factors with the energetic efficiency of swimming.

Leaping clear of the water (porpoising) during respiration is a behavioural strategy that theoretically reduces the cost of high-speed swimming in dolphins (Au & Weihs 1980; Blake 1983). This activity allows the animal to avoid high levels of drag associated with surface swimming. For the adult dolphins in this study, minimum predicted porpoising speeds (v_p) ranged from 2.4 to 2.89 m · s^{-1}, and were higher than the range for v_{mr}. Contrary to the theoretical predictions, the dolphins in this study did not porpoise at specific speeds. Behaviour during respiration ranged from minimum surfacing to full leaps while the animals travelled at identical speeds.

Clearly, individual variation is an important factor affecting the preferred swimming speeds of bottlenose dolphins.

Aerobic costs during diving

The pool of oxygen available to support aerobic processes in the diving mammal can be assessed from reserves in the lungs, muscles and blood (Kooyman 1985, 1989). The calculations for cetaceans are identical to those for pinnipeds except for parameters specific to bottlenose dolphins: these include haematological characteristics, larger diving lung volume (0.75 × total lung capacity), larger muscle mass (36% × body mass), and lower myoglobin concentration (2.54 g% wet weight) for dolphins. (See Fig. 2 legend.) The distribution of oxygen stores for a 145 kg bottlenose dolphin is presented in Fig. 2. The lungs account for 1.18 l (25%) of the O_2 reserve. Blood stores and locomotor muscles each hold approximately 1.78 l (37–38%) of oxygen. Together these reserves provide 32.7 mlO_2 per kg of body mass. These results are similar to values for dolphins reported by Kooyman (1989), although muscle reserves are 10% higher and blood reserves are 7% lower in the earlier calculation. The discrepancy is attributed to the myoglobin concentration used in the calculations, which results in a 9% difference in total oxygen stores. When compared to other taxa of marine mammals, the total available oxygen reserve of bottlenose dolphins more closely resembles levels reported for otariids (*Callorhinus ursinus, Zalophus californianus*) than those for phocids (*Phoca vitulina, Leptonychotes*

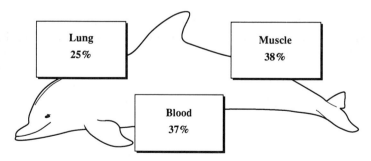

Total oxygen store = 32.7 ml O_2 /kg body weight

Fig. 2. Lung, blood and muscle oxygen stores for an adult 145 kg bottlenose dolphin. Calculations are based on Kooyman (1989). Assumptions used in the calculations are: total lung volume (TLV) from Stahl (1967), diving lung volume = 0.75 TLV with 15% extractable oxygen in the lung of quiescent, breath-holding dolphins (Goforth 1986); blood volume = 80 ml/kg body mass (Ridgway & Harrison 1986) and blood oxygen carrying capacity = 21 mlO_2/ 100 ml blood (Lenfant 1969); arterial and venous blood volumes and extractions from Kooyman (1989) and present study; locomotor muscle mass = 36% body mass (Goforth 1986) with a myoglobin concentration of 2.54 g Mb/100 g wet weight (Castellini 1981); myoglobin oxygen affinity = 1.34 mlO_2/g Mb (Kooyman 1989).

Table 1. Aerobic breath-hold duration for resting and swimming bottlenose dolphins[a]

Speed $(m \cdot s^{-1})$	Metabolic rate $(mlO_2 \cdot kg^{-1} \cdot min^{-1})$	Aerobic breath-hold duration (min)
Rest	7.39 ± 0.91	4.42 ± 0.48
1.0	7.73^b	4.23
2.1	8.07 ± 0.33	4.05 ± 0.16
2.9	24.67 ± 2.41	1.33 ± 0.12

[a] Durations were calculated from a total oxygen reserve of 32.7 $mlO_2 \cdot kg^{-1}$ (Fig. 2) and metabolic rates measured for each level of activity. Metabolic rates are from Williams, Friedl, Fong et al. (1992) and Williams, Friedl & Haun (in press); $n = 11$ for resting animals, $n = 4$ for swimming dolphins. Breath-hold durations for swimming animals are equivalent to the aerobic dive limit defined by Kooyman (1985).

[b] Estimated.

weddelli). Phocids maintain total oxygen reserves that are 21–79% higher than those calculated for dolphins (Kooyman 1989).

Total available oxygen plays an important role in defining the limits of submersion for marine mammals. The aerobic dive limit (ADL), as defined by Kooyman (1985), describes the maximum breath-hold that can be supported by available oxygen stores without a post-dive rise in blood lactic acid. It can be calculated by dividing the oxygen store by the metabolic rate, and therefore is dependent on the animal's level of activity. For pinnipeds ranging in size from 40 to 450 kg, the calculated ADL correlates well with observations of diving behaviour (Kooyman et al. 1983; Gentry et al. 1986). More than 90% of routine dives by these animals fall at or below the ADL.

Maximum aerobic breath-hold durations or ADL for bottlenose dolphins range from 4.1 to 4.4 min for animals at rest or swimming at speeds below v_{mr} (Table 1). The curvilinear rise in metabolic rate with higher swimming speeds (Williams, Friedl, Fong et al. 1992) results in a significant decrease (at $p < 0.05$) in ADL at 2.9 m · s^{-1}. Slow swimming speeds apparently provide a metabolic advantage for prolonging total aerobic submergence time. If dolphins follow the relationship for ADL and diving behaviour reported for pinnipeds, these results indicate that routine dive durations should be comparatively short for bottlenose dolphins. As found for the total oxygen store, the predicted routine dive duration of dolphins (< 4.5 min) more closely approximates the diving limitations of otariids than those of phocids.

Limits to aerobic breath-hold

Although the aerobic dive limit provides valuable information about oxygen stores, the management of these stores has been difficult to measure. To understand the processes by which dolphins utilize oxygen reserves, we examined changes in blood gases during breath-hold. The purpose of this study was to compare the pattern of oxygen depletion in the blood to the aerobic breath-hold limit predicted on the basis of resting metabolism (Table 1). A 163 kg bottlenose dolphin was trained to hold a padded biteplate in its mouth while resting quietly, ventral side up, on the water surface. In this position, the blowhole was submerged and the animal was required to hold its breath until an acoustic signal was given by a trainer. Sequential blood samples were taken from a fluke vessel throughout the breath-hold and were analysed for oxygen and carbon dioxide content (158 pH/Blood Gas Analyzer, Corning). The heart rate of the animal remained above 30 beats per minute, approximating resting apneustic bradycardia (Williams, Friedl, Fong *et al.* 1992).

The changes we observed in blood P_{O_2} during trained breath-holds (Fig. 3) were similar in pattern to end-tidal respiratory gases of submerged bottlenose dolphins (Ridgway, Scronce & Kanwisher 1969) and to venous samples of Weddell seals during forced submergence (Qvist unpubl. obs. in Kooyman 1989). Blood P_{O_2} of the resting dolphin ranged from 42.6 mmHg to 52.5 mmHg (5.7–7.0 kPa) and reflected the duration of breath-hold preceding the sample. During the trained breath-hold, P_{O_2} of the blood was

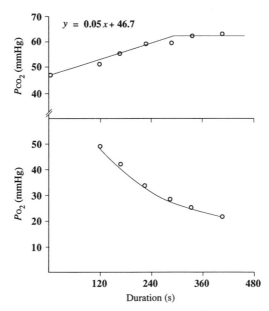

Fig. 3. Changes in blood gases during a 6.5 min breath-hold for a quiescent, adult dolphin. Points represent values for sequential blood samples. Lines are least squares regressions through the data points. The break point in the P_{CO_2} relationship was determined from the intersection of linear ($n = 4$, $r = 0.98$) and plateau ($n = 3$) regressions. A resting value for blood P_{O_2} was unavailable for analysis.

49.0 mmHg (6.5 kPa) after 2 min of submersion, and it decreased to 21.4 mmHg (2.8 kPa) after 6.5 min of submersion. Blood carbon dioxide increased linearly during the first 4.5 min of the breath-hold and averaged 61.8 ± 1.8 mmHg ($n = 3$) for the remainder of the period. The predicted aerobic limit for breath-hold for a 145 kg resting dolphin, 4.42 min (Table 1), correlated well with the plateau in P_{CO_2}. Conversely, blood P_{O_2} was markedly lower, but did not reach a consistent minimum level at the predicted breath-hold limit. This may reflect physiological responses to conserve oxygen stores as the breath-hold limit is approached, and is supported by changes in blood lactate concentration. Lactate levels of the dolphin increased from 1.0 mmol $\cdot$ l^{-1} at rest to 2.7 mmol $\cdot$ l^{-1} following a 6.5-min breath-hold, indicating that the ADL was surpassed.

Elevated metabolic demands associated with swimming movements will undoubtedly change the rates of blood P_{O_2} utilization and carbon dioxide production during diving. However, there appears to be a reasonable correlation between the reduction of oxygen stores, elevation of carbon dioxide and predicted aerobic breath-hold limits. In view of this correlation, we can begin to predict limits for aerobic diving depth and duration based on available oxygen stores, metabolic rate, and swimming speed.

Simple calculations allow us to determine the minimum duration of a dive if we assign specific swimming speeds and depths (Fig. 4). These calculations assume a straight transit between the water surface and maximum depth (no 'bottom or search time'). Swimming speed is averaged from ascent and descent rates. For example, a dive to 200 m (total swimming distance = 400 m) at an average speed of 2.0 m $\cdot$ s^{-1} would require 3.3 min to complete. If we examine several combinations of depth and swimming speed, the result is a series of curves (Fig. 4). The advantage of high swimming speeds during diving quickly becomes apparent. At slow swimming speeds, there is a prohibitive increase in the duration of the dive as depth increases. A dolphin swimming at 1.0 m $\cdot$ s^{-1} would take approximately 3 min to return from a 100 m dive and 10 min to complete a 300 m dive. The same dives conducted at 4.0 m $\cdot$ s^{-1} would take 0.8 min and 2.5 min, respectively.

Although high swimming speeds reduce dive duration, they incur high energetic costs. The balance between swim speed, energetic cost and dive depth is illustrated by comparing maximum aerobic breath-hold duration of dolphins (Table 1) to the calculated minimum dive duration (Fig. 4). By superimposing the curves, we are able to make predictions about the ability of dolphins to support aerobic dives. The combination of diving depths and swimming speeds occurring below the dashed line can be supported aerobically. Thus, a 145 kg bottlenose dolphin is able to rely on aerobic processes when diving to 100 m and swimming between 1.0 and 3.0 m $\cdot$ s^{-1}. If depth is increased to 200 m aerobic dives are limited to a narrow range of

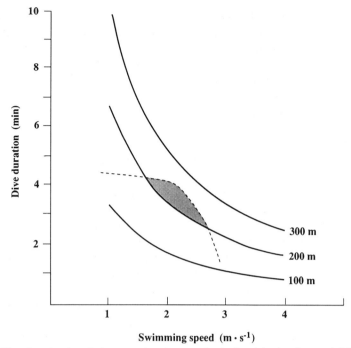

Fig. 4. Dive duration in relation to swimming speed calculated for bottlenose dolphins. The results for three depths are compared. The dashed line represents the aerobic dive limits for a 145 kg dolphin (from Table 1). Aerobic dives to 200 m are limited to the swimming speed range 1.60–2.65 m.s^{-1} (shaded area).

speeds that correspond to minimum transport costs and v_{mr}. Slower speeds to this depth prolong the duration of submergence and cause the available oxygen stores to be exceeded; at faster speeds, high transport costs deplete oxygen stores before the dive can be completed. These calculations indicate that bottlenose dolphins could perform straight dives approximating 250 m without employing specialized physiological mechanisms for diving, as long as swimming speed was maintained near 2.0 m · s^{-1}. Beyond these limits, anaerobic processes and metabolic adjustments specific for conserving oxygen, particularly in hypoxia-sensitive organs, would be required.

Balancing dive duration and speed in trained dolphins

The preceding sections defined the theoretical limits for swimming speed during aerobic diving for bottlenose dolphins. To test these theoretical predictions empirically, we examined routine speeds used by wild and trained animals during diving.

Few studies have evaluated the relationship between speed, dive depth and duration of aerobic submergence in bottlenose dolphins. Detailed observations of diving behaviour in wild *Tursiops* have been limited primarily to coastal populations. Shane (1990) reported mean dive times of 20–25 s for bottlenose dolphins in waters of less than 7 m off the western coast of southern Florida. Würsig & Würsig (1979) found nearshore *Tursiops* of Golfo San Jose, Argentina, in waters shallower than 39 m. These animals demonstrated mean dive times of less than 22 s (Würsig 1978). Longer dive durations were reported for a juvenile male bottlenose dolphin off Pembrokeshire, Wales. The mean dive duration of this animal was 55.3 s; maximum duration was less than 150 s (Lockyer & Morris 1987). These durations are well below the aerobic dive limit calculated for bottlenose dolphins. Most probably behavioural rather than physiological factors contributed to the short durations. Thus, the evaluation of elite diving performance in wild dolphins must be postponed until data concerning the diving behaviour of pelagic animals become available.

Currently, the best information on diving performance in bottlenose dolphins comes from experiments with trained animals (Ridgway *et al.* 1969; Ridgway & Howard 1979; Ridgway & Harrison 1986). In general, the experiments required trained bottlenose dolphins to dive on command to a submerged target in an open ocean situation. Dive depth, descent, ascent, and surface intervals as well as physiological parameters were measured for each dive or series of dives. Following a similar protocol, we trained four bottlenose dolphins (mean body weight = 179 kg) to dive to a submerged marker placed at 202 m. At this depth, metabolic support of the dive could be aerobic or anaerobic depending on swimming speed (Fig. 4). The pattern of movement during submergence was monitored by a microprocessor-based time-depth recorder (Wildlife Computers, Woodinville, WA) attached to a strap around the pectoral fin of each animal.

The results from these studies indicated that dolphins selected swimming speeds that were coincident with the lowest transport costs and provided maximum range. Swimming speeds calculated from ascent and descent rates ranged from 1.6 to 2.2 m · s^{-1} for depths ranging from 100 m to 535 m (Table 2). Trained dolphins in our study maintained an average speed of

Table 2. Dive duration and swimming speed measured for bottlenose dolphins[a]

Dive depth (m)	Duration (min)	Speed (m · s^{-1})	Reference
100	1.5	2.2	Ridgway & Howard (1979)
202	3.3	1.9	Present study
390	8.0	1.6	Ridgway & Harrison (1986)
535	8.0	2.2	Ridgway & Harrison (1986)

[a] Depth was determined by training animals to swim directly to and from submerged markers.

approximately 1.90 m · s^{-1} when diving to 202 m. At this speed, the dolphins were able to maximize aerobic submergence time through cost-efficient locomotion. A similar strategy was employed by dolphins performing shallower and deeper dives. The mean speed of diving dolphins, 1.98 ± 0.29 m · s^{-1}, was not significantly different from the mean speed of coastal dolphins during transit swimming.

Summary

The range of speeds providing maximum range and minimum transport costs is 1.67–2.27 m · s^{-1} for bottlenose dolphins. These speeds are used routinely by wild dolphins. Speed during diving depends on total available oxygen and swimming metabolic rate rather than on depth *per se*. Total available oxygen for a 145 kg dolphin was 32.7 mlO_2 · kg^{-1} body weight. These stores support aerobic breath-hold durations of 4 to 4.5 min during rest and swimming speeds below v_{mr}. These limitations restrict aerobic dives to depths shallower than 250 m. Blood gases during breath-hold and behaviour during trained dives indicate that bottlenose dolphins maximize submergence time by swimming at cost-efficient speeds approaching 2.0 m · s^{-1}.

These calculations and conclusions are limited to bottlenose dolphins, *Tursiops truncatus*. Although it is tempting to make predictions about other species, cetaceans, like their terrestrial counterparts, display a wide range of performance capabilities, anatomical characteristics and physiological limitations. Therefore, the relationships among energetic cost, swimming speed and dive duration for other whales must await further investigation.

Acknowledgements

This study was supported by an ASEE-ONT fellowship to T. M. Williams and includes procedures from AMBS project #SR02301. All experimental procedures were evaluated and approved according to animal welfare regulations specified under NIH guidelines. We thank G. Kooyman and R. Gentry for comments on drafts of the manuscript and insightful discussions of the results. The authors also thank the many people who assisted with the dolphins. Veterinary support was provided by P. Nachtigall, M. Magee, E. Huber and E. Rawitz. K. Keller was a valuable blood laboratory assistant. Creative training of the dolphins was provided by R. Yamada, M. Rothe, S. Shippee, L. Fogg and T. Sullivan.

References

Au, D. & Weihs, D. (1980). At high speeds dolphins save energy by leaping. *Nature, Lond.* **284**: 548–550.

Blake, R. W. (1983). Energetics of leaping in dolphins and other aquatic animals. *J. mar. biol. Ass. U.K.* **63**: 61–70.

Castellini, M. A. (1981). *Biochemical adaptations for diving in marine mammals.* PhD thesis: Univ. of California, San Diego.

Castellini, M. A., Murphy, B. J., Fedak, M., Ronald, K., Gofton, N. & Hochachka, P. W. (1985). Potentially conflicting metabolic demands of diving and exercise in seals. *J. appl. Physiol.* **58**: 392–399.

Fish, F. E. & Hui, C. A. (1991). Dolphin swimming—a review. *Mammal Rev.* **21**: 181–195.

Gentry, R. L., Costa, D. P., Croxall, J. P., David, J. H. M., Davis, R. W., Kooyman, G. L., Majluf, P., McCann, T. S. & Trillmich, F. (1986). Synthesis and conclusions. In *Fur seals: maternal strategies on land and at sea*: 220–264. (Eds Gentry, R. L. & Kooyman, G. L.). Princeton University Press, New Jersey.

Goforth, H. W. (1986). *Glycogenolytic responses and force production characteristics of a bottlenose dolphin* (Tursiops truncatus) *while exercising against a force transducer.* PhD thesis: Univ. of California, Los Angeles.

Hertel, H. (1966). *Structure, form, movement.* Reinhold Publishing Corporation, New York.

Hochachka, P. W. (1986). Balancing conflicting metabolic demands of exercise and diving. *Fed. Proc.* **45**: 2948–2952.

Kooyman, G. L. (1985). Physiology without restraint in diving mammals. *Mar. Mamm. Sci.* **1**: 166–178.

Kooyman, G. L. (1989). *Diverse divers: physiology and behavior.* Springer-Verlag, Berlin. (*Zoophysiology* **23**).

Kooyman, G. L., Castellini, M. A., Davis, R. W. & Maue, R. A. (1983). Aerobic diving limits of immature Weddell seals. *J. comp. Physiol.* **151**: 171–174.

Lenfant, C. (1969). Physiological properties of blood of marine mammals. In *The biology of marine mammals.* 95–116. (Ed. Andersen, H. T.). Academic Press, New York & London.

Lockyer, C. & Morris, R. (1987). Observations on diving behaviour and swimming speeds in a wild juvenile *Tursiops truncatus. Aquat. Mamm.* **13**: 31–35.

Ridgway, S. H. & Harrison, R. J. (1986). Diving dolphins. In *Research on dolphins*: 33–58. (Eds Bryden, M. M. & Harrison, R.). Clarendon Press, Oxford.

Ridgway, S. H. & Howard, R. (1979). Dolphin lung collapse and intramuscular circulation during free diving: evidence from nitrogen washout. *Science* **206**: 1182–1183.

Ridgway, S. H., Scronce, B. L. & Kanwisher, J. (1969). Respiration and deep diving in the bottlenose porpoise. *Science* **166**: 1651–1654.

Shane, S. H. (1990). Behavior and ecology of the bottlenose dolphin at Sanibel Island, Florida. In *The bottlenose dolphin*: 245–265. (Eds Leatherwood, S. & Reeves, R. R.). Academic Press, San Diego, New York etc.

Stahl, W. R. (1967). Scaling of respiratory variables in mammals. *J. appl. Physiol.* **22**: 453–460.

Williams, T. M. (1987). Approaches for the study of exercise physiology and hydrodynamics in marine mammals. In *Approaches to marine mammal energetics*: 127–145. (Eds Huntley, A. C., Costa, D. P., Worthy, G. A. J. & Castellini, M. A.). Allen Press, Kansas. (*Spec. Publs Soc. mar. Mammal.* No. 1.)

Williams, T. M. (1989). Swimming by sea otters: adaptations for low energetic cost locomotion. *J. comp. Physiol. A* **164**: 815–824.

Williams, T. M., Friedl, W. A., Fong, M. L., Yamada, R. M., Sedivy, P. & Haun, J. E. (1992). Travel at low energetic cost by swimming and wave-riding bottlenose dolphins. *Nature, Lond.* **355**: 821–823.

Williams, T. M., Friedl, W. A. & Haun, J. E. (In press). The physiology of bottlenose dolphins (*Tursiops truncatus*): heart rate, metabolic rate and plasma lactate concentration during exercise. *J. exp. Biol.*

Williams, T. M. & Kooyman, G. L. (1985). Swimming performance and hydrodynamic characteristics of harbor seals *Phoca vitulina*. *Physiol. Zool.* **58**: 576–589.

Würsig, B. (1978). Occurrence and group organization of Atlantic bottlenose porpoises (*Tursiops truncatus*) in an Argentine bay. *Biol. Bull. mar. biol. Lab. Woods Hole* **154**: 349–359.

Würsig, B. & Würsig, M. (1979). Behavior and ecology of the bottlenose dolphin, *Tursiops truncatus*, in the south Atlantic. *Fish. Bull. U.S. Fish Wildl. Serv.* **77**: 399–412.

Index